RadTool Nuclear Medicine MCQs

Medhat Sam Gabriel • Bital Savir-Baruch
Editors

RadTool Nuclear Medicine MCQs

Board Exam Preparation

Springer

Editors
Medhat Sam Gabriel
Division of Nuclear Medicine
Department of Radiology
Loyola University Medical Center
Maywood, IL
USA

Bital Savir-Baruch
Division of Nuclear Medicine
Department of Radiology
Loyola University Medical Center
Maywood, IL
USA

ISBN 978-3-030-69280-3 ISBN 978-3-030-69281-0 (eBook)
https://doi.org/10.1007/978-3-030-69281-0

This Springer imprint is published by the registered company Springer Nature Switzerland AG
The registered company address is: Gewerbestrasse 11, 6330 Cham, Switzerland

Preface

The field of Nuclear Medicine and Molecular Imaging has many similarities to Radiology in general although in many respects it is different and unique. The practice of Nuclear Medicine does not only evaluate the anatomy but extends to assess the underlying physiological and cellular abnormalities. Understanding the physiology of the organ or system of interest while simultaneously looking at the anatomy adds another dimension in the interpretation of images. The challenge comes from integrating both sets of data to reach a meaningful conclusion. Working with nuclear medicine and radiology residents for many years taught me that this is not an easy task. In reality, this integration is more complicated than it looks and some residents struggle with this concept for a long time.

Many textbooks explain the combined anatomy and physiology of diseases but may be difficult to read or comprehend. Putting these issues in a thought-provoking format and dividing them into smaller pieces of information with motivating questions would hopefully make them easier to understand. This was the ultimate goal of this book that is written primarily for the residents and would benefit radiologists who read nuclear scans. I hope that those residents and radiologists will find this book useful and helpful in making their understanding of nuclear scans easier and the interpretation more specific.

This multiple-choice question book is the second book in the Nuclear Medicine Flash Facts series and offers more than 400 multiple-choice questions of high-yield topics in the field of nuclear medicine. It covers basic science and clinical nuclear medicine with designated chapters on physics, radiopharmacy, nuclear oncology, cardiology, endocrinology, pulmonology, musculoskeletal, genitourinary, and the central nervous system. The book is written for the radiology and nuclear medicine residents in training but extends to benefit radiologists who practice nuclear medicine or preparing for the recertification examination.

Chicago, Illinois

Medhat Sam Gabriel
Bital Savir-Baruch

Genitourinary System 299
Medhat Sam Gabriel, Bital Savir-Baruch, Robert H. Wagner, Bruch J. Barron, Davide Bova, Aamna Hassan, Stephanie Rodriguez-Santiago, and Yekaterina Kucerova

Central Nervous System 319
Medhat Sam Gabriel, Bital Savir-Baruch, Robert H. Wagner, Bruch J. Barron, Davide Bova, Aamna Hassan, Stephanie Rodriguez-Santiago, and Yekaterina Kucerova

Radionuclide Therapy 327
Medhat Sam Gabriel, Bital Savir-Baruch, Robert H. Wagner, Bruch J. Barron, Davide Bova, Aamna Hassan, Stephanie Rodriguez-Santiago, and Yekaterina Kucerova

Miscellaneous 337
Medhat Sam Gabriel, Bital Savir-Baruch, Robert H. Wagner, Bruch J. Barron, Davide Bova, Aamna Hassan, Stephanie Rodriguez-Santiago, and Yekaterina Kucerova

List of Contributors

Editors

Medhat Sam Gabriel Division of Nuclear Medicine, Department of Radiology, Loyola University Medical Center, Maywood, IL, USA

Bital Savir-Baruch Division of Nuclear Medicine, Department of Radiology, Loyola University Medical Center, Maywood, IL, USA

Authors

Bruch J. Barron Division of Nuclear Medicine and Molecular Imaging, Department of Radiology and Imaging Sciences, Atlanta, GA, USA

Davide Bova Division of Nuclear Medicine, Department of Radiology, Loyola University Medical Center, Maywood, IL, USA

James Halama Department of Nuclear Medicine, Loyola University Medical Center, Maywood, IL, USA

Aamna Hassan Department of Nuclear Medicine, Shaukat Khanum Memorial Cancer Hospital and Research Center, Lahore, Punjab, Pakistan

Stephen Karesh Department of Nuclear Medicine, Loyola University Medical Center, Maywood, IL, USA

Yekaterina Kucerova Division of Nuclear Medicine, Department of Radiology, Loyola University Medical Center, Maywood, IL, USA

Department of Radiology, Cook County Healthy, Chicago, IL, USA

Stephanie Rodriguez-Santiago Department of Nuclear Medicine, C.W. Bill Young VA Medical Center, Bay Pines, FL, USA

Robert H. Wagner Division of Nuclear Medicine, Department of Radiology, Loyola University Medical Center, Maywood, IL, USA

About the Author

Medhat Sam Gabriel is a Professor of Radiology at the Stritch School of Medicine of Loyola University in Chicago. He joined Loyola in 2006 and served as the Director of the Nuclear Medicine Residency Program from 2009 to 2020. Dr. Gabriel is certified by the American Board of Nuclear Medicine (ABNM). He is a distinguished teacher and gave many lecture and review courses in nuclear medicine and nuclear cardiology in multiple institutions and states. He was also invited as a visiting professor to give lectures at the University of San Juan, Puerto Rico. He received his Medical Degree from Assiut University, Faculty of Medicine in Egypt. Dr. Gabriel received the Award of Teacher of the Year for Radiology Resident in 2012–2013. He serves in the Central Chapter of the Society of Nuclear Medicine and Molecular Imaging (CCSNMMI). He was the President of CCSNMMI in 2017–2018. He is an active member of ABNM, ACNM, RSNA, and SNMMI.

Bital Savir-Baruch is an Associate Professor of Radiology in the Nuclear Medicine Division of Loyola University Health System and the Loyola University Stritch School of Medicine in Chicago. Dr. Savir is board-certified in Nuclear Medicine and Nuclear Cardiology and received her medical degree from Semmelweis University, Budapest, Hungary. She completed a research fellowship and Nuclear Medicine residency at Emory University Hospital, Atlanta, GA. At Loyola University Medical Center, she serves as a clinical physician, educator, researcher, and mentor for medical students and residents. She is an active member of ACNM, SNMMI, RSNA, and ACR societies lecturing frequently and has several leadership positions. Dr. Savir is the author of *RadTool Nuclear Medicine Flash Facts*, the first book in the flash fact educational book series. Her main research interest lies in the innovation of new techniques to guide invasive procedures and investigate novel radiotracers and imaging devices to detect and manage neoplastic diseases.

Part I
Basic Science

3. When the gamma camera with a pinhole collimator moves away from the patient, what happens to the count rate?

   (a) Increases
   (b) Does not change
   (c) Decreases
   (d) Depends on the organ imaged

   Correct answer: (c) decreases.
   The count rate decreases, following the inverse square law. If the distance from the pinhole to the patient is doubled the count rate will decrease by a factor of 1/4. It is not organ dependence.

4. The sinogram of a selected slice in SPECT may be used to:

   (a) Eliminate the bullseye artifact.
   (b) Display significant non-uniformities in the gamma camera images.
   (c) Assign the appropriate filter for image reconstruction.
   (d) Detect patient motion

   Correct answer: (d) detect patient motion.
   The sinogram of a selected SPECT slice is extremely sensitive for detecting patient motion. For a given transaxial slice, the sinogram shows the projected patient anatomy (radionuclide activity distribution) onto the gamma camera crystal as the detector rotates during the acquisition. The projection of the anatomy is expected to follow a sinusoidal path across the camera crystal. Patient motion is seen whenever the projected anatomy position deviates from the predicted path. Motion correction software is designed to apply corrections to any such deviations.

5. If images from a gamma camera with a useful field-of-view of 384 mm in diameter are digitized to 128 × 128 image matrices, what would be the image pixel cross-sectional area?

   (a) 0.11 mm^2
   (b) 0.34 mm^2
   (c) 3 mm^2
   (d) 9 mm^2

   Correct answer: **(d) 9 mm^2.**
   The 384 mm diameter of the gamma camera field-of-view is subdivided into 128 pixels horizontally and vertically. Assuming that the pixels are square, the cross-sectional area of each pixel is then:

$$\text{cross}-\text{sectional area} = (\text{horizontal pixel width}) \times (\text{vertical pixel width})$$

$$= (384\,\text{mm} / 128\,\text{pixels}) \times (384\,\text{mm} / 128\,\text{pixels})$$

$$= (3\,\text{mm}) \times (3\,\text{mm})$$

$$= 9\,\text{mm}^2$$

6. Which type of radiation detector is the radionuclide dose calibrator?

   (a) Scintillation
   (b) Geiger–Mueller
   (c) Gas ionization
   (d) Liquid scintillation counter

Correct answer: (c) Gas ionization.
The dose calibrator detector has a cylinder into which the radionuclide is inserted. The air (gas) within the cylinder becomes ionized by the emitted gamma radiation. The ionization rate, measured in Roentgens/min (R/min), is used to calculate the dose by applying the known specific gamma ray constant for the radionuclide tested. The specific gamma ray constant has units of R/min/mCi. The scintillation detector is used in gamma cameras. The Geiger–Mueller detector is a highly sensitized gas ion chamber used for survey radiative contamination. The liquid scintillation counter is used for detecting beta emitting radioisotopes.

7. Which instrument is used for locating radioactive contamination?

(a) Well counter
(b) Gas ion chamber
(c) Geiger counter
(d) Proportional counter

Correct answer: (c) Geiger Counter.
The Geiger counter is a highly sensitized gas ion chamber and has a handheld probe sensitive to low levels of radiation that can be used to survey across working areas such as tabletops, workbenches, and floors. The well counter has a NaI crystal with a hole drilled within it in order to insert test tube samples containing known or unknown radionuclides. The gas ion chamber is in the dose calibrator, and proportional counters are to measure neutron radiation.

8. If the count rate from a 50 kBq ^{99m}Tc source on an uptake probe is 30,000 counts/min (CPM), what is the counting efficiency of the source?

(a) 0.1%
(b) 0.6%
(c) 1.0%
(d) 6.0%

Correct answer: (c) 1%. A 50 kBq source has 50,000 decays/s 30,000 CPM means that 500 counts/s of gamma photons were measured. Therefore, the counting efficiency = 500 counts gamma/s divided by 50,000 decays/s = 0.01, or 1%.

9. The intrinsic efficiency in detecting gamma rays in a gamma camera is increased by

(a) Adding to the crystal thickness
(b) Swapping NaI for a crystal that has shorter fluorescent emissions
(c) Increasing the number of photomultiplier tubes in the gamma camera
(d) Moving the detector away from the radioactive source

Correct answer: (a) adding to the crystal thickness. The intrinsic efficiency in the gamma camera is defined as the fraction of gamma rays that reach the camera crystal that are counted and recorded in the image. This fraction is always increased by adding to the thickness of the crystal. Changing to a crystal that emits its scintillations faster does not change intrinsic efficiency. Such crystals have less dead-time and are able to process higher count rates. Increasing the number of PMTs improves spatial resolution only. Moving the detector away from the source is a geometric factor that leads to loss of count rate because of detector source distance.

10. When performing a thyroid uptake measurement with I-123 with a probe detector, what are the leg counts used for?

(a) Correct for extra-thyroidal I-123 in the patient's neck
(b) Determine the non-thyroidal uptake of I-123

(c) Determine the background correction for the standard reference count
(d) Correct for attenuation of I-123 photons

Correct answer: (a) correct for extra-thyroidal I-123 in the patient's neck. The uptake probe when pointed at the patient's leg measures count rate of I-123 in the blood of the leg. The leg's diameter is similar to the diameter of the patient's neck, and thus this count rate is used to estimate extra-thyroidal I-123 blood activity in the patient's neck. The leg counts would provide insufficient information to determine the non-thyroidal uptake of I-123. The room background count rate in the absence of all sources is used to correct the standard reference count rate. Attenuation correction is not applied in this measurement. Attenuation is estimated by counting the reference standard inserted 1 cm inside a cylindrical plastic phantom to simulate the depth of the patient's thyroid in the neck.

11. What is the purpose of the diverging-hole collimator?

(a) Enlarge the gamma camera field-of-view
(b) Magnify patient images
(c) Decrease gamma camera sensitivity
(d) Image high-energy isotopes

Correct answer: (a) enlarge the gamma camera field-of-view. The diverging-hole collimator has holes that are angled outward away from the crystal center, increases the gamma camera field-of-view. The consequence is that the patient image is minified. The camera system has a higher count rate and more sensitivity because more of the patient is imaged. There are diverging collimators designed for high and low-energy isotope. The only difference is the thickness of the lead septa between the holes.

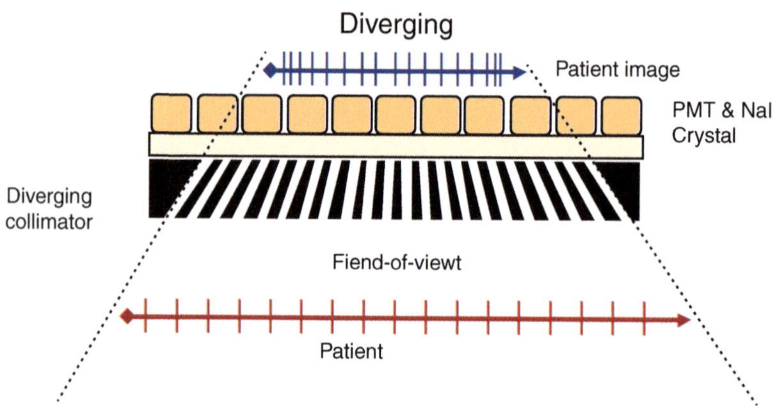

12. Which of the following is the single most important parameter in maintaining gamma camera uniformity?

(a) Imaging a low-energy isotope
(b) Balancing the PMT gains
(c) Imaging at high count rates
(d) Using a wide energy window

Correct answer: (b) balancing the PMT gains. An array of photomultiplier tubes (PMTs) is used to measure the scintillation light emitted from the camera NaI(Tl) crystal following gamma ray interactions. To have a uniform response, the PMT gain of each tube must be matched to one another. Imaging a low-energy radioisotope or imaging with a wide energy window does not necessarily yield good camera uniformity. Imaging at high count rates generally leads to saturation in the PMT signal output leading to a loss of uniformity.

13. What happens to the count rate when imaging a patient using a parallel-hole collimator?

   (a) The count rate remains the same as the distance between the camera collimator and patient increases.
   (b) The count rate decreases as the distance between the camera collimator and patient increases.
   (c) The count rate does not change as the time between radioisotope injection and the time of imaging increases.
   (d) The count rate increases when choosing to make the energy window smaller.

   Correct answer: (a) remains the same as the distance between the camera collimator and patient increases. This is a special case. The parallel-hole geometry allows only gamma rays that are parallel to one another to be recorded. The distance from the origin does not matter. The count rate always decreases when choosing a smaller energy window regardless of the type of collimator used.

14. What is the best combination to achieve a high spatial resolution collimator?

   (a) Long holes of small diameter
   (b) Long holes with thin lead septa
   (c) Long square holes
   (d) Short holes of large diameter

   Correct answer: (a) long holes of small diameter. Spatial resolution achieved with a gamma camera is best when using long-channel, small diameter holes. Such holes are narrowly focused on activity that is distributed in the patient. The lead septa thickness is adjusted thicker for high-energy gamma rays and does not directly affect the resolution. Designing square holes may lead to higher sensitivity but not necessarily spatial resolution.

15. Which collimator is required for whole-body Ga-67 imaging?

   (a) Low-energy parallel-hole
   (b) Medium-energy parallel-hole
   (c) High-energy parallel-hole
   (d) Pinhole

   Correct answer: (b) Medium-energy parallel-hole. The gamma rays emitted by Ga-67 are of 90, 180, and 300 keV energy. The highest energy of 300 keV by definition falls within the medium-energy classification of collimators. The pinhole collimator is rarely, if ever, used for imaging Ga-67. There are two types of pinhole collimators, low-energy and high-energy. The high-energy is used for I-131 imaging.

16. Imaging Tl-201 requires the use of which collimator?

   (a) low-energy general purpose collimator
   (b) medium-energy collimator
   (c) high-energy collimator
   (d) pinhole collimator

   Correct answer: (a) low-energy general purpose collimator. Tl-201 emits a 70 keV characteristic X-ray and 167 keV gamma ray. By definition, both fall within the low-energy classification of collimators. There are two types of pinhole collimators, low-energy and high-energy. The low-energy pinhole collimator would be used in such situations.

17. How does scatter affect the image contrast?

    (a) By filling in of counts in background regions
    (b) By increasing the noise in background regions
    (c) By blurring the hot, high count density areas
    (d) By decreasing the hot, high count density areas, due to dead-time count losses

    Correct answer: (a) filling of counts in background regions. Scatter adds a background haze to the gamma camera images. Gamma rays from within the patient that scatter are no longer seen as directly coming from the radionuclide source, but nearby in the background regions. The noise is directly determined by the number of recorded counts, and as such scatter in the background will decrease noise. The phenomenon of blurring is restricted to the discussion of loss of spatial resolution. The dead-time at high count rates leads to count losses equally for scatter and non-scatter counts.

18. What is the purpose of using body-contouring during SPECT acquisitions in gamma cameras equipped with parallel-hole collimators?

    (a) Increase the sensitivity.
    (b) Increase the spatial resolution.
    (c) Decrease the effects of attenuation.
    (d) Simplify the reconstruction algorithm.

    Correct answer: (b) increase the spatial resolution. The spatial resolution in SPECT is best when the gamma detectors stay close to the patient during gamma camera rotation. Choosing a body-contouring method achieves this goal. The count rate (sensitivity) of parallel-hole collimators does not change at detector-patient distances and therefore will not change during body-contouring. Attenuation is determined by the patient thickness and not by detector-camera distance. The extra procedure for performing body-contouring will only complicate the reconstruction algorithm.

19. Why is it that the Chang attenuation correction algorithm does not work well in the chest?

    (a) Because of cardiac and respiratory motion.
    (b) Because only a 180 acquisition arc is used.
    (c) Because the tissue density of the lungs is one-third that of soft tissue.
    (d) Because of noncircular motion SPECT acquisition.

    Correct answer: (c) the tissue density of the lungs is one-third that of soft tissue. The Chang attenuation correction algorithm assumes a uniform tissue density across the entire patient cross-section. This assumption is reasonable for cross-sections in the abdomen and brain, but not in the chest. The Chang algorithm does not require that more than 180° acquisition arc. The algorithm is applied post reconstruction. Noncircular motion SPECT and corrections for cardiac and respiratory motion are applied during the initial image reconstruction.

20. What type of SPECT artifact results from nonuniformities seen in gamma camera flood images?

    (a) Streaks in the transverse images passing through the patient volume.
    (b) Concentric rings of alternating high-low count density starting at the center-of-rotation.
    (c) High count density rim artifacts at the periphery of internal organs.
    (d) High count density rim at reconstructed edge of transverse image.

    Correct answer: (b) Concentric rings of alternating high-low count density starting at the center-of-rotation. Streaks radiating through the SPECT slices are caused by counting statistics and inconsistencies in measurement in projects of the patient activity distribution (high count density bladder seen with high sensitivity AP and PA, but not laterally). High count density at the rim of internal organs is related to attenuation artifacts and high count density at the periphery of the reconstruction is related to truncation of parts of the patient during camera rotation.

21. If there is misalignment of the gamma camera center-of-rotation (COR) with the image matrix center, how would the reconstructed SPECT images be affected?

    (a) They will exhibit overall loss in image resolution.
    (b) They will become excessively noisy.
    (c) They will be split in half and mirrored.
    (d) They will show a hot rim artifact.

    Correct answer: (a) exhibit overall loss in image resolution. Misalignment of the center-of-rotation causes the reconstruction program to incorrectly place the count profiles during back projection and leads to a blurring of the image and loss of spatial resolution. If severe enough, point sources appear artefactually as donuts. There is no loss of counts because of the misalignment and therefore the image noise will be unaffected. Split in half and mirrored is potentially possible, but only in severe cases of improper calibration. The hot rim is strictly related to other phenomenon of attenuation and truncation.

22. Which filter is routinely used in SPECT image reconstructions to reduce the image noise?

    (a) Butterworth
    (b) Hanning
    (c) Ramp
    (d) Weiner

    Correct Answer: (a) Butterworth. Although the Hanning filter also reduces noise, it has not been routinely the choice. Both ramp and Weiner filters are used to enhance image contrast at the expense of increasing image noise.

23. The positron emitting radioisotopes C-11, N-13, O-15, and F-18 are commonly used in PET imaging. What is the order of these positron emitters in terms of increasing half-life?

    (a) C-11, O-15, N-13, F-18
    (b) O-15, N-13, C-11, F-18
    (c) N-13, O-15, C-11, F-18
    (d) O-15, C-11, N-13, F-18

    Correct answer: (b) O-15, N-13, C-11, F-18
    The half-lives of these positron emitters are

O-15	2 min
N-13	10 min
C-11	20 min
F-18	110 min

24. If the technologist increases the cutoff frequency of the Butterworth filter, the resultant images:

    (a) are smoother and of less contrast
    (b) are noisier and of higher contrast

(c) exhibit a Moiré pattern
(d) are higher in count density

Correct answer: (b) are noisier and of higher contrast. The higher cutoff frequency allows high-frequency information to remain in the reconstructed image which preserves more of the contrast but also leaves more noise. The Moiré pattern is a consequence of poor sampling of the highest of frequency content, ordinarily not observed in SPECT images. The overall count density remains unchanged during filtering.

25. What is the purpose for imaging a SPECT phantom?

(a) to perform a center-of-rotation calibration.
(b) to perform a sensitivity calibration.
(c) to determine dead-time corrections for SPECT.
(d) It is a quality assurance test of the SPECT system.

Correct answer: (d) as a quality assurance test of the SPECT system. Center-of-rotation measurements are made with small point sources on the SPECT image pallet. Although the sensitivity (count rate in CPM) can be measured with the SPECT phantom, the result provides no useful results. Meaningful sensitivity is measured by imaging a small point source of activity on the imaging table. High count rates are not encountered in SPECT imaging, and dead-time measurements are not required.

26. If too few counts are acquired during a SPECT acquisition, what may be done improve the reconstructed images?

(a) Choosing a reconstruction filter with a lower cutoff frequency
(b) Apply a resolution enhancement reconstruction filter
(c) Sum adjacent projection images from the acquisition prior to image reconstruction
(d) Use the FBP algorithm

Correct Answer: (a) Choosing a reconstruction filter with a lower cutoff frequency. An insufficient number of counts in a SPECT acquisition leads to reconstructed images that are very noisy. Smoothing the images by choosing a lower cutoff frequency in the SPECT reconstruction filter will reduce the noise in the image. Applying a resolution enhancement filter will further amplify the noise. Summing adjacent projection images offer no benefit. Iterative reconstruction algorithms offer much improved reconstruction with low counts than filtered backprojection (FBP).

27. What does the coincidence detection technique used in PET eliminate the need for?

(a) lead collimation
(b) energy discrimination
(c) photon attenuation corrections
(d) FBP reconstruction

Correct Answer: (a) lead collimation. Lead collimation in gamma cameras localizes radioactivity within a patient by collimating (channeling) gamma rays reaching the NaI crystal. The coincidence detection of the two opposite traveling gamma rays from positron annihilations within the patient allows for localization of radioactivity without the use of lead collimation. Energy discrimination, photon attenuation, and FBP reconstruction offer no role in collimation.

28. What is a limiting factor in achieving high spatial resolution in PET imaging?

(a) The required detector crystal density
(b) Limited organ uptake of PET radiopharmaceuticals
(c) Positron range
(d) Low annihilation radiation energy

Correct Answer: (c) positron range. Positrons travel a short distance (positron range) before annihilation occurs. The PET scanner measures the location of the annihilation and not the origin of the positron. This difference limits the spatial resolution that can be achieved. High crystal density is needed for PET that actually leads to improvements in spatial resolution. The annihilation radiation energy of 511 keV is considered to be high-energy.

29. What does time-of-flight in PET measure?

(a) The random fraction of counts by knowing the delay time
(b) The coincidence timing window
(c) Blood flow transit in the PET imager field-of-view
(d) The location of positron annihilations that occur between a pair of PET detectors

Correct Answer: (d) the location of positron annihilations that occur between a pair of PET detectors. The difference in arrival time between the two annihilation gamma rays determines the location of the annihilation event between detector pairs. The coincidence timing window is a predefined parameter used to identify the occurrence of coincident detection events. A delay time added inline to one of the detector pairs is used to measure random count rates and is unrelated to time-of-flight measurement. PET imaging makes no measurement of blood flow transit time.

30. The true coincidence counts in PET imaging correspond to the number of __________ that occur between a pair of PET detector crystals.

(a) positron-electron annihilations
(b) annihilation gamma rays
(c) nuclear decays
(d) positrons emitted

Correct Answer: (a) positron-electron annihilations. The number of positrons emitted could be considered an answer except that some of the positrons emitted travel outside the boundary of the line drawn between a pair of detectors. Many annihilation gamma rays are detected but only a small fraction (1/50) are actually in coincident with another in an opposite PET detector. The true count rate is linearly proportional to the nuclear decay rate. A sensitivity calibration is needed to convert the true count rate to nuclear decay rate.

31. In PET, how is a line-of-response (LOR) defined?

(a) A line drawn between a PET detector pairs
(b) A line drawn between detector rings
(c) The line tangent to a coincident event pair
(d) Time-of-flight events

Correct Answer: (a) A line drawn between a PET detector pairs. The LOR is not defined for detector rings, but detector pairs only which may extend between detector rings. A line tangent to an event pair or time-of-flight has no relationship to the definition of the LOR.

32. How is coincidence detection of annihilation photons in 2D PET limited? Identify the incorrect statement:

 (a) to within each detector ring and one adjacent detector ring
 (b) to within specified axial slices
 (c) between detector pairs within one ring
 (d) LORs in between installed lead septa

 Correct Answer: (c) between detector pairs within one ring. The 2D method uses coincidence counts to reconstruct slices within the direct plane of a detector ring, and for slices situated halfway between adjacent detector rings. The reconstructed axial slices correspond to defined by the lines-of-response of the accepted coincidence detector pairs. The lead septa are installed to limit the detection of annihilation photons arising outside the defined slices.

33. How does the random coincident count rate vary with patient activity at scan time?

 (a) It decreases with patient activity
 (b) It increases with patient activity
 (c) It increases with the square of the patient activity
 (d) It is independent of patient activity

 Correct Answer: (c) It increases with the square of the patient activity. The random coincident count rate can be calculated by multiplying the single photon count rate on each of a detector pair times the coincidence timing window. As the patient activity increases the single count rate increase linearly on each of the detectors, the product corresponds to the square of the patient activity.

34. Which expression is used to calculate the standardized uptake value (SUV)?

 (a) tissue concentration of FDG per unit mass
 (b) tissue concentration of FDG per unit injected activity per unit mass
 (c) tissue concentration of FDG per unit injected activity/h
 (d) tissue concentration of FDG per unit injected activity

 Correct Answer: (b) tissue concentration of FDG per unit injected activity per unit mass. Another way of writing the definition is (tissue FDG in Bq/mL)/injected Activity in Bq/mass in g).

35. Under what conditions is the standard uptake value of tumors most accurate?

 (a) Tumors that are metabolically active
 (b) Tumors whose diameter is greater than three times the resolution of the PET scanner
 (c) In PET scanners after the coincidence timing window has been increased
 (d) PET imaging of large patients

 Correct Answer: (b) Tumors whose diameter is greater than three times the resolution of the PET scanner. The SUV accuracy is strongly dependent on partial volume effects. The effect is eliminated in centers of tumors whose diameter is three times the resolution of the PET scanner. A metabolically active tumor is not necessarily large. The SUV of small active tumors will be underrepresented due to partial volume effects. Imaging of large patients leads to quantitative inaccuracies overall not necessarily just tumors. Increasing the coincidence timing window will increase random counts leading to quantitative inaccuracies.

36. What does the sensitivity calibration of a PET scanner determine?

(a) true coincidence count rate in CPM/mL
(b) true coincidence count rate in CPM/Bq/mL
(c) attenuation correction factor
(d) dead-time correction curve

Correct Answer: (b) true coincidence count rate in CPM/Bq/mL. The sensitivity calibration is used to convert the true coincident count rate measured in count/min (CPM) to units of Bq/mL required to calculate the SUV. It is obtained by imaging a cylinder of water that contains a known concentration of activity in Bq/mL. An attenuation correction factor is needed for accurate sensitivity calibration but is obtained from the corresponding CT image data. Dead-time corrections are applied in the sensitivity calibration but are measured separately by starting with high levels of activity in the phantom and imaging multiple times as the activity in the phantom decays.

37. What area in a PET facility yields the highest radiation exposures to personnel?

(a) hot lab
(b) PET scanner Room
(c) bathroom
(d) uptake room

Correct Answer: (d) uptake room. The uptake room is the room in which patients stay during the time of the radiopharmaceutical uptake which may be as long as 90 min. Patients spend less time in the scanner room. Typically less than 30 min. Also, there is significant radioactive decay during the uptake. Patients excrete PET radioisotope in the urine but spend a small amount of time in the bathroom. In most PET facilities patients are injected in the uptake room. If injected in the hot lab, the time present in the room is short. Injection of patients always poses a high radiation exposure to the technologist. With proper shielding of the syringe reduces the exposure.

38. If an unaccounted fraction of an FDG dose is infiltrated into the patient's arm during injection, how is the SUV affected?

(a) SUV is higher than expected
(b) SUV is lower than expected
(c) SUV is unaffected
(d) SUV is seen to be elevated in tumors

Correct Answer: (b) SUV is lower than expected. The FDG uptake in tissues is directly related to the concentration of circulating FDG. If the concentration is lower than expected because of dose infiltration, the resultant SUV as a whole will be lower. To be accurate in this circumstance the injected dose must be corrected for the infiltration fraction.

39. How does the intrinsic radioactivity of lutetium in LSO (lutetium oxyorthosilicate) PET crystals affect PET images?

(a) Increases the patient radiation dose
(b) Adds scatter to the PET images
(c) Causes streak artifacts across the PET images
(d) Adds a background of random coincidences in the PET images

Correct Answer: (d) Adds a background of random coincidences in the PET images. Internal radioactive emissions increase the single detector count rate which in turn leads to increased random coincidence counts. The random counts add uniformly to the background in PET images. The internal emissions will not contribute to the recording of scattered coincidence counts and will not contribute to additional image artifacts.

40. If the lean body mass were used to calculate the SUV instead of the total body mass, how is the SUV affected?

(a) SUV is higher
(b) SUV is lower
(c) SUV is unaffected and remains the same
(d) SUV is higher in the fat

Correct Answer: (b) SUV is lower. The SUV is the ratio (tissue FDG concentration/injected activity/mass). Lean body mass is always smaller than total body mass and therefore the SUV will be lower in value.

41. If the residual FDG activity remaining in the syringe after patient IV injection was not accounted for, how is the SUV affected?

(a) SUV is higher
(b) SUV is lower
(c) SUV is unaffected and remains the same
(d) SUV becomes negative

Correct Answer: (b) SUV is lower. The SUV is the ratio (tissue FDG concentration/injected activity/mass). The injected activity in the formula is the net activity administered after subtracting the residual from the assayed activity. An incorrect larger value of injected activity will result in a lower SUV.

42. If for some reason the clock in the PET scanner did not fall back by 1 h on the day that daylight savings time ended (the PET scanner clock is 1 h ahead), how is the resultant SUV affected?

(a) SUV is higher
(b) SUV is lower
(c) SUV is unaffected and remains the same
(d) SUV becomes negative

Correct Answer: (a) SUV is higher. The clock in the PET scanner not only stamps the time of image acquisition but is also used to correct the tissue FDG concentration for radioactive decay. The extra hour of decay time (false because of the incorrect PET scanner clock) leads to an increased decay factor of 1.46 ($2^{60/110}$) and thus all SUVs will be higher by this factor.

43. If the patient does not void just prior to scanning, how is the resultant overall SUV affected?

(a) SUV is higher
(b) SUV is lower
(c) SUV is unaffected and remains the same
(d) SUV becomes negative

Correct Answer: (c) SUV is unaffected and remains the same. Emptying the bladder just prior to scanning should not alter the uptake of FDG in the body tissues, and thus the SUV remains the same. The patient's bladder is emptied in order to diminish reconstruction artifacts in the pelvis. These artifacts could both elevate or decrease the SUV in the region of the pelvis.

44. If scatter was not corrected for during image reconstruction, how the resultant overall SUV affected?

(a) SUV is higher
(b) SUV is lower

(c) SUV is unaffected and remains the same
(d) SUV is zero

Correct Answer: (a) SUV is higher. The body tissue FDG concentration is higher because coincidence counts from scatter were not subtracted in reconstruction, and therefore the higher SUV.

45. If dead-time correction was not applied during image reconstruction, how is the resultant overall SUV affected?

(a) SUV is higher
(b) SUV is lower
(c) SUV is unaffected and remains the same
(d) SUV is zero

Correct Answer: (b) SUV is lower. The body tissue FDG concentration is lower because counts lost during the acquisition because of dead-time were not recovered (corrected for) in the reconstruction, and therefore the lower SUV.

46. If time-of-flight is enabled, how is the overall resultant SUV affected?

(a) SUV is higher
(b) SUV is lower
(c) SUV is unaffected and remains the same
(d) SUV in tumors is lower

Correct Answer: (c) SUV is unaffected and remains the same. The time-of-flight information is used in the image reconstruction to improve the quality of the reconstructed images, and should not change the overall, average SUV. SUV of tumors may be slightly higher because of the improved reconstruction.

47. In the gamma camera, how is the location of photon interaction in the detector crystal determined?

(a) The location is calculated by triangulation of the PMT signals
(b) The location is calculated by summing the PMT signals
(c) The location is calculated by position weighted sum of the PMT signals
(d) The location is calculated by averaging the PMT signals

Correct Answer: (a) The location is calculated **by position weighted sum of the PMT signals.** The centroid of the scintillation emissions in the NaI crystal corresponding to the location of the photon interaction is calculated by a position weighted sum of PMT signals in the camera.

48. What is the best option to increase gamma camera detector spatial resolution?

(a) decrease the NaI(Tl) crystal thickness
(b) increase the NaI(Tl) crystal thickness
(c) decrease the number of photomultiplier tubes
(d) increase the energy resolution of the gamma camera

Correct Answer: (a) decrease the NaI(Tl) crystal thickness. In thinner crystals, the centroid of scintillation emissions is closer to the array of PMT's which in turn leads to a more precise calculation of the photon interaction location. An increase in the number of photomultiplier tubes would improve spatial resolution. Change in the energy resolution will not alter the spatial resolution. Increasing the energy resolution reduces scatter leading to images with higher contrast.

49. How does the intrinsic spatial resolution of the gamma camera vary?

 (a) It increases with increasing photon energy
 (b) It decreases with increasing photon energy
 (c) It increases with energy resolution
 (d) It increases with increasing dead-time count losses

Correct Answer: (a) It increases with increasing photon energy. PMT signals are stronger with high-energy photons and therefore the position weighted sum calculation to locate photon interactions is more precise. The primary example is that the intrinsic spatial resolution of Tl-201 with 70 keV photons is less than that of Tc-99m with 140 keV photons. There is no change with energy resolution. Increasing dead-time count losses by using leads to a decrease in spatial resolution.

50. Of the collimator types listed below, which has the highest sensitivity measured in CPM/uCi:

 (a) pinhole
 (b) parallel-hole
 (c) converging
 (d) diverging

Correct Answer: (c) converging. For sources within the collimator field-of-view, the converging collimator has the highest number of holes focused at the radioactive source and therefore the highest sensitivity. The diverging-hole collimator is opposite whose holes are focused outward to expand the camera field-of-view. The parallel-hole collimator sensitivity is seen to be in between that of the converging and diverging-hole collimators. The pinhole collimator with only a single aperture hole has the least sensitivity.

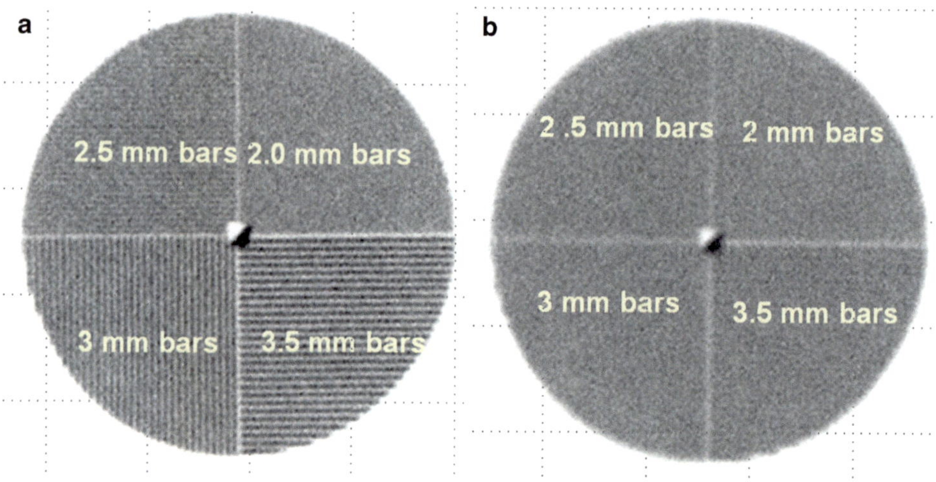

51. A bar phantom was imaged twice with a ^{57}Co flood source on the same gamma camera. Under what condition would explain the loss in detail observed in (b) as compared to (a)?

 (a) The collimator was removed for (b)
 (b) A whole-body scan mode was used for (b)
 (c) The bar phantom in (b) was 10 cm above the gamma camera collimator
 (d) The image matrix was too large for (b)

Correct answer: (C) The bar phantom in (b) was 10 cm above the gamma camera collimator. Image resolution with the gamma camera is best with sources at the collimator surface.

Imaging the bar phantom with an air gap of just 10 cm leads to a total loss of bar detail as compared to the phantom properly placed at the collimator surface. (a) would be a possible choice as an answer except for being able to observe the spacing between the four bar quadrants. The spacing would be totally absent as well in the image. Also, the collimator would always be present when using a ^{57}Co source. For (b) the image resolution is expected to be the same in whole-body mode. For (d) the largest 3.5 mm bars would be observed with even the coarsest available matrix of 64 × 64 pixels.

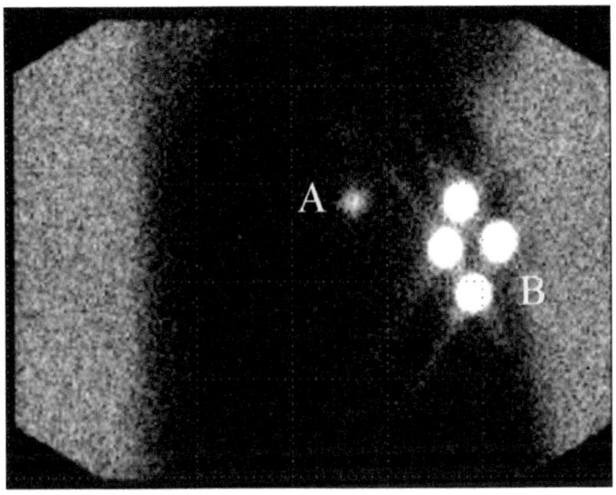

Sentinel Lymph node study image

52. For the sentinel lymph node study image shown above, where was ^{99m}Tc-SC injected?

    (a) Location A
    (b) Location B
    (c) Arm vein
    (d) Breast parenchyma

    Correct answer: (b) Location B. The ^{99m}Tc-SC was injected just under the skin at four locations surrounding a tumor. The injected sites have much higher concentrations of activity and are seen with high count density in the images. Location A is the site of the sentinel lymph node in this study.

53. For the sentinel lymph node study image shown above, the outline of the patient is obtained by placing a radioactive source behind the patient. What source is used?

    (a) ^{99m}Tc point source
    (b) ^{99m}Tc scatter source
    (c) ^{57}Co button source
    (d) ^{57}Co flood source

    Correct answer: (d) Co-57 flood source. A ^{57}Co flood source illuminates the entire gamma crystal. When the flood source is placed behind the patient, the patient outline is observed as it absorbs the ^{57}Co gamma rays. A ^{99m}Tc point source will not illuminate the entire crystal and a ^{99m}Tc scatter does not exist.

54. For the sentinel lymph node study image shown above, the streaks that radiate outward from the high count density spots in the image are:

(a) artifacts from penetration of the lead septa of the gamma camera collimator by ^{99m}Tc gamma rays.
(b) artifacts from a high count rate.
(c) lymph channels through which the ^{99m}Tc-SC migrates.
(d) artifacts from ^{99m}Tc point source gamma ray scattering.

Correct answer: (a) artifacts from penetration of the lead septa of the gamma camera collimator by ^{99m}Tc point source gamma rays. Even gamma rays from ^{99m}Tc, which are considered low-energy, penetrate the collimator septa. The streaks arise from preferential penetration in directions of the thinnest septa. The artifactual pattern corresponds to the collimator holes being arranged in a hexagonal pattern.

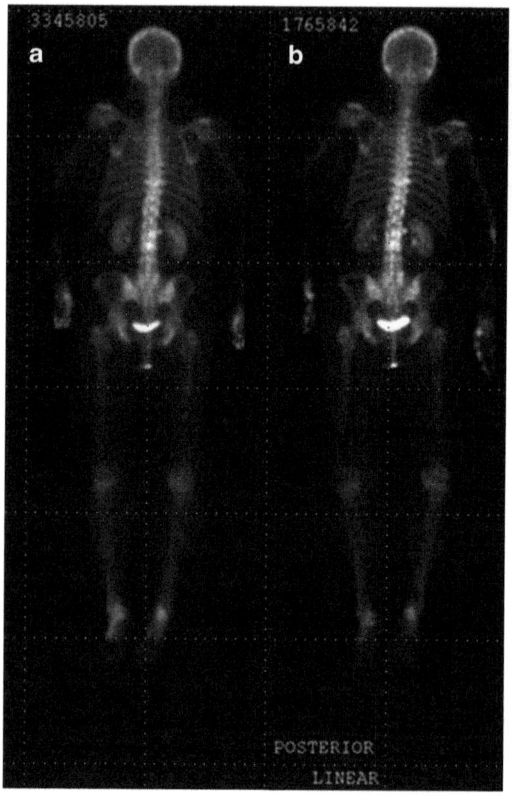

55. For the whole-body ^{99m}Tc-MDP bone scan study shown above, the same patient was scanned twice at the same speed with two different collimators. What collimator was used for A?

(a) LEAP
(b) LEHR
(c) MEGP
(d) Pinhole

Correct answer: (a) LEAP. The low-energy all-purpose (LEAP) collimator has high sensitivity and low resolution. Note there is less observed detail in A as compared to B, yet more counts were acquired. The numbers at the top of the two images correspond to the total counts acquired for each image. The MEGP (medium-energy general purpose) collimator is not specified for imaging ^{99m}Tc. Also, the pinhole collimator would not be specified for use for whole-body imaging. Image B is done by LEHR (Low-energy high resolution) collimator.

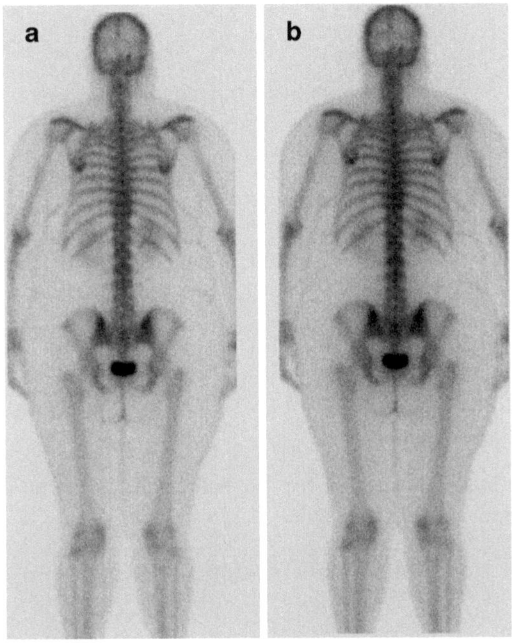

56. For the whole-body ^{99m}Tc-MDP bone scan study shown above, the same patient was scanned twice under two different conditions. Image B as compared to A exhibits a higher background count density? What is a possible reason?

(a) The gamma camera detector was further away from the patient in B than in A
(b) A high sensitivity collimator was used for B
(c) A pinhole collimator was used for B
(d) The gamma camera energy acquisition window for A is 20% and for B is 30%

Correct answer: (d) The gamma camera energy acquisition window for A is 20% and for B is 30%. The higher background counts are from the acceptance of more scatter in the image in B as a result of using a larger energy window for imaging. Also, when comparing the two images detail and edge sharpness do not differ and therefore differing resolution is not the case, eliminating answers (**a**) and (**b**). A pinhole collimator would not be used for whole-body bone imaging.

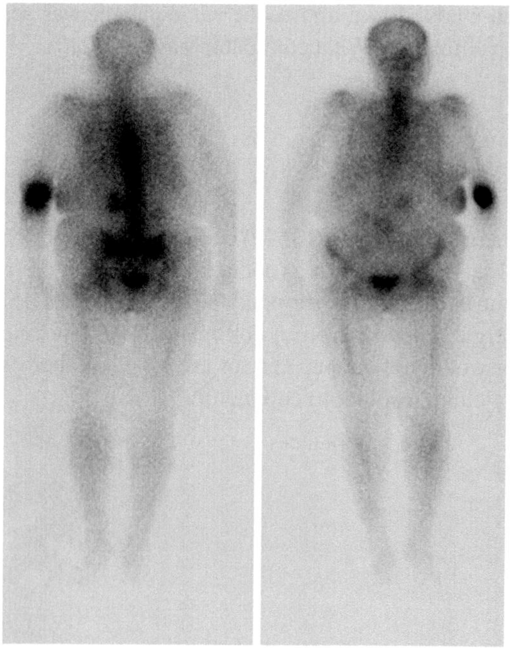

57. For the whole-body ^{99m}Tc-MDP bone scan shown above, what is the reason for the non-visualization of the bones?

(a) A ^{57}Co energy window was used for the scan
(b) Free ^{99m}Tc-MDP
(c) Patient scanned with a high-energy collimator
(d) All of the activity remains at the injection site

Correct answer: (a) A ^{57}Co energy window was used for the scan. All the counts in the image represent down scatter of ^{99m}Tc gamma rays.

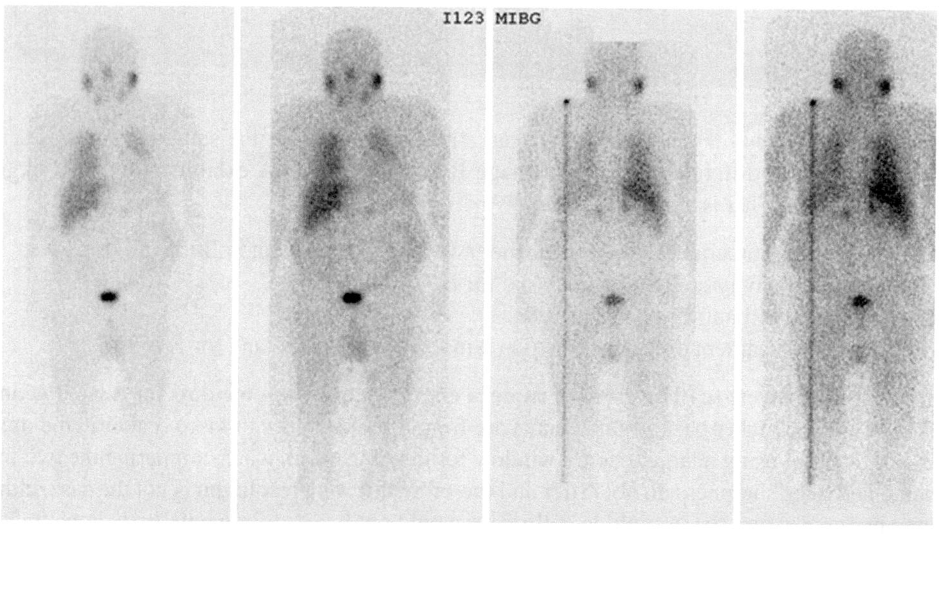

58. For the whole-body ^{123}I-MIBG scan shown above, the cause of the vertical linear artifact seen on the posterior scan image is from contamination with ^{99m}Tc located on the

(a) patient shoulder
(b) imaging table
(c) anterior detector head
(d) posterior detector head

Correct answer: (d) posterior detector head. The contamination must be on the posterior detector head as it is seen to move during the scan. The artifact is not seen at all on the anterior detector since the counts are diminished by absorption through the patient and seen at a much lower spatial resolution.

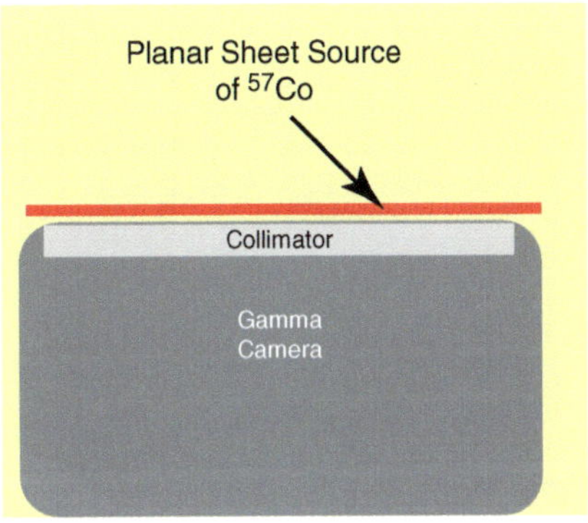

59. The schematic shown above demonstrates which quality control procedure?

(a) performing a SPECT uniformity calibration
(b) performing an intrinsic uniformity calibration
(c) acquiring an intrinsic flood image
(d) acquiring an extrinsic flood image

Correct answer: (d) acquiring an extrinsic flood image. The planar ^{57}Co flood source is shown to be placed directly above the collimator in order to acquire an extrinsic flood image (extrinsic because the flood image is acquired with an installed collimator). Uniformity calibrations are not routine quality control procedures and are performed when extrinsic or intrinsic flood images show nonuniformities.

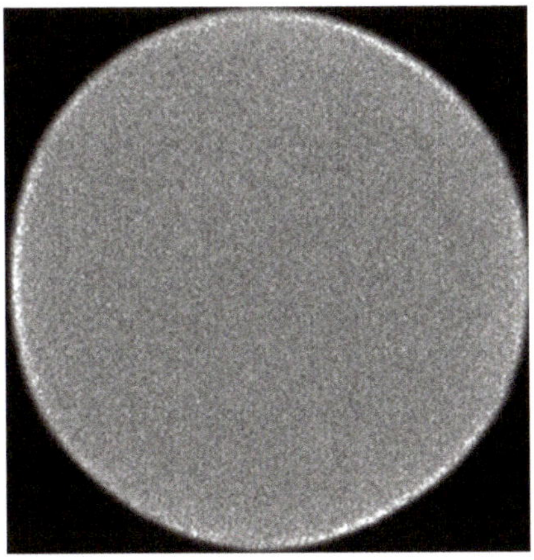

64. The crescent artifact observed in the above intrinsic flood image results from:

(a) Broken crystal
(b) Attenuation of the ^{99m}Tc point source
(c) A second source of ^{99m}Tc contamination within the gamma camera room
(d) Detector nonlinearity

Correct answer: (c) A second source of ^{99m}Tc contamination within the gamma camera room. The second source in the camera room, perhaps located at the wall, was partially shielded by the detector housing leading to the crescent-shaped artifact observed in the flood image.

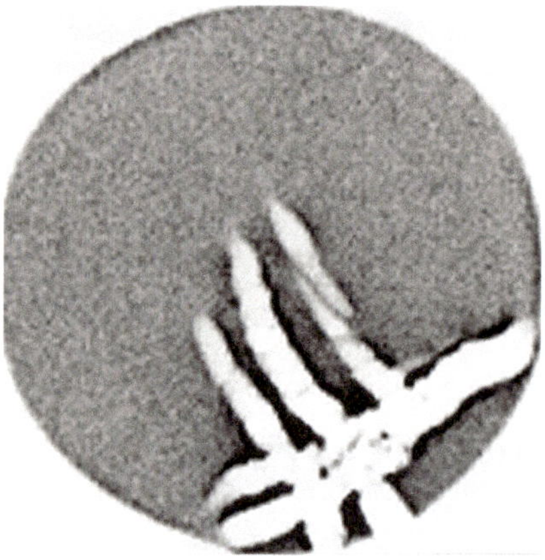

65. The artifact observed in the above flood image results from:

(a) Decoupling of the PMTs
(b) Hydration within the crystal
(c) Fractures within the crystal
(d) Collimator damage

Correct answer: (c) Fractures within the crystal. When fractures are present, the scintillations are reflected by the fracture interfaces leading to high count densities all along the fractures. This is unique to such fractures.

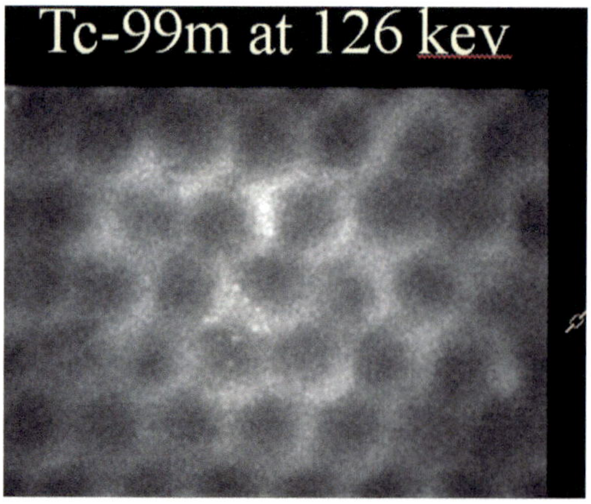

66. The white spots observed in the above flood image of a ^{99m}Tc point source acquired at a low-energy window of 126 keV are caused by:

(a) Decoupling of the PMTs
(b) Hydration within the crystal
(c) Fractures within a broken crystal
(d) Damaged collimator holes

Correct answer: (b) Hydration within the crystal. The white spots correspond to small holes within the hermetic seal separating the NaI crystal from the room air. The crystal in these locations has changed its characteristics.

67. The structure observed in the above extrinsic flood image with a ^{57}Co sheet source results from:

(a) Decoupling of the PMTs
(b) Damaged collimator holes
(c) Fracture within a broken crystal
(d) Hydration within the crystal

Correct answer: (b) Damaged collimator holes. The extrinsic flood image is acquired with an installed collimator. Highly defined structure as seen is caused by a crushed section of collimator holes.

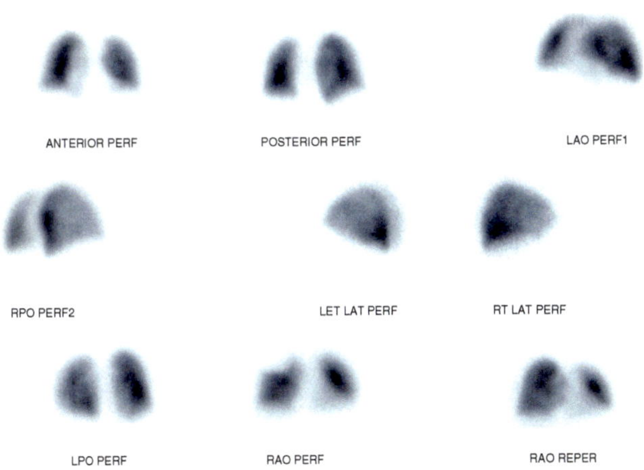

68. The acquisition sequence of images in this lung scan is from left-to-right to top-to-bottom row. The defect observed in the RAO PERF image:

(a) Results from a complete loss of a detector PMT
(b) Results from decreased perfusion
(c) Results from a pacemaker implanted in the chest
(d) Results from detector nonuniformity

Correct answer: (a) Results from a complete loss of a detector PMT. No other defects are observed in any of the prior images. The PMT in this location burnt out just prior to acquiring the RAO PERF image. The RAO REPEAT image was taken on another gamma camera to demonstrate the loss of the PMT tube.

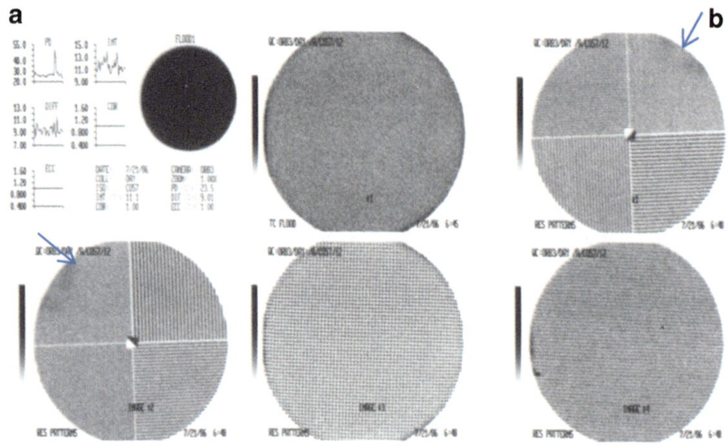

69. The extra count density observed in the quadrant bar phantom images above at locations A and B results from:

(a) Contamination on the quadrant bar phantom with ^{99m}Tc
(b) Contamination on the camera detector with ^{99m}Tc
(c) Nonuniformity of the camera detector
(d) Improper position of the ^{99m}Tc flood source

Correct answer: (b) Contamination with ^{99m}Tc on the quadrant bar phantom. The area of contamination rotates with the phantom.

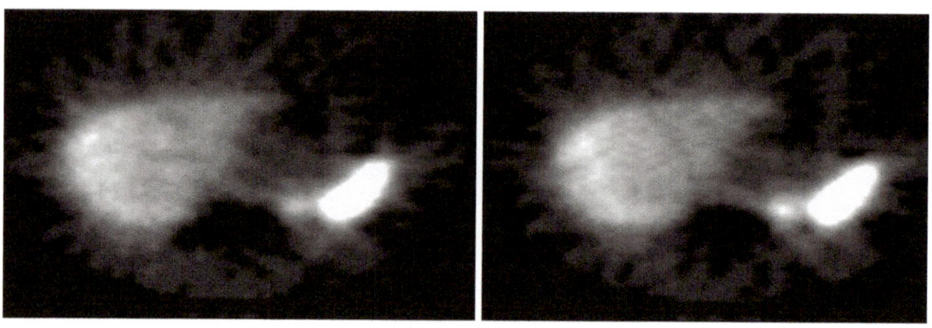

70. The higher count density observed at the periphery of the liver in these two Tc-99 m sulfur colloid liver SPECT images corresponds to:

(a) greater uptake of the sulfur colloid at lateral portions of the liver
(b) greater sensitivity to the sulfur colloid activity at the lateral portions of the liver because of photon attenuation
(c) Applying a high-resolution SPECT reconstruction filter
(d) A center-of-rotation misalignment

Correct answer: (b) greater sensitivity to the sulfur colloid activity at the lateral portions of the liver because of photon attenuation. A uniform distribution of sulfur colloid is expected throughout the liver. SPECT reconstruction filters affect the whole image not any one segment. A center-of-rotation misalignment would result in a blurring of the whole image.

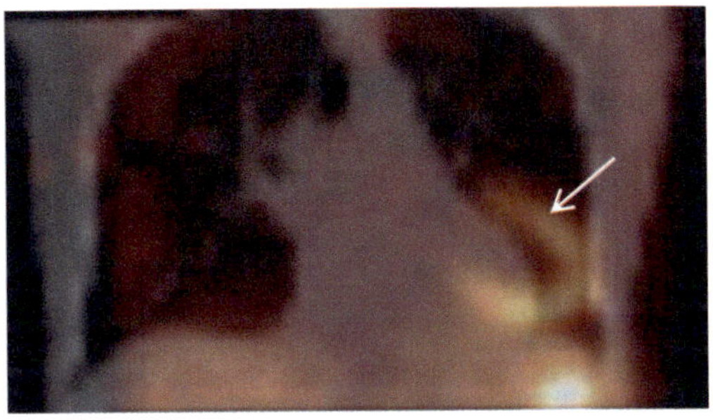

71. What is the reason for the decreased perfusion observed in the lateral wall of this myocardial perfusion SPECT study done with CT attenuation correction?

(a) Respiratory motion
(b) The lateral wall of the heart is attenuated by the lung tissue
(c) The high activity concentration in the gall bladder leads to artefactual lowering the counts in adjacent slices
(d) The lateral displacement of the heart into the lungs leads to a lowering of attenuation correction factors.

Correct answer: **(d) the lateral displacement of the heart into the lungs leads to a lowering of attenuation correction factors.** The lateral displacement occurred between the CT and SPECT acquisitions. The lung tissue attenuates the ^{99m}Tc photons less than soft tissue in the mediastinum.

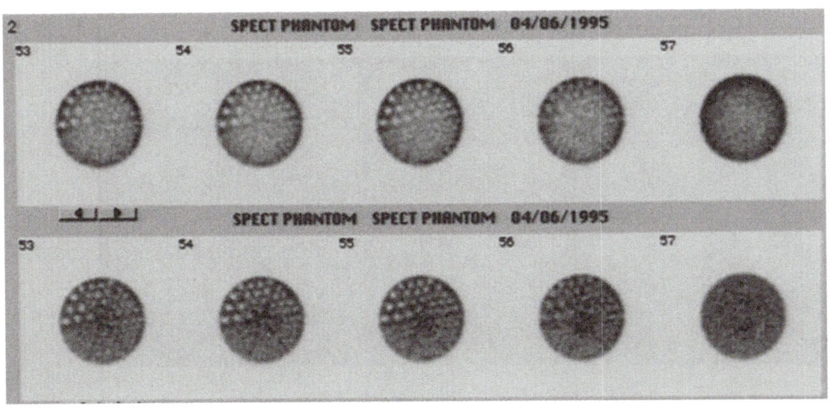

72. The SPECT phantom above was reconstructed twice. What algorithm was applied to reconstruct the second row of images?

(a) Chang
(b) OSEM
(c) Linogram
(d) COR

Correct answer: (a) Chang. Attenuation correction was added to the second row of images. Note, the decreased count density in the reconstructed slices to the top row is a result of photon attenuation. The Chang algorithm applies a uniform tissue density attenuation coefficient to each of the slices. OSEM is an iterative reconstruction algorithm. Linogram is an image derived to detect patient motions. COR is short for center-of-rotation.

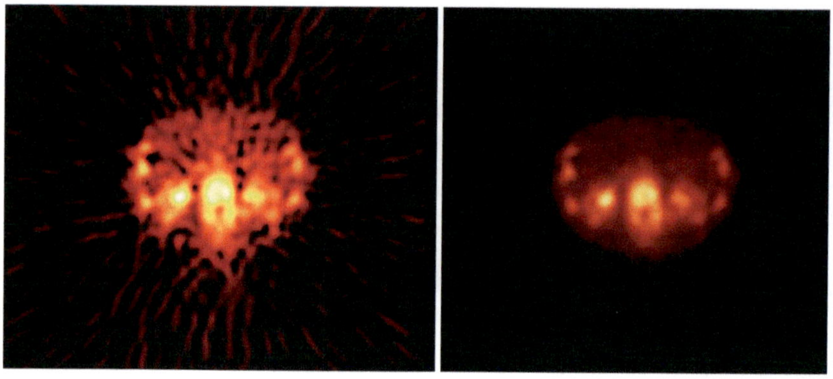

73. The same transaxial slice in a ^{99m}Tc-MDP bone SPECT is reconstructed in two ways. What do the radial streaks observed in the image on the left result from?

(a) partial volume effects
(b) not using attenuation correction
(c) use of filtered backprojection reconstruction algorithm
(d) gamma camera detector nonuniformities

Correct answer: (c) use of filtered backprojection SPECT reconstruction algorithm. Tomographic image reconstruction proceeds by backprojecting counts along projection rays within the patient volume. Statistical noise in the acquisition prevents the backprojection rays to be eliminated in the reconstruction. The image on the right is reconstructed by an iterative algorithm that was developed to minimize the artifacts of the noise.

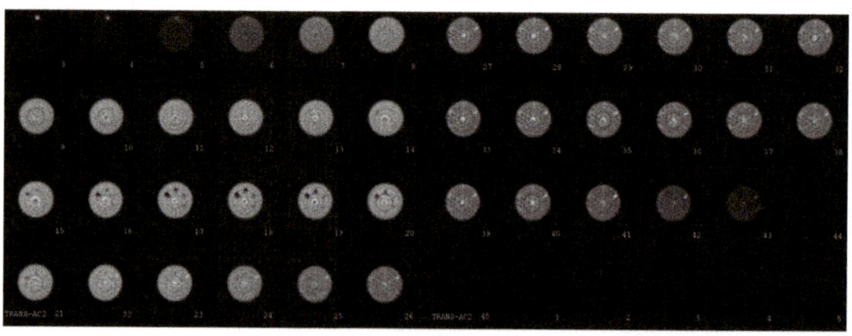

74. Reconstructed transaxial slice slices of a SPECT phantom. What are the concentric ring artifacts due to?

(a) nonuniformity of the gamma camera detectors
(b) improperly applying attenuation correction
(c) too many iterations during reconstruction
(d) incomplete mixing of the radionuclide inside the phantom

Correct answer: (a) nonuniformity of the gamma camera detectors. This is the classic concentric ring artifact observed whenever there is nonuniformity in one or both gamma camera detectors. These nonuniformities are observed in routine flood images. It is necessary to recalibrate the gamma camera detectors for uniformity using a high count flood image. Attenuation correction and iterative reconstructions will not introduce ring artifacts. Incomplete mixing would result in artifact associated with the presence and absence of activity within the SPECT phantom volume.

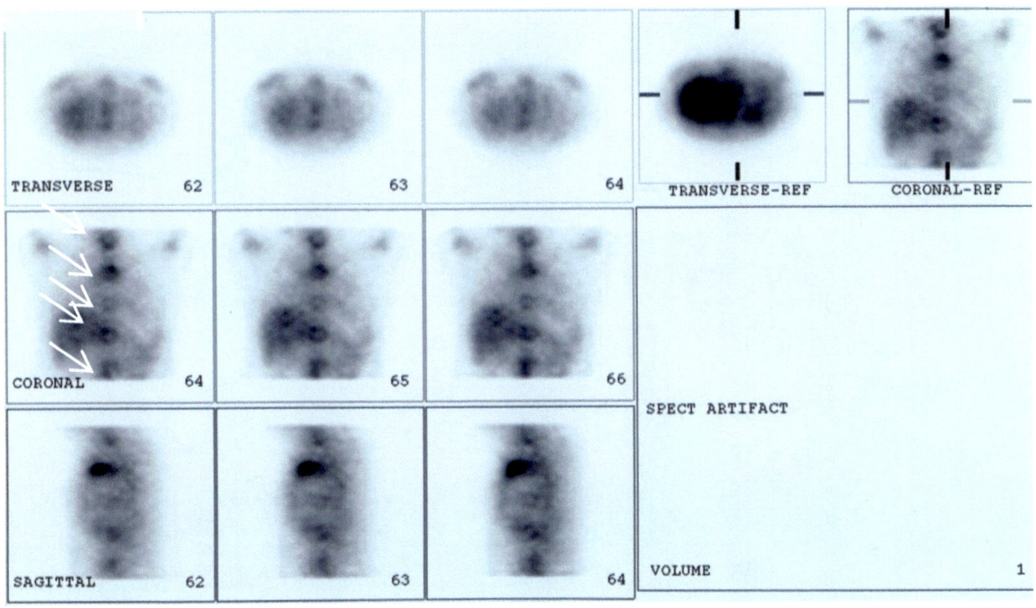

75. Above are reconstructed SPECT slices of a Ga-67 citrate study. What the 5 centrally located artifacts result from?

(a) nonuniformity of the gamma camera detectors
(b) improperly applying attenuation correction
(c) too many iterations during reconstruction
(d) imaging too early after injection

Correct answer: (a) nonuniformity of the gamma camera detectors. This is the ring artifact observed whenever there is nonuniformity in one or both gamma camera detectors, but in this case only the center of the concentric ring is observed and is seen in many slices superior to inferior. The corresponding flood image is below:

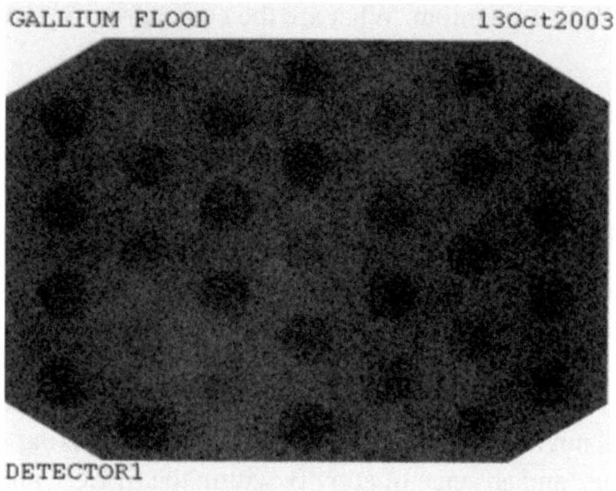

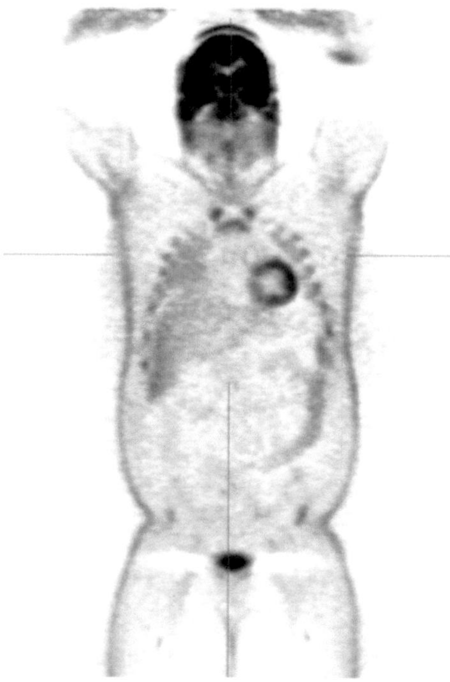

76. The apparent high uptake of FDG in the skin of the above PET scan is observed when?

(a) no attenuation correction is applied
(b) too many random counts are present
(c) counts from scatter are very high
(d) count rate is very low

Correct answer: (a) no attenuation correction is applied. This is a classic image observed whenever attenuation correction has not been applied. The PET scanner is most sensitive to positron annihilations at the periphery of the patient and therefore the observed high uptake of FDG in the skin. FDG uptake internally is ordinarily higher but the counts are attenuated and therefore appear with lower count density.

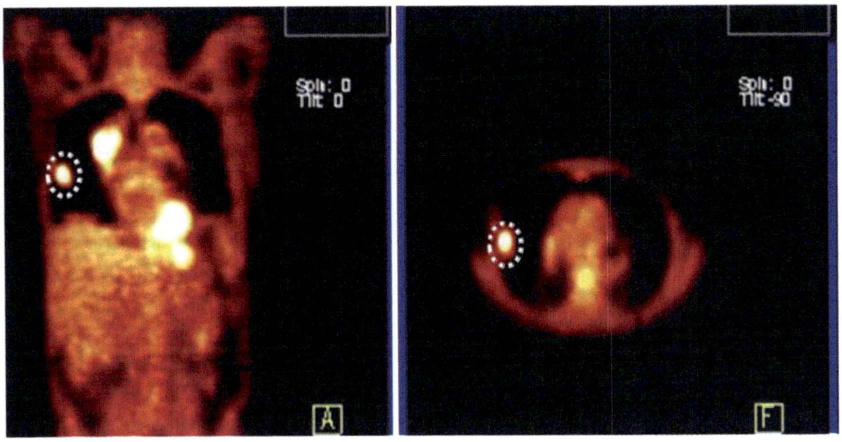

77. The region drawn over the suspected lesion in the FDG PET study above:

(a) is drawn too large as the reported average SUV includes tracer uptake in the normal lung
(b) is drawn with boundaries outside the lesion to ensure that the highest SUV within the lesion is reported
(c) is properly drawn, but the reported SUV is inaccurately high because of the high contrast adjustment
(d) is drawn with boundaries outside the lesion in order to properly identify the lesion boundaries.

Correct answer: (b) is drawn with boundaries outside the lesion to ensure that the highest SUV within the lesion is reported. The objective is to report the highest SUV within lesions. Drawing a large region ensures that the maximum within the lesion is found. Changing the image contrast does not alter the SUV.

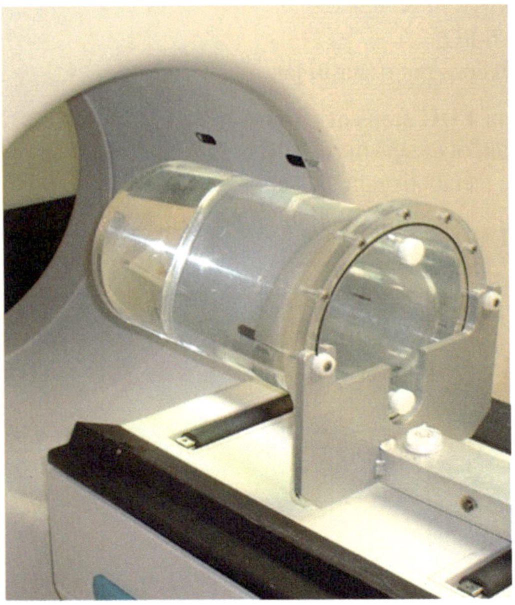

78. The cylindrical phantom above is filled with water mixed with ^{18}F-FDG. The volume of water and the concentration of the ^{18}F-FDG are known. What does this phantom used for?

    (a) validate the sensitivity calibration of the PET scanner.
    (b) test the PET scanner spatial resolution
    (c) measure the lowest significant SUV
    (d) measure the coincidence timing window

    Correct answer: (a) validate the sensitivity calibration of the PET scanner. The reconstruction algorithm in its final step converts the coincidence counts into activity concentration in units of Bq/mL. When the sensitivity calibration is correct, the SUV of a uniform mixture ^{18}F-FDG in the phantom is expected to be 1.0.

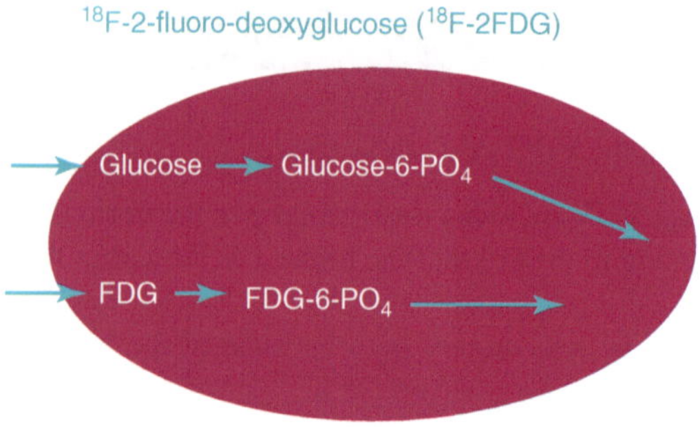

79. The above diagram describes the initial stage in the metabolism of FDG in cells. Identify the correct statement:

    (a) The metabolism of FDG-6-PO_4 continues in the glycolytic pathway like Glucose-6-PO_4, except at a much slower rate.
    (b) The metabolism of FDG to FDG-6-PO_4 occurs at a rate that is much slower than glucose to Glucose-6-PO_4.
    (c) The metabolism of FDG stops at FDG-6-PO_4.
    (d) The metabolism of FDG to FDG-6-PO_4 does not occur in fat tissues.

    Correct answer: (c) The metabolism of FDG stops at FDG-6-PO4. The FDG-6-PO_4 is no longer a substrate for glycolysis and also cannot escape the cell. The rate of the build-up of FDG-6-PO_4 is proportional to the rate of glucose metabolism in the cell.

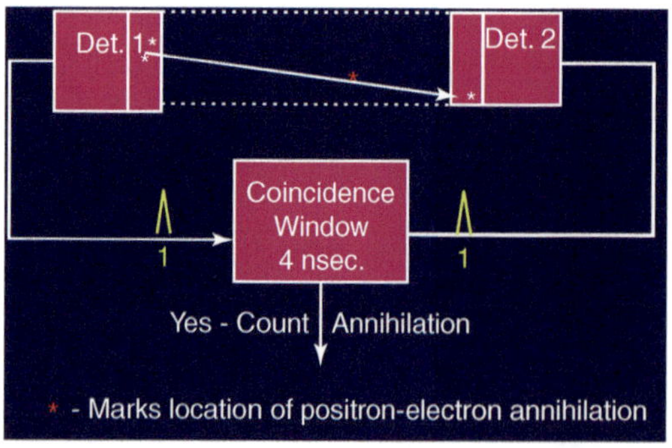

80. The above diagram describes the method of coincidence detection of the two oppositely traveling positron-electron annihilation gamma photons by a pair of gamma ray detectors. Identify the purpose of the Coincidence Window.

(a) The coincidence window allows for counting annihilation events anywhere within the volume between the detector pairs.
(b) The coincidence window prevents counting of annihilation events purely at random.
(c) The coincidence window prevents counting of coincidence events whenever one or both of the annihilation gamma photons undergo scattering.
(d) The coincidence window decreases the coincidence count loss under conditions of high dead-time.

Correct answer: (a) The coincidence window allows for counting annihilation events anywhere within the volume between the detector pairs. Patients will fill the volume between detector pairs in a PET detector ring. The 4 nanoseconds (ns) coincidence window allows for up to a difference of 4 ns in arrival time between annihilation gamma photon pairs. Without a coincidence window, the system would only be sensitive to annihilations that occur exactly halfway between a detector pair. The answer in (b) is incorrect because the use of the coincidence timing window actually causes random counts to be recorded. Using a window of 8 ns would record twice as many random counts as for a 4 ns window. For (c) the energy window set for each detector is used to prevent scatter from being recorded, and (d) is incorrect altogether.

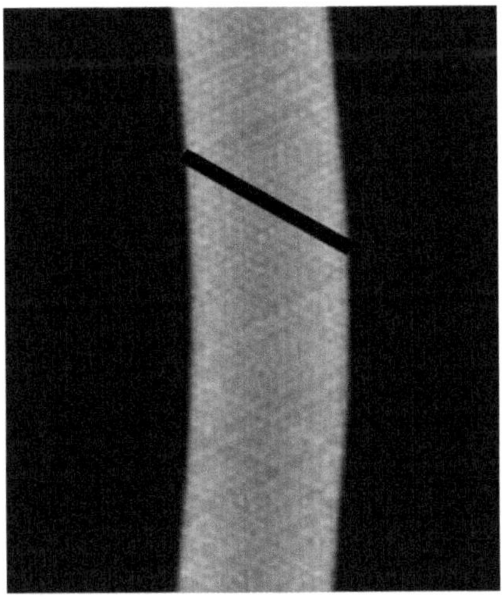

81. The above image is a sinogram from the acquisition of a cylinder of water with ^{18}F-FDG. What is the reason for the cross-hatched structure seen in the sinogram?

(a) Differences in the arrival time of positron-electron annihilation gamma photon pairs
(b) Variation in the dead-time of detector pairs within a PET ring
(c) Variation from statistical nature of counting coincidence events.
(d) Variation in the sensitivity between detector pairs within a PET detector ring.

Correct answer: (d) Variation in the sensitivity between detector pairs within a PET detector ring. Stored in the sinogram are the coincidence counts recorded for detector pairs within a single PET detector ring. The orientations of the detector pairs encircle the object leading to the visible pattern. Often these are identified as the lines-of-response for all the detector pairs.

For (b) there could be some variation in dead-time, but in the case of cylinder this would not be the case. For (c), statistical count variations show as the grainy pattern in the sinogram and not as a regular pattern.

82. The above image is a sinogram from the acquisition of a cylinder of water with ^{18}F-FDG. What is the cause of the black stripe in the sinogram?

    (a) A malfunctioning module of detectors in the PET ring.
    (b) Total loss of coincidence counts due to dead-time in this portion of the acquisition
    (c) A lead strip located near the center of the water phantom.
    (d) The coincidence window was off in this portion of the acquisition.

    Correct answer: (a) A malfunctioning module of detectors in the PET ring. There is a total loss of coincidence counts in the affected detector module. All detectors are in operation during acquisition and not just one portion of the acquisition. A lead object in the phantom would be seen from top-to-bottom in the sinogram.

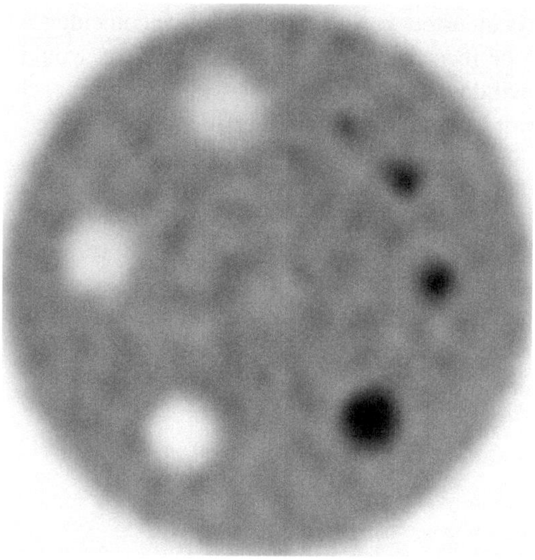

83. The above image is a reconstructed image of a PET phantom containing 3 cold cylinders with no activity each at 25 mm in diameter and 4 hot cylinders of sizes 25, 16, 12, and 8 mm diameter filled with FDG at a concentration of 2.5 times higher than the background activity. What is the expected SUV of the background and that of the largest hot cylinder?

    (a) Unable to get from the image, must extract the numbers with an ROI.
    (b) 0.5, 2.5
    (c) 1.0, 2.5
    (d) 2.0, 5.0

    Correct answer: (c) 1.0, 2.5. In a phantom, the injected FDG activity is uniformly distributed throughout the background with no uptake and with an expected SUV = 1.0. The cylinders have 2.5 times the background, and therefore the expected SUV = 2.5.

84. The above image is a reconstructed image of a PET phantom containing 3 cylinders with no activity each at 25 mm in diameter and 4 cylinders of sizes 25, 16, 12, and 8 mm diameter filled with FDG at a concentration of 2.5 times higher than the background cylinder activity. Compared to the 25 mm cylinder, it appears that the FDG concentration is decreasingly smaller in the 16, 12, and 8 mm cylinders, respectively. Select the best answer to this observation:

(a) Partial volume effect
(b) Loss of spatial resolution for the smaller cylinders
(c) The background is bleeding into the smaller cylinders
(d) The scatter is higher in smaller cylinders

Correct answer: (a) Partial volume effect. The PET scanner has limited spatial resolution and uses 4 mm voxels. The activity concentration in voxels at the edges of the cylinder is an average of the FDG concentration in the background and in the cylinder. The weighting of the background concentration in the average increases as the cylinders decrease in diameter and therefore the appearance of decreasing SUV.

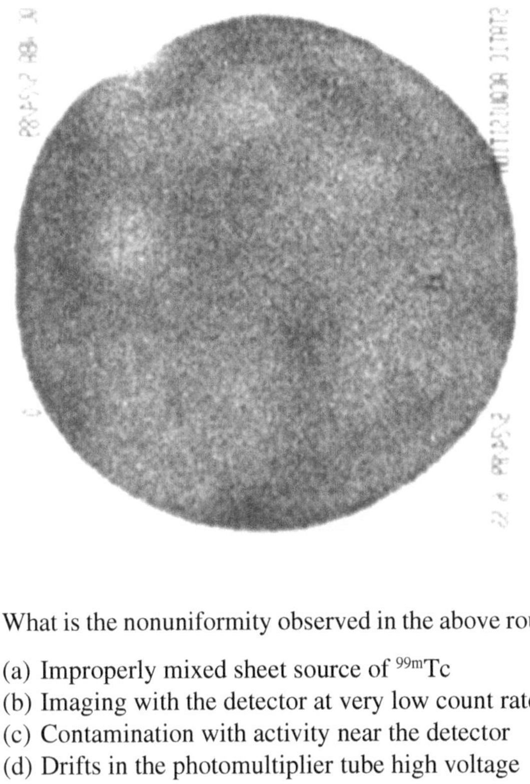

85. What is the nonuniformity observed in the above routine intrinsic flood image likely the result of?

(a) Improperly mixed sheet source of ^{99m}Tc
(b) Imaging with the detector at very low count rates
(c) Contamination with activity near the detector
(d) Drifts in the photomultiplier tube high voltage

Correct answer: (d) Drifts in the photomultiplier tube high voltage. The high voltage set across the photomultiplier tubes must be matched and held constant. A high voltage tuning process is used to achieve excellent uniformity in the flood images. Imaging at very low count rates does not affect uniformity. Contamination with activity near the detector in all likelihood would yield a much more uniform flood image. A sheet source of ^{99m}Tc is not used.

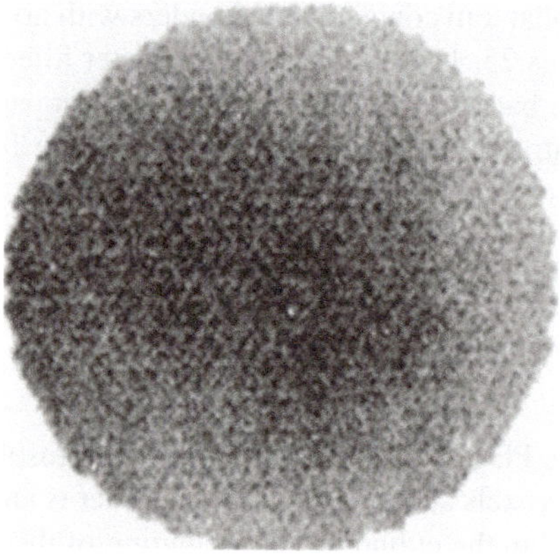

86. What is the nonuniformity observed in the above routine flood image caused by?

(a) Improperly manufactured sheet source of ^{57}Co
(b) Flood image with a high-energy collimator
(c) Photon counting statistics
(d) Drift in the photomultiplier tube high voltage

Correct answer: (b) Flood image with a high-energy collimator. The flood image was acquired with a high-energy collimator and ^{57}Co sheet source. The high-energy collimator has 4–5 mm septa and holes similar in size. This pattern becomes visible in the flood image. A poorly manufactured ^{57}Co sheet source would produce hot and cold spots in the flood image but not in a regular pattern. The nonuniformity due to counting statistics would not yield a regular pattern. Drifts in photomultiplier tube high voltage results in blotches of nonuniformities matched to the photomultiplier tube locations.

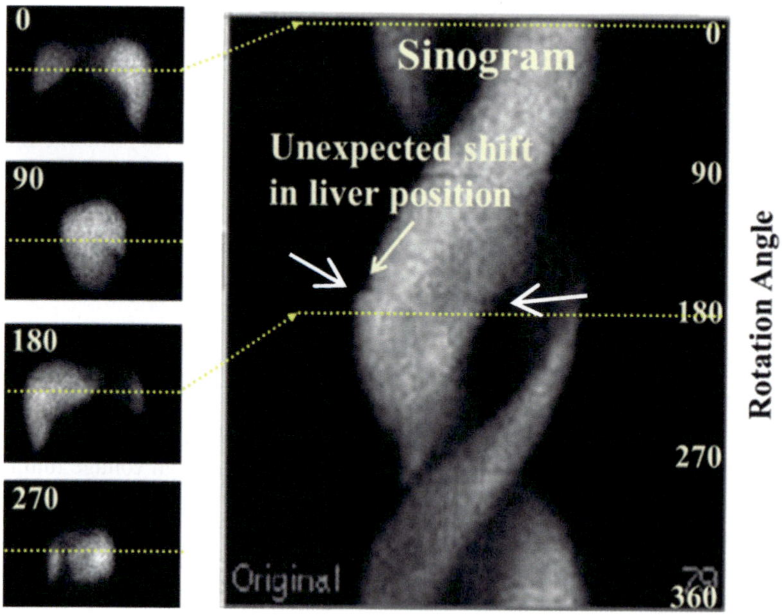

87. The above figure shows raw image data and a sinogram for a SPECT acquisition of ^{99m}Tc-sulfur colloid in the liver and spleen. The top half of the sinogram is projection data from detector 1 at acquisition angles of 0–180, and the bottom half is from detector 2 at acquisition angles of 180–360 degrees. What is the cause of the unexpected shift in the position of the liver shown?

(a) Center-of-rotation misregistration between the two detector heads
(b) The patient moved at the halfway point of the SPECT acquisition
(c) Mismatch in the pixel size of the detectors
(d) Multiple-window misregistration in the detectors

Correct answer: (a) Center-of-rotation misregistration between the two detector heads. It is required that both detectors be registered to the same axes. This is accomplished by performing a center-of-rotation calibration. The calibration requires SPECT imaging of small ^{99m}Tc sources placed on the imaging pallet. Patient motion would be observed as discontinuities in the projected liver in detector rotations between 0–180 and 180–360. A mismatch in the detector pixels sizes would result in the mismatch in the width of the liver projections between the detectors on the sinogram. Multiple-window misregistration is an energy misregistration artifact when multiple-energy isotopes are imaged.

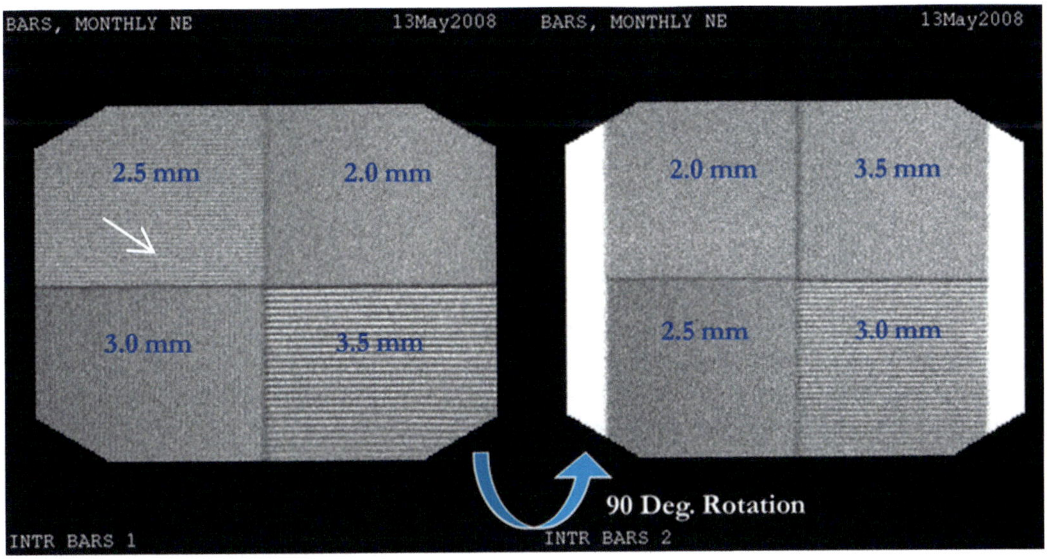

88. Quadrant bar images of ^{99m}Tc shown above were acquired on the same gamma camera detector with the second image (INTR BARS 2) obtained after the bar phantom had been rotated counterclockwise by 90°. The width of the bars is shown in the images. What is the cause of the observed difference in the spatial resolution between these images?

(a) Problem with detector spatial linearity
(b) The higher count rate at the left and right edges in the second image causes a loss in spatial resolution
(c) The spatial resolution in the × (horizontal direction) is absent
(d) Multiple-window misregistration

Correct answer: (c) The spatial resolution in the × (horizontal direction) is absent. Note that on the first image the 3.0 mm bars are absent but seen after 90° rotation in image 2. Likewise, the 3.5 mm bars are seen in image 1 but not in image 2 after 90° rotation. Both cases lead to the conclusion that the spatial resolution along the *x*-axis is absent. Spatial linearity affects the linear shape of the bars, but not their visibility. The higher count rate in image 2 could possibly affect spatial resolution but a loss in resolution is also observed in image 1. Multiple-window misregistration is an energy misregistration artifact when multiple-energy isotopes are imaged.

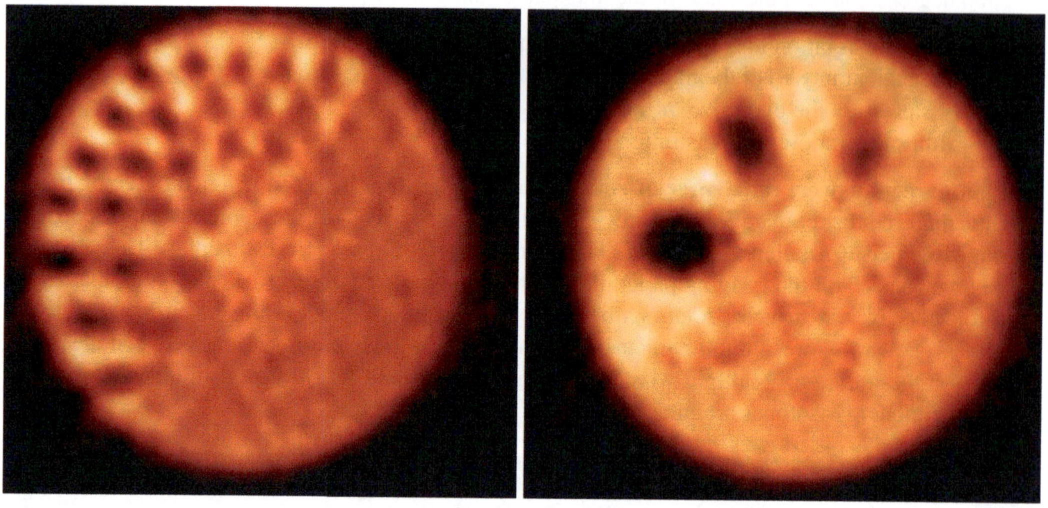

89. Why do cylinders and spheres observed in the SPECT phantom slices above appear elliptical in shape?

(a) Use of the filtered backprojection reconstruction algorithm
(b) Center-of-rotation error
(c) Presence of gamma camera detector nonuniformities
(d) Result of a cardiac 180° acquisition orbit.

Correct answer: (d) Result of a cardiac 180° acquisition orbit. Theoretically only a 180° acquisition orbit is needed, but as such leads to the artifacts observed. Attenuation and resolution changes at depth are not fully measured with a limited 180 acquisition leading to the artifact. A full 360° acquisition will result in properly circular shaped objects within the phantom. For myocardial perfusion imaging however the heart lies very anterior in the chest and a 180° orbit has been shown to provide a better result. Note also for cardiac imaging that the acquisition angle covers an arc from −135 to 45° centered over the heart. The direction of the elliptical artifacts is oriented as such. Error in the center-of-rotation will result in an overall blurring of the objects. Gamma camera detector nonuniformities lead to ring artifacts. The filtered backprojection algorithm adds streak artifacts to the images but does alter the shapes of reconstructed objects.

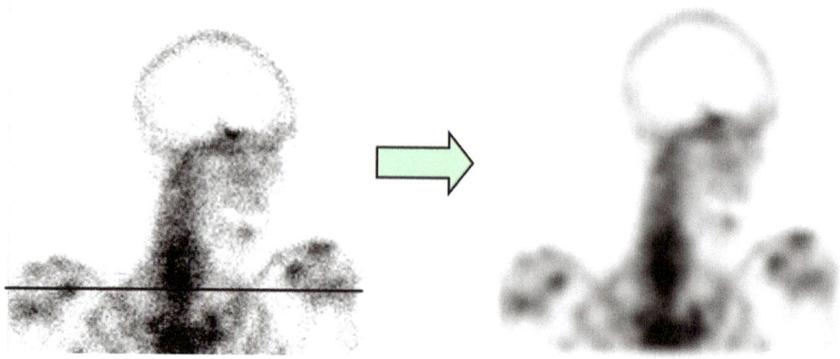

90. The original gamma camera image on the left (ignoring the horizontal line) was processed with what type of filter to achieve the image on the right?

(a) Ramp filter
(b) Butterworth filter
(c) Gaussian filter
(d) Wiener filter

Correct answer: (b) Butterworth filter. The Butterworth filter is a low-pass frequency filter that leads to smoothing of the image. Its benefit is to reduce the grainy noise appearance. The ramp, Gaussian, and Wiener filters are high-pass frequency filters used to enhance and sharpen the image detail.

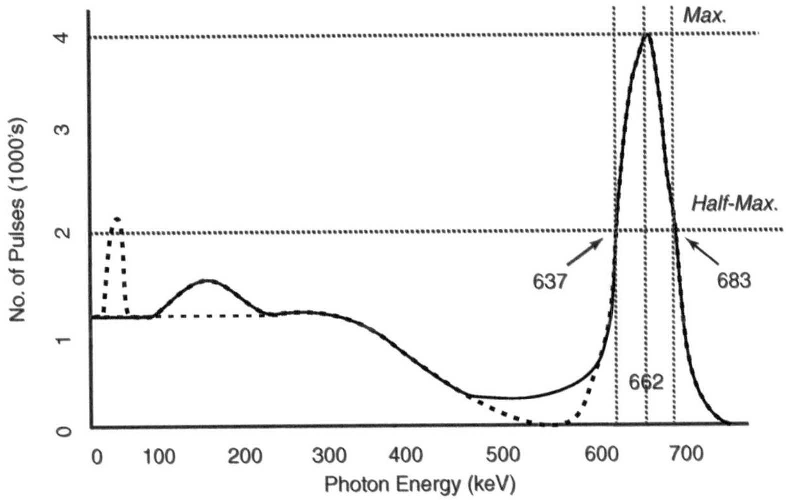

91. What is the ful0l-width-half maximum (FWHM) as a percentage of the photopeak energy of ^{137}Cs in the energy spectrum shown above?

(a) 1%
(b) 7%
(c) 20%
(d) Insufficient information to determine

Correct answer: (b) 7%. The FWHM = 100% (683–637)/662 = 7%. The FWHM is a measure of the energy resolution of the gamma detector and its ability to separate gamma photons from scattered photons.

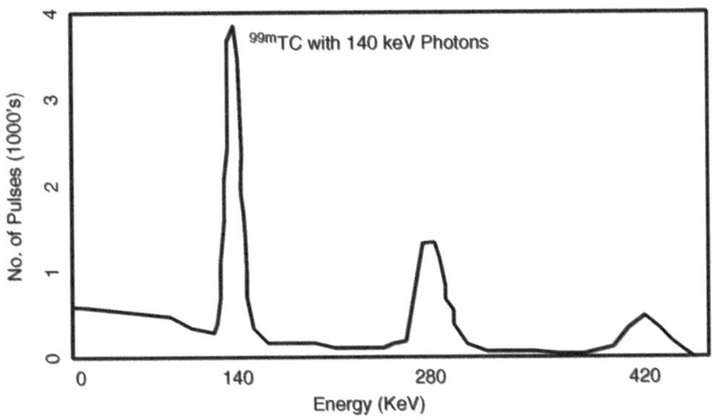

92. Under what condition does the energy peak centered at 280 keV in the energy spectrum of ^{99m}Tc shown above occur?

(a) Very high count rate
(b) Very high scattering
(c) Very low count rate
(d) Defective hydrated NaI detector crystal

Correct answer: (a) Very high count rate. At very high count rates, the likelihood of simultaneous photon interactions becomes prevalent. Two simultaneous interactions of 140 keV gamma photons show in a second sum peak at 280 keV. Note, also, that the 420 keV sum peak corresponds to the simultaneous interaction of three gamma photons of ^{99m}Tc. These are identified as sum peaks. For proper count rate measurements, the sum peaks should be absent and is the case at much lower count rates. Scattering would not contribute to these sum peaks, but do contribute to the recording of energy events in between the 140, 280, and 420 keV sum peaks. Defective crystals lead to an overall decrease in the energy signals detected.

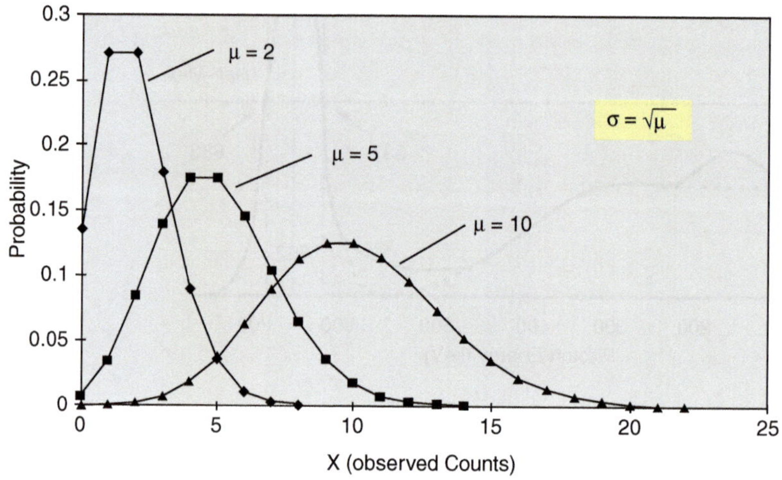

93. What type of probability distribution function do the curves above demonstrate?

(a) Gaussian
(b) binomial
(c) chi square
(d) Poisson

Correct answer: (d) Poisson. The Poisson distribution describes the statistical variability in counting gamma rays from a radioisotope. Its shape varies, skewed to the right, depending on the distribution mean value. The standard deviation of the distribution is always the square root of the distribution mean value. As the mean becomes larger the Poisson distribution becomes more symmetrical approaching the characteristics of the Gaussian distribution.

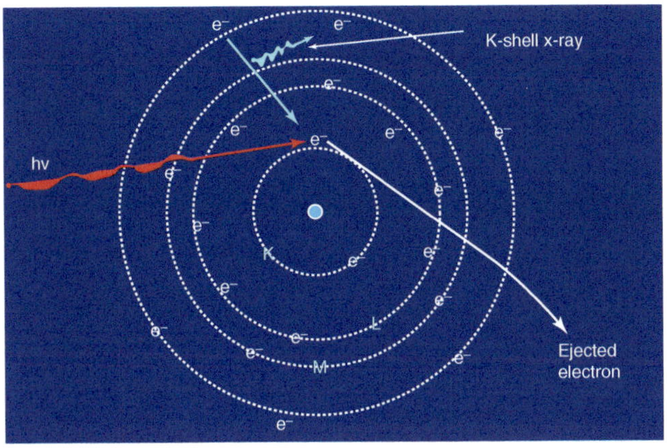

94. The diagram above illustrates which photon (hν) interaction in matter?

(a) Thompson scattering
(b) Photoelectric
(c) Compton scattering
(d) Bremsstrahlung

Correct answer: (b) Photoelectric. The photoelectric interaction targets k-shell electrons of atoms in the absorbing medium. All the photon energy is transferred to the k-shell electron and it is ejected. An outer shell electron in the atom moves in to fill the k-shell vacancy causing a k-shell characteristic X-ray to be emitted.

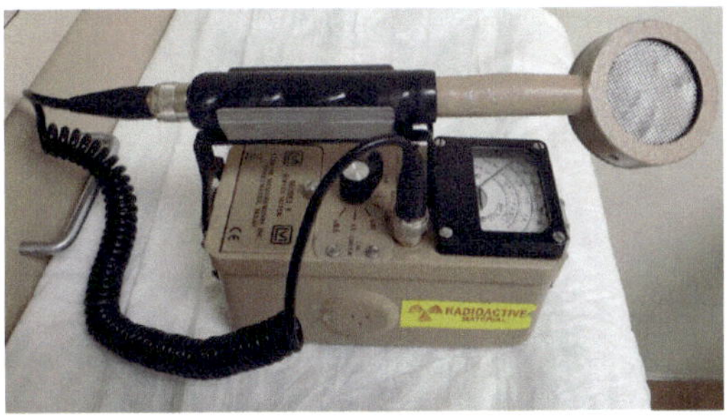

95. What type of radiation detector is the picture above?

(a) Gamma counter
(b) Dose calibrator
(c) Beta counter
(d) Geiger counter

Correct answer: (d) Geiger counter. The Geiger counter has a handheld wand that is used to search a work area for radioactive contamination. It is sensitive to both gamma and beta emissions.

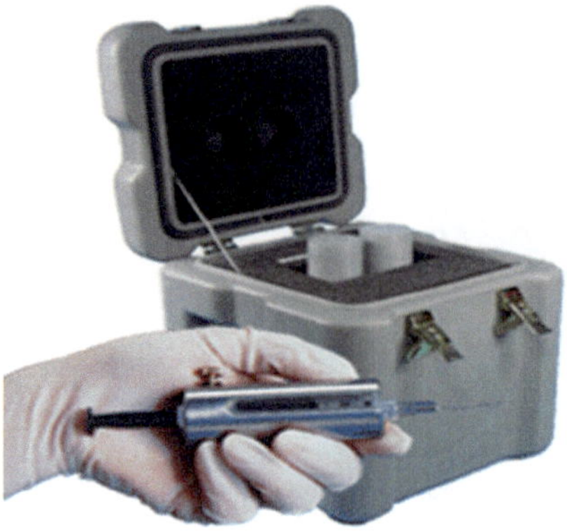

96. What shielding material in the syringe shield pictured above is preferred for handling PET radioisotopes?

    (a) lead
    (b) tungsten
    (c) plastic
    (d) gold

    Correct answer: (b) tungsten. Tungsten has a density of 18 g/cc, as compared to lead with 11 g/cc. Therefore, for the same thickness of the shield, tungsten is more effective than lead. Lead is a much more effective shield for lower energy gammas used in nuclear medicine. Plastic is an effective shield for purely beta emitting radioisotopes. Gold is also a very effective gamma shield but too expensive to use commonly.

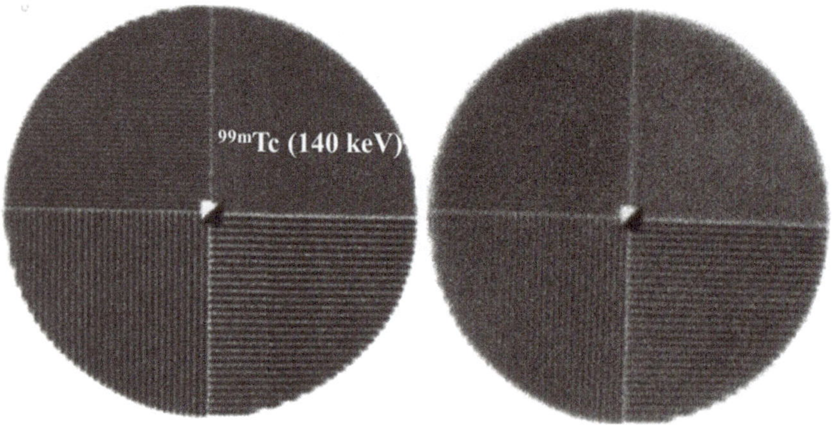

97. The image on the left is an intrinsic quadrant bar phantom image using ^{99m}Tc. The image on the right is using the same gamma camera and quadrant bar phantom, only the radioisotope was changed? What radioisotope was used?

    (a) ^{201}Tl
    (b) ^{131}I
    (c) ^{18}F
    (d) ^{111}In

    Correct answer: (a) ^{201}Tl. The primary ^{201}Tl photon emission is 70 keV characteristic X-ray. The energy signals generated in the gamma camera crystal are half that of 140 keV photons. As a result, the position circuitry in the gamma camera is less accurate and leads to images of less spatial resolution. This is one reason for moving away from Tl-201 for myocardial perfusion imaging. The second reason is a larger scatter fraction for low-energy photons. The third reason is higher radiation dose to the patient. All the other radioisotopes listed have gamma photon emission larger than 140 keV. There is an expected increase in spatial resolution, intrinsically. However, needed medium and high-energy collimators, ultimately decrease the spatial resolution when imaging patients.

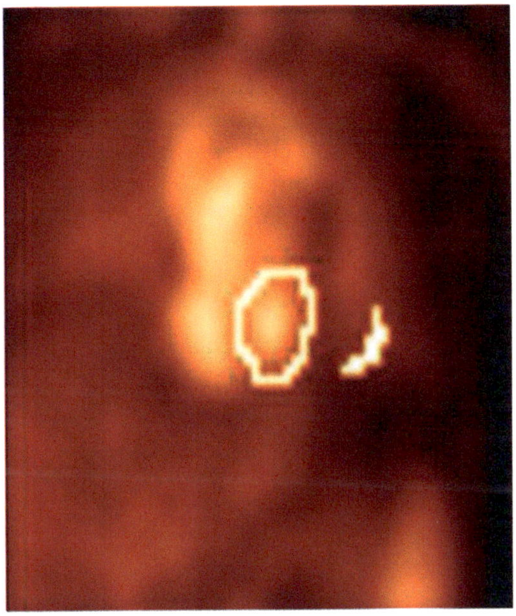

98. The left ventricle and background ROIs are shown in the above image in processing a MUGA study. The background ROI is incorrectly placed over the area of the descending aorta. How does this placement affect the ejection fraction calculation?

(a) The ejection fraction is not affected
(b) The ejection fraction is artifactually low
(c) The ejection fraction is artifactually high
(d) The ejection fraction becomes negative

Correct answer: (c) The ejection fraction is artifactually high. The ejection fraction = 100%*(end diastolic counts – end systolic counts)/(end diastolic counts – background counts). The background counts are overestimated by placing the background ROI over a vascular structure, such as the descending aorta, and therefore the ejection is artifactually high.

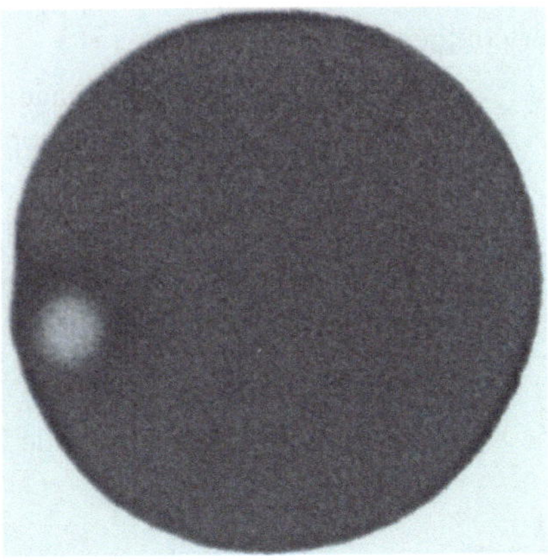

99. What step(s) should be taken to correct the artifact observed in the above flood image with ^{99m}Tc?

(a) Remove the gamma camera from service and inform the service engineer of a nonfunctioning photomultiplier tube
(b) Perform a uniformity calibration for ^{99m}Tc
(c) Check to see if the artifact remains after installing a low-energy collimator
(d) Increase the strength of the flood source and acquire a new flood and check to see if the artifact remains.

Correct answer: (a) Remove the gamma camera from service and inform the service engineer of a nonfunctioning photomultiplier tube. This artifact is definitely the result of a nonfunctioning photomultiplier. The photomultiplier tube can only be removed and replaced by a service engineer. A uniformity calibration may improve the artifact but will not remove it. The artifact will remain after installing a collimator or changing the strength of the ^{99m}Tc source.

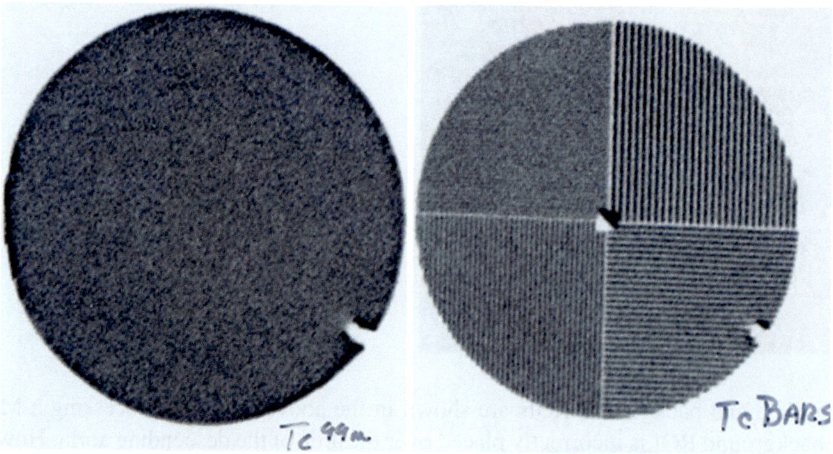

100. What is the cause of the artifact observed in the intrinsic QC images above?

(a) Nonfunctioning photomultiplier tube
(b) Fractured crystal
(c) Damaged collimator
(d) Dense absorbing object set on top of the crystal

Correct answer: (b) Fractured crystal. This artifact is the result of a crystal fracture limited to the crystal edge. It was created by a piece of hardware (screw) left on top of the crystal while a collimator was being installed. The screw was impaled into the crystal. An artifact from a damaged collimator or of a dense object on the camera would look similar except for the absence of the high count density observed all along the periphery of the artifact.

Radiopharmacy

Stephen Karesh and Medhat Sam Gabriel

Radiopharmacy

Part 1: Physics of Nuclear Medicine: Atomic Structure, Radioactive Decay, Interaction of Ionizing Radiation with Matter

1. An atomic nucleus contains 39 protons and 50 neutrons. Its mass number (A) is

   (a) 11
   (b) 39
   (c) 50
   (d) 89

   Correct answer is (d). The mass number is determined by the addition of the number of protons and neutrons in the nucleus. Collectively they are referred to as nucleons.

2. In standard notation, one of the isotopes of bromine is Br_{35}. How many neutrons does this nucleus contain?

   (a) 35
   (b) 38
   (c) 73
   (d) 108

   Correct answer is (a). The number of neutrons is determined by subtracting the number of protons from the mass number. N = A – Z

S. Karesh
Department of Nuclear Medicine, Loyola University Medical Center, Maywood, IL, USA

M. S. Gabriel (✉)
Division of Nuclear Medicine, Department of Radiology, Loyola University Medical Center, Maywood, IL, USA
e-mail: sgabriel@lumc.edu

M. S. Gabriel, B. Savir-Baruch (eds.), *RadTool Nuclear Medicine MCQs*,
https://doi.org/10.1007/978-3-030-69281-0_2

(c) isomeric transition
(d) photoelectric effect

Correct answer is (c). Decay of Tc-99m to Tc-99g (ground state) is a perfect example of isomeric transition. An excited Tc-99m atom de-excites itself and emits a gamma-ray. The A, Z, and N numbers remain unchanged; only the energy level has been reduced. The ground state has a different half-life (2.2×10^5 years) than the metastable state (6 h).

13. After an atom has decayed by giving off a negative beta particle, the remaining atom is:

(a) An atom of a new element having an atomic number one higher than the old and with no or little change in mass number.
(b) An atom of a new element having an atomic number one higher than the old and with a significant lowering of mass number.
(c) An atom of a new element having an atomic number one less than the old and with no or little change in mass number.
(d) Unchanged, except it has now become stable.

Correct answer is (a). In beta-minus decay, the Z number of the daughter always increases by one unit to counterbalance the −1 charge on the emitted electron. There is only a negligible change in mass. The generic equation representing this decay is shown below:

A		A		0				
X	----→	Y	+	e	+	ν	+	Energy
N		Z	N-1	Z+1		-1		

14. Internal conversion is most similar diagrammatically to:

(a) isomeric transition
(b) compton effect
(c) pair production
(d) photoelectric effect

Correct answer is (d). In reference to the diagram, both internal conversion (Fig. 1) and photoelectric effect (Fig. 2) involve the removal of an electron from an orbital in a totally inelastic collision. There is no residual gamma-ray energy. The vacancy is filled by a higher-shell electron, releasing energy in the form of characteristic X-rays. If additional electrons are knocked out of their orbits by these secondary radiations, they are called *Auger electrons*. Isomeric transition is a decay mode and not an interaction with matter; pair production involves the conversion of a gamma-ray with E > 1.02 MeV into a pair of electrons, e^+ and e^-. Compton Effect is a partially elastic collision with a residual gamma-ray that can interact with matter.

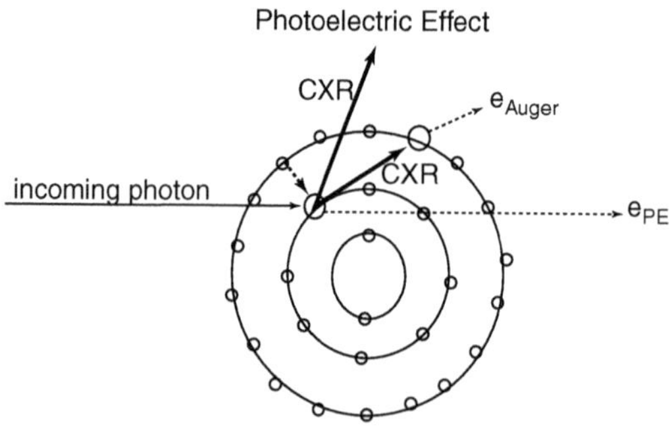

Fig. 1 Auger electrons are ejected by the secondaryprocess rather than the primary process

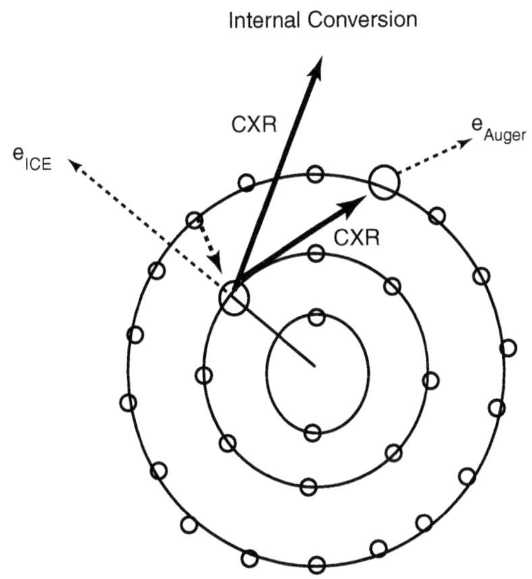

Fig. 2 Auger electrons are ejected by the secondaryprocess rather than the primary process

15. The formula $E_e = E_\gamma - BE$ describes which of the following interactions with matter?

    (a) compton effect
    (b) photoelectric effect
    (c) pair production
    (d) isomeric transition
    (e) none of the above

    Correct answer is (b). This formula best describes the photoelectric effect, in which the energy of the gamma-ray that strikes the electron is converted into the kinetic energy of the ejected photoelectron. The formula means that the energy of the electron equals the energy of the incident photon—the binding energy holding the electron in its orbit. This is considered to be a totally inelastic collision.

16. Pair production takes place

    (a) only in the presence of nuclei
    (b) only in a vacuum
    (c) only in solids
    (d) none of the above

    Correct answer is (a). Pair production only takes place in the presence of nuclei. It is an interaction in matter in which the incident gamma-ray with energy >1.02 MeV must impinge on a nuclear field. While it is more likely to occur in solids and less likely in gases, all that is required is the presence of a nucleus. Pair production will not take place in a vacuum.

17. The literature value for the β-energy of Ca^{45} is 0.255 MeV. This represents

    (a) the mean energy
    (b) the median energy
    (c) the maximum energy
    (d) the minimum energy

Correct answer is (c). By convention, physicists consider the literature value for energy of β^- emitting nuclides to be the maximum energy.

18. A true statement regarding β^+ emitters is:

(a) Positrons are imaged following their interaction in a sodium iodide crystal
(b) β^+ emitting nuclides have previously undergone pair production
(c) When a β^+ at its rest mass interacts with matter, it loses its energy by photoelectric and Compton effects
(d) The process that competes with β^+ emission is called electron capture

Correct answer is (d). Choice (a) is incorrect because positrons do not interact in a NaI crystal, but rather in the patient. Therefore, they are not imaged; their annihilation energy is imaged. Choice (b) is incorrect because beta-plus emitters do not undergo pair production; only high-energy photons do. Choice (c) is incorrect because a positron at rest mass interacts with matter by annihilating an electron.

19. Which of the following processes occurs spontaneously in an effort to reduce the N/P ratio?

(a) β^- emission
(b) β^+ emission
(c) α^{2+} emission
(d) α^{2-} emission

Correct answer is (a). Beta-minus emission is the only mode of decay in which the Z number of the daughter increases relative to the parent. Choices (b) and (c) are incorrect because in those decay modes, the Z number decreases. Choice d is incorrect because there is no such thing as an α^{2-} particle.

20. Which of the following is an example of specific activity?

(a) mg/Ci
(b) Ci/μmole
(c) counts/μmole
(d) none of the above

Correct answer is (b). Specific activity is defined as **A/m** (Activity/unit mass). Numerator must be directly relatable to disintegrations. Examples are Ci/μmole or Bq/mole. None of the other answers is expressed as **A/m**.

21. Which of the following accurately defines a millicurie?

(a) 3.7×10^7 counts/s
(b) 3.7×10^7 counts/min
(c) 3.7×10^7 disintegrations/s
(d) 3.7×10^7 disintegrations/min

Correct answer is (c). A mCi is 10^{-3} Ci; a Ci is defined as 3.7×10^{10} disintegrations/s so a mCi = 3.7×10^7 disintegrations/s.

22. Which of the following is correct when F-18 FDG decays?

(a) The atomic number (Z) increases.
(b) An electron is emitted.

(c) The mass of the daughter is higher than the parent.
(d) 0.511 MeV photons are emitted.

Correct answer is (d). F-18 FDG decays by positron emission. 511 keV photons are emitted in all positron decays. The Z number of the daughter **decreases** by one unit in this decay mode and a neutrino is emitted, not an electron. The mass of the daughter is always **less** than that of the parent.

23. Rank the following radionuclides in order of increasing gamma ray energy.

1. I-123
2. I-125
3. I-131
4. Tc-99 m

(a) I-123, I-125, I-131, Tc-99m
(b) I-123, Tc-99m, I-125, I-131
(c) I-125, Tc-99m, I-123, I-131
(d) Tc-99m, I-125, I-123, I-131,

Correct answer is (c). I-125 (28, 35 keV) < Tc-99m (140 keV) < I-123 (159 keV) < I-131 (364 keV).

Part 2: Mathematical Problem Solving

1. A vial contains 100 mCi of an unidentified radionuclide, at 10 a.m. on Thursday. At 10 a.m. on Friday, the activity has decayed to 25 mCi. What is the half-life of this radionuclide?

(a) 24 h
(b) 12 h
(c) 4 h
(d) 100 h

Correct answer is (b). Simply, 100 mCi to 50 mCi to 25 mCi requires two half-lives. If 24 h = 2 half-lives, then the half-life = 12 h,
$A_t = A_0 \times 0.5^{(\text{elapsed time/half-life})}$, therefore, $25 = 100 \times 0.5^{(24\ \text{h/t})}$ and
$5 = 0.5^{(24/t)}$. Taking the log of both sides,
$\log 0.25 = 24/t \times \log 0.5$, therefore, $t = \frac{24 \times \log 0.5}{\log 0.25}$ so half-life = 12 h

2. A vial contains a mixture of 20 mCi of I^{124} (half-life 4 days) and 6 mCi of I-131 (half-life 8 days). What will be the activity in the vial 8 days from now?

(a) 3 mCi
(b) 5 mCi
(c) 8 mCi
(d) 13 mCi

Correct answer is (c).
Activity = $(20 \times 0.5^{8/4}) + (6 \times 0.5^{8/8})$
= 5 + 3 = 8 mCi
so **total activity = 8 mCi**

3. For a T_{phys} = 1 year and a T_{biol} = 10 years, what is the T_{eff}

   (a) about 10 years
   (b) about 9 years
   (c) about 4 years
   (d) about 0.9 year

   Correct answer is (d).
   $I/t_{eff} = 1/t_{phys} + 1/t_{biol}$
   $= 1/1 + 1/10 = 1.10$ therefore
   $t_{eff} = 1/1.10$ So
   $\mathbf{t_{eff} = 0.909}$
   Also, by definition, T_{eff} must be less than T_{biol} or T_{phys}, and therefore the only correct option is (d) (less than the physical ½ life of 1 year).

4. A collimated beam with 3 half-value layers of lead in front of it gives a reading of 10 mR/h. If 1 half-value layer is removed, what would be the intensity reading?

   (a) 5 mR/h
   (b) 20 mR/h
   (c) 30 mR/h
   (d) 40 mR/h

   Correct answer is (b). Any time one half-value layer is removed from between the source and the detector, the intensity reading doubles. Since I_0 is 10 mR/h, the new **reading will be 20 mR/h**

5. A collimated beam has initial intensity I_1 = 10 mR/h. What would be the intensity I_2 if 3 half-value layers of lead (density 13.6 g/cm^3) were placed between source and detector?

   (a) 3.33 mR/h
   (b) 1.25 mR/h
   (c) 0.7 mR/h
   (d) 0.1 mR/h

   Correct answer is (b). $I = I_0 \times 0.5^3 = 10$ mR/h $\times$ 0.125 so **I = 1.25 mR/h.**

6. Refer to the diagram and assume an uncollimated point source at point A. Calculate the value for I_2

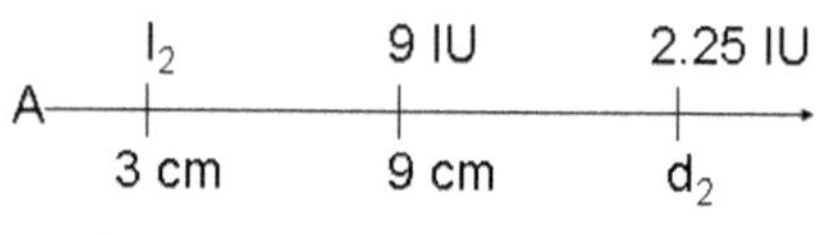

   (a) 9 IU
   (b) 81 IU
   (c) 27 IU
   (d) 243 IU

   Correct answer is (b). By inverse square law, 9 IU $\times$ 9^2 cm^2 = $I_2 \times 3^2$ cm^2 so $\mathbf{I_2}$ **= 81 mR/h**

7. Refer to the same diagram and calculate the value for d_2.

   (a) 16 cm
   (b) 18 cm
   (c) 22.5 cm
   (d) 324 cm

Correct answer is (b). By inverse square law, 9 IU × 9^2 cm^2 = 2.25 × d^2 so **d = 18 cm**
Please note that this is not inverse law; it is an inverse square law

8. A1 Ci source of Tc-99m was left to decay in the hot lab overnight. The next time the technologist checked it, the activity was only 62.5 mCi. How many half-lives have elapsed?

(a) three
(b) four
(c) five
(d) can not be determined from data provided

Correct answer is (b).
Fraction remaining = 62.5/1000 = 0.0625 = 0.5^n where n = # of half-lives. Taking the log of both sides, log 0.0625 = n × log 0.5 and
n = log 0.0625/log 0.5
therefore **n = 4**
Logically, the activity would decrease from 1000 to 500 to 250 to 125 to 62.5 mCi. That implies that 4 half-lives have elapsed.

9. For a T_{phys} = 56 days and a T_{eff} = 14 days what is T_{biol}

(a) about 14 days
(b) about 19 days
(c) about 28 days
(d) about 42 days

Correct answer is (d).
$1/t_{biol} = 1/t_{eff} - 1/t_{phys}$
= 1/14–1/56
= 0.0714–0.0178 = 0.0536
therefore, t_{biol} = 1/0.0536 and
t_{biol} = 18.66 days or about 19 days

10. If t_{phys} of Ga-67 is 3 days and t_{eff} of Ga-citrate = 1.5 days, what is t_{biol}?

(a) 1.5 days
(b) 3 days
(c) 4.5 days
(d) 6 days

Correct answer is (b). This is a special case in which t_{eff} = ½ t_{phys}. In this case, t_{biol} will equal t_{phys} (3 days).
Mathematically,
$1/t_{biol} = 1/t_{eff} - 1/t_{phys}$ so
$1/t_{biol}$= 1/1.5–1/3 = 2/3–1/3 =1/3
Therefore, **t_{biol} = 3 days**

11. At 10 cm from a point source, the intensity is 1000 R/h. At what distance from the source would the intensity be reduced to 1000 mR/h?

(a) 1000 cm
(b) 500 cm
(c) 100 cm
(d) 50 cm

Correct answer is (c). By inverse square law, $1000\ R/h \times 10^2\ cm^2 = 1\ R/h \times d^2$ therefore $d^2 = 10{,}000\ cm^2$ and **d = 100 cm**

12. For Tc-99 m, what fraction remains after 18 h?

(a) 6/18
(b) $(6/18)^2$
(c) $(0.5)^{6/18}$
(d) $(0.5)^{18/6}$

Correct answer is (d).
Fraction remaining = $0.5^{\#\text{ of half-lives}} = 0.5^{(18/6)} = (0.5)^3 = 0.125$

13. The first half-value layer (HVL) for a polychromatic beam is 3 mm. of Aluminum. The second HVL is ______________ the first.

(a) markedly less than
(b) slightly less than
(c) equal to
(d) greater than

Correct answer is (d). For a polychromatic beam, the first half-value layer reduces the original intensity reading to ½ its value. In doing so, it eliminates essentially all the low-energy emissions, but only a small fraction of the high energy emissions. The net effect is to raise the average energy of the beam. This is called *beam hardening*. The second half-value layer therefore, has to be thicker than the first since the average energy of the beam has increased

14. A source has an intensity of 4 mR/h. A barrier 4.5 cm thick is placed between the source and the detector and the reading drops to 0.5 mR/h. What is the half-value layer of the absorber?

(a) 0.5 cm
(b) 1.0 cm
(c) 1.5 cm
(d) 2 cm

Correct answer is (b).
$I = I_0 \times 0.5^{\#\text{ of HVL's}}$
$= I_0 \times 0.5^{\text{thickness/HVL}}$ therefore
$0.5\ mR/h = 4.0\ mR/h \times 0.5^{4.5cm/HVL}$ and
$0.125 = 0.5^{4.5cm/HVL}$. Taking log of both sides,
log 0.125 = (4.5 cm/HVL) × log 0.5). Rearranging,
HVL = 4.5 cm × log 0.5/log 0.125 and
HVL = 1.5 cm
Simply, for the intensity to decrease from 4 to 0.5 mR/h, it needs to go through 3 HLVs (4 → 2 → 1 → 0.5). The barrier is 4.5 cm thick; so, the HLV is 4.5/3=1.5 cm

15. If a certain ^{131}I compound has a t_{biol} of 24 days, what is the effective half-life?

(a) 8 days
(b) 7 days
(c) 6 days
(d) 5 days

Correct answer is (c).
$I/t_{eff} = 1/t_{phys} + 1/t_{biol}$
$= 1/24 + 1/8 = 4/26$ or $1/6$.
Therefore, $t_{eff} = 6$ days.

16. A source of 5 mCi of Tc-99 m is placed in front of a detector. If the detector efficiency is 40%, how many counts/s can be detected?

(a) 3.7×10^8 counts
(b) 7.4×10^8 counts
(c) 3.7×10^7 counts
(d) 7.4×10^7 counts

Correct answer is (d).
1 mCi gives rise to 3.7×10^7 dis/s
The source activity = $5 \times (3.7 \times 10^7$ dis/s)
Counts detected are only 40% or $0.4 \times 5 \times (3.7 \times 10^7$ counts/s) = $2 \times (3.7 \times 10^7$ counts/s) or 7.4×10^7 counts/s

17. Two camera heads are alike except that one has a 10-inch diameter scintillation crystal and the other has a 15-inch diameter detector. The collimators for the detectors are identical in all parameters (length, focal length, etc). The counting rate obtained with the 15-inch head from an extended source compared to that obtained with the 10-inch head with comparable collimation is:

(a) 1.5 times as great
(b) 0.667 times as great
(c) 2.25 times as great
(d) the same

Correct answer is (c).
Ratio of $\frac{\text{counts from 15 inch crystal}}{\text{counts from 10 inch crystal}} = \frac{15^2}{10^2}$
therefore **ratio = 2.25**

18. A radioactive source has a dose rate of 40 mR/h at a distance of one foot. At what distance from this source would a technician working 40 h/week for 50 weeks/year receive a total dose of 5000 mR?

(a) 10 feet
(b) 6 feet
(c) 4 feet
(d) 2 feet

Correct answer is (c).
40 h/week × 50 weeks/year = 2000 h/year. Dose rate received is 5000 mR/2000 h = 2.5 mR/h. By inverse square law,
$40 \text{ mR/h} \times (1 \text{ ft})^2 = 2.5 \text{ mR/h} \times d^2$.
$d^2 = 16 \text{ ft}^2$ and therefore
$d = 4$ ft

Part 3: Statistics: Concepts and Problems

1. Which of the following tests are performed periodically on a radioisotope dose calibrator?

   (a) accuracy, precision, constancy
   (b) precision, reliability, constancy
   (c) accuracy, constancy, reliability
   (d) accuracy, constancy, linearity

   Correct answer is (d). These are the three tests prescribed by law. Accuracy determines the machine's ability to yield correct value for a series of standards; constancy is really a measure of the machine's precision; linearity determines its ability to properly calibrate doses from the microcurie range to the millicurie range.

2. A 3 min observation is made from which the count rate is found to be 300 counts/min. The standard deviation of this count rate is:

   (a) 30
   (b) 20
   (c) 15
   (d) 10

   Correct answer is (d).
   300 c/min × 3 min = 900 cts collected.

   $\sigma = \pm \sqrt{900} = 30$ cts
   since counted for 3 min,
   σ of the count rate = 30/3 so
   σ = 10 c/min

Part 4: Equilibrium and Generators

1. The Tc-99 m eluted from the Mo-99/Tc-99 m generator is in the chemical form of:

   (a) Na pertechnetate $(TcO_4)^{-1}$
   (b) HR Tc, e.g., $TcO(OH)_2.(H_2O)$
   (c) Technetium hydroxide, e.g., $Tc(OH)_4$
   (d) none of the above

   Correct answer is (a). Na pertechnetate is the most highly oxidized chemical form of Tc, in which the oxidation number is 7^+. Choice **(b)** is incorrect—Hydrolyzed reduced Tc is an insoluble colloid localizing in the RES. Also, Technetium hydroxide represents another insoluble form.

2. The parent/daughter relationship in a Mo/Tc generator is a typical example of

   (a) no stability
   (b) secular equilibrium
   (c) transient equilibrium
   (d) no equilibrium

   Correct answer is (c). Transient equilibrium is reached when the parent's half-life is approximately 10 times that of the daughter. The half-life of Mo-99 is 67 h or about 11 times that of

Tc-99m (6 h). The Mo/Tc generator is considered an ideal example of transient equilibrium. The requirement for secular equilibrium, on the other hand, is that the half-life of the parent is far greater than the half-life of the daughter, e.g., 100:1 or more.

3. Which one of the following is true for **transient equilibrium**:

   (a) at equilibrium, the physical half-life of the daughter equals the physical half-life of the parent
   (b) at equilibrium, the physical half-life of the daughter is 10 times as great as the physical half-life of the parent
   (c) at equilibrium, the apparent half-life of the parent equals the physical half-life of the daughter
   (d) at equilibrium, the apparent half-life of the daughter equals the physical half-life of the parent

   Correct answer is (d). The daughter can not decay until it is formed. Since the rate of formation of the daughter is slow, in the equilibrium mixture, the daughter will appear to decay with the parent's half-life. When separated from the mixture (e.g., when the generator is eluted), the daughter decays with its characteristic half-life.

4. The Mo/Tc generator is eluted with which of the following?

   (a) 0.001 M hydrochloric acid
   (b) sterile distilled water
   (c) normal saline
   (d) 5% glucose

   Correct answer is (c). Elution of the Mo/Tc generator is done by using normal physiological saline (0.9% NaCl). It is compatible with blood and the eluate requires no pH adjustment.

5. After a typical Mo-99/Tc-99m generator elution, the ingrowing Tc-99m reaches a maximum

   (a) at about 2 h
   (b) between 2 and 6 h
   (c) between 6 and 12 h
   (d) between 12 and 24 h

   Correct answer is (d). The actual time required to reach maximum activity level in the Mo/Tc generator is 23 h. This is an ideal equilibrium time since the generator is typically eluted every 24 h, right after equilibrium is reached. This maximizes the yield of pertechnetate removed from the generator.

6. In an equilibrium mixture of parent Mo-99 and daughter Tc-99m, the ratio of Tc-99m activity to Mo-99 activity could reach:

   (a) about 2
   (b) about 1.7
   (c) about 1.4
   (d) about 1.05

   Correct answer is (c). The activity ratio of daughter to parent in transient equilibrium is typically 0.8–1.05. The number of daughter atoms cannot exceed that of the parent, but the activity level can do so, especially if the half-lives are close to one another.

5. A bone scan reveals bone to soft tissue ratio of 5:1 and lesion to bone ratio of 5:1. What is the ratio of lesion to soft tissue?

   (a) 1:1
   (b) 5:1
   (c) 10:1
   (d) 25:1

   Correct answer is (d). $\frac{5\ \text{bone}}{1\ \text{soft tissue}} \times \frac{5\ \text{lesion}}{1\ \text{bone}} = \frac{25\ \text{lesion}}{1\ \text{soft tissue}}$

6. In adult population, what % of the injected dose of Tc-99 m bone agents localizes in bone and what % is excreted through the kidneys? Assume normal renal function.

   (a) 30%/70%
   (b) 50%/50%
   (c) 70%/30%
   (d) 90%/10%

   Correct answer is (b). Approximately 45–50% localizes in bones, 45–50% is cleared through the kidneys, and the remainder constitutes activity in soft tissue. In the pediatric population, about 70% will localize in the skeleton.

7. A patient scheduled for an RAIU is taking 100 μg of Synthroid (Levothytoxine) daily. How long must the patient be off Synthroid to obtain a valid RAIU?

   (a) 1 day
   (b) 1 week
   (c) 2 weeks
   (d) 4 weeks

   Correct answer is (d). RAIU is invalid if Synthroid has been administered 4 weeks prior to the procedure.

Part 6: Quality Control and Regulatory Issues

1. A female technologist enters your office and announces to you that she is pregnant and gives you the estimated date of conception. Which one of the following statements is correct?

   (a) She is now a "DECLARED PREGNANT WORKER"
   (b) She did not provide all required information for being categorized as a "DECLARED PREGNANT WORKER"
   (c) The technologist provided all required information for being categorized as a "DECLARED PREGNANT WORKER," but it was not in writing
   (d) The category of "DECLARED PREGNANT WORKER" ends 6 months after it becomes official

   Correct answer (c). According to the current NRC regulations, notification of her status **must be** in writing. It terminates when she is no longer pregnant.

2. What are the dose limits to the embryo/fetus for a woman who meets the requirements of a declared pregnant worker?

   (a) Dose to the embryo/fetus over the entire period of gestation (9 months) is 0.05 Rem
   (b) Dose to the embryo/fetus over the entire period of gestation (9 months) is 0.5 Rem

(c) Dose to the embryo/fetus over the entire period of gestation (9 months) is 5.0 Rem
(d) Dose to the embryo/fetus over the entire period of gestation (9 months) is 50 Rem

Correct answer is (b). According to the current NRC regulations, the dose to the embryo/fetus over the entire period of gestation (9 months) is 0.5 rem and should be delivered at a fairly uniform rate over the entire gestation period, not in a few large doses.

3. For an **occupationally exposed individual** for 1 year, the maximum permissible whole-body dose, the extremity dose, internal organ dose, and the dose to the lens of the eye are, respectively

(a) 5 Rem, 50 Rem, 50 Rem, and 15 Rem
(b) 5 Rem, 50 Rem, 15 Rem, and 50 Rem
(c) 5 Rem, 15 Rem, 50 Rem, and
(d) 15 Rem, 50 Rem, 5 Rem, and 15 Rem

Correct answer is (a) According to NRC Regulation 20.1201(a), the licensee shall control the occupational dose to individual adults to:
5.0 Rem Total Effective Dose Equivalent per year (whole-body dose).
50.0 Rem/year dose to extremities
50.0 Rem Total Organ Dose Equivalent per year (individual organ)
15.0 Rem/year to the lens of the eye

4. How often MUST an employer inform his radiation workers of their radiation exposure levels?

(a) weekly
(b) monthly
(c) quarterly
(d) annually

Correct answer is (d). Although film badge readings are routinely posted on the bulletin board on a monthly basis, according to the current NRC regulations, every employer is responsible for informing each employee on an **annual basis** of his cumulative radiation dose.

5. Institutions are permitted to designate one sink in the laboratory as a "HOT SINK" for disposal of low-activity radioactive waste. The limit for activity is:

(a) 1 Ci per year for all isotopes
(b) 10 Ci per year for all isotopes
(c) 100 mCi per year for all isotopes
(d) No limit

Correct answer is (a). Disposal into sanitary sewerage is for readily soluble material or readily dispersible biological material. Each liquid waste disposal must be documented by the radioisotope, amount, date, and initials of the person involved. Total annual quantity must not exceed 1 Ci for all isotopes for all users on one license

6. All administered doses must be within what percentage of the prescribed dose?

(a) 5%
(b) 10%
(c) 15%
(d) 20%

Correct answer is (d). The prior 10% rule no longer applies. In 2006, the rule was changed to 20%. If a prescribed dose must be changed due to patient size or another valid reason, that dose is modifiable only by a physician and the change must be in writing.

7. A patient was treated with 14 mCi of I-131 NaI for hyperthyroidism. On further review, the prescribed dose in the written directive was 10 mCi. Which of the following statements is correct?

   (a) This is not a recordable or reportable event because I-131 NaI is a relatively safe drug
   (b) This is not a recordable or reportable event because the patient did not receive more than 50% of the prescribed dose
   (c) This is a recordable event because the patient received more than 20% but less than 50% of the prescribed dose
   (d) This is a reportable event because the patient received more than 20% of the prescribed dose

   Correct answer is (d). This is a reportable event. For therapeutic radiopharmaceuticals, the administered dose must be within 20% of the prescribed dose. In this case, the patient received a 40% overdose.

   Action required for Therapeutic Reportable Events: Document details and notify the NRC/state in writing of the event, giving pertinent details and plans for preventing recurrence. Do NOT supply any patient-specific information, e.g., name, SS #.

8. How often hand monitoring and treadmill monitoring are required?

   (a) Twice daily
   (b) Daily
   (c) Twice weekly
   (d) Weekly

   Correct answer is (b). Hand monitoring is required daily for most licensees. Some technologists monitor after every patient or every time they leave the laboratory. Treadmill monitoring is required on a daily basis. Results of these surveys must be recorded in an appropriate logbook.

9. How often must a dose calibrator be checked for accuracy?

   (a) Daily
   (b) Monthly
   (c) Quarterly
   (d) Annually

   Correct answer is (d). This test should be performed at installation and annually. It is designed to show that the calibrator is giving correct readings throughout the entire energy scale that are likely to encounter. Low, medium, and high energy standards (usually Co-57, Ba-133 or Cs-137, and Co-60, respectively) are measured in the dose calibrator using appropriate settings. Standard and measured values are compared.

10. How often must a dose calibrator be checked for constancy?

    (a) Daily
    (b) Monthly
    (c) Quarterly
    (d) Annually

    Correct answer is (a). This test, performed **at installation and daily**, measures instrument precision and is designed to show that a long-lived source, usually 30 y Cs-137, yields reproducible readings on a daily basis on all isotope settings we are likely to use. The Cs-137 source is placed in the dose calibrator. Activity is then measured on the Cs-137 setting and all other settings used on a daily basis. Values are recorded in the dose calibrator logbook and are compared with recent values to determine if the instrument is maintaining constancy on a daily basis.

11. How often must a dose calibrator be checked for linearity?

(a) Daily
(b) Monthly
(c) Quarterly
(d) Annually

Correct answer is (c). This test, performed at installation and quarterly, is designed to prove that the dose calibrator readout is linear for sources varying from the μCi range through the mCi range. A high activity Tc-99m source (300 mCi) is measured at T_0 and at predetermined time intervals up to 48 h. Expected and actual measurements are compared (and may be analyzed graphically) to determine if the instrument is linear throughout the activity range we are likely to encounter.

12. How often must a dose calibrator be checked for geometry?

(a) Monthly
(b) Quarterly
(c) Annually
(d) At Installation

Correct answer is (d). This test is performed at the installation of the dose calibrator or after moving the instrument. It is designed to show that correct readings can be obtained regardless of the sample size or geometry. 0.5 mL of Tc-99 m in a 10 mL syringe (activity 25 mCi) is measured in the dose calibrator and the value obtained is recorded. The activity is then diluted with water to 1 mL, 2 mL, 3 mL, 5 mL, and 10 mL. At each of these points, a reading is taken and the value recorded. Data are then evaluated to determine the effect of sample geometry on the dose calibrator reading. If an instrument is geometry-dependent, it may be necessary to routinely correct readings obtained when using a calibrator.

13. Which of the following is an example of a radionuclide impurity?

(a) Mo-99 in the generator eluate
(b) I-123 Iodide in a preparation of I-123 mIBG
(c) Aluminum ion breakthrough
(d) Hydrolyzed reduced Tc in a preparation of Tc-99m MDP

Correct answer is (a). A radionuclide impurity implies that the incorrect radionuclide is present. For example, Mo-99 should never be present in the eluate containing Tc-99m pertechnetate. Examples of radiochemical impurity are I-123 Iodide in a preparation of I-123 MIBG and hydrolyzed reduced Tc in a preparation of Tc-99m MDP. Aluminum ion breakthrough is an example of a chemical impurity. It implies that no radionuclide is present (note the absence of prefix "radio"). Soluble aluminum ion is not harmful to patients but interferes with the labeling of bone agents, sulfur colloid, and red cells. All these impurities will alter radiation dose and affect image quality.

14. What is the upper limit of Al^{3+} breakthrough in a generator eluate?

(a) 1 ppm
(b) 5 ppm
(c) 10 ppm
(d) 20 ppm

Correct answer is (c). Exchange diffusion takes place when fluoride and hydroxide ions on hydroxyapatite structure of bone exchange places and fluoroapatite is formed. No other examples of exchange diffusion are available.

Part 8: Radionuclide Therapy

1. A hyperthyroid patient has a 50% uptake and was treated with 10 mCi of I-131 NaI solution. The radiation dose to the thyroid is approximately:

   (a) 1000 Rads
   (b) 5000 Rads
   (c) 10,000 Rads
   (d) 50,000 Rads

   Correct answer is (c). The package insert for I-131 NaI indicates that a dose of 11,000 Rads is delivered to the thyroid gland of a patient with a 50% uptake who was treated with 10 mCi of I-131.

2. A hospitalized patient for I-131 treatment of thyroid may be discharged at a reading of:

   (a) 5 mR/h at body surface
   (b) 7 mR/h at body surface
   (c) 5 mR/h at 1 m from patient's chest
   (d) 7 mR/h at 1 m from patient's chest

   Correct answer is (d). According to the current NRC Regulations, the patient may be released if a reading of <7 mR/h at 1 m from patient's chest is obtained.

3. Which of the following therapies involves the use of an unsealed source in the body?

   (a) Brachytherapy
   (b) Teletherapy
   (c) Radiopharmaceutical Therapy
   (d) Neutron irradiation

   Correct answer is (c). Radiopharmaceutical Therapy is the only therapy listed involving the use of an unsealed source in the body. Brachytherapy involves sealed sources and the others involve the use of external beams.

4. Which of the following therapeutic radioisotopes places the dose administrator at greatest risk?
   (a) I-131
   (b) P-32
   (c) Sr-89
   (d) Y-90

   Correct answer is (a). I-131 is a beta/gamma emitter; all others are pure beta emitters. The gamma rays emitted by I-131, therefore, place the person administering the dose at much greater risk than any of the other isotopes.

5. Typical prescribed dose for treating Toxic Multinodular Goiter is:

   (a) 5–10 mCi
   (b) 15–20 mCi
   (c) 25–30 mCi
   (d) 35–40 mCi

   Correct answer is (c). 25–30 mCi are typically prescribed dose for treating toxic MNG.

6. If the first dose of I-131 NaI administered for treating hyperthyroidism fails to adequately treat the patient, the second dose typically equals to:

   (a) 50% of the first dose
   (b) Same as the first dose
   (c) 150% of the first dose
   (d) Twice as much as the first dose

   Correct answer is (c). If the first dose of I-131 fails to cure hyperthyroidism, the second dose is usually about 150% of the first dose.

7. Which one of the following statements is true?

   (a) The biological distribution of Sr-89 is essentially identical to that of Tc-99 m MDP
   (b) Sr-89 is an alpha and beta emitter
   (c) Therapy with Sr-89 is palliative and slightly increases the patient's life expectancy
   (d) Sr-89 is retained in metastases and in normal bone tissue to the same degree

   Correct answer is (a). Sr-89 therapy is palliative in the sense that it reduces pain and improves the quality of the patient's life; it cannot cure the primary disease. It's biological distribution is similar to MDP. It is a beta emitter only.

8. A typical injected dose of Sr-89 chloride is

   (a) 0.4 mCi
   (b) 4 mCi
   (c) 0.4 mCi/kg
   (d) 4 mCi/kg

   Correct answer is (b). The recommended dose in the package insert is 4 mCi, regardless of the patient's weight. Samarium-153 and Radium-223 are based on weight.

9. Which ONE of the following statements regarding both Sr-89 and I-131 is TRUE?

   (a) Both decay by β^+ emission
   (b) Both decay by β^- emission
   (c) Both emit γ photons along with the β particles
   (d) Both require hospitalization until the patient reading is <7 mR/h at 1 m from the patient's chest

   Correct answer is (b). Both decay by β^- emission. (c) is false—Sr-89 emits no gammas; (d) is false—The7 mR/h at 1 m limit applies to gamma ray emitters, not alpha or beta emitters.

Part 9: Radiation Dosimetry

1. Marie Curie's Professor carried in an inside pocket of his coat a glass vial containing radium-226; shortly thereafter, the first radiation burn was documented. This skin burn was caused by

   (a) The potent 4.8 MeV α particles emitted by this isotope
   (b) The gamma rays accompanying α decay
   (c) Cerenkov radiation
   (d) 0.005 keV X-rays

   Correct answer is (b). High energy gammas cause essentially all the damage; the high energy alphas cannot penetrate the wall of the glass wall; Cerenkov radiation is irrelevant; and 5 keV X-rays have very poor penetrability and are unable to confer a significant dose.

2. Irradiation of the juvenile thyroid has been associated with an increased risk of developing

   (a) thyroid carcinoma and parathyroid adenomas
   (b) hypothyroidism
   (c) hyperthyroidism
   (d) aplastic anemia

   Correct answer is (a). Epidemiological studies have correlated irradiation of the juvenile thyroid with an increased incidence of thyroid carcinoma and parathyroid adenomas

3. The ALARA Concept, implemented in the late 1970s, has contributed to

   (a) A reduction in the cost of radiopharmaceuticals
   (b) An overall reduction in dose to radiation workers.
   (c) Increased compliance with continuing education requirements
   (d) Increased availability of new radiopharmaceuticals

   Correct answer is (b). The ALARA concept was developed to insure that radiation absorbed dose for Radiation Workers would be as Low as Reasonably Achievable. The net result was an overall reduction in dose to radiation workers.

4. In any given year, a radiation worker may receive a whole-body dose of

   (a) 1 R
   (b) 5 R
   (c) 10 R
   (d) 50 R

   Correct answer is (b). As Radiation Workers, we are permitted a Maximum Permissible Dose of 5 R whole body per year.

5. In any given year, family members of a radiation worker may receive a *whole-body dose* of

   (a) 0.1 R
   (b) 0.5 R
   (c) 1 R
   (d) 5 R

   Correct answer is (b). Families of Radiation Workers, as well as members of the general public, are permitted a Maximum Permissible Dose of 0.5 R whole body per year.

Part 10: Computers in Nuclear Medicine

1. The number of bits in a kilobyte is

   (a) 256
   (b) 1000
   (c) 1024
   (d) 8192

   Correct answer is (d) 8192. The number of bits in a kilobyte is determined as follows:
   # bits/kB = (1024 bytes/kB) × (8 bits/byte)
   = 8192 bits

2. The decimal equivalent of the binary number 01101 is

   (a) 5
   (b) 7
   (c) 13
   (d) 65

   Correct answer is (c) 13. The decimal equivalent for the binary number 01101 is determined as follows:
   decimal value = $0 \times 2^4 + 1 \times 2^3 + 1 \times 2^2 + 0 \times 2^1 + 1 \times 2^0 = 0 + 8 + 4 + 0 + 1 = 13$

3. The binary equivalent of the decimal number 21 is

   (a) 001101
   (b) 100111
   (c) 010101
   (d) 011101

   Correct answer is (c) 010101. The binary equivalent for the decimal number 21 is determined by successive approximation starting at the most significant digit as follows:
   Place 5: $2^5 = 32 > 21$; therefore, bit set to 0 with remainder of 21
   Place 4: $2^4 = 16 < 21$; therefore, bit set to 1 with remainder of 5
   Place 3: $2^3 = 8 > 5$; therefore, bit set to 0 with remainder of 5
   Place 2: $2^2 = 4 < 5$; therefore, bit set to 1 with remainder of 1
   Place 1: $2^1 = 2 > 1$; therefore, bit set to 0 with remainder of 1
   Place 0: $2^0 = 1 = 1$; therefore, bit set to 1 with remainder of 0; finished!
   Resultant binary number is 010101

4. An X-ray film digitizer produces 2048 × 2048 × 8 bit digital images. The number of gray levels captured by the digitizer is

   (a) 8
   (b) 256
   (c) 4194304
   (d) 33554432

   Correct answer is (b) 256. The matrix size in this image is 2048 × 2048 pixels. The pixel depth in this image is 8 bits. The number of gray levels in the image is given by the pixel depth and is determined as follows:
   $2^{(\text{pixel depth})} = 2^8 = 256$.

5. A 512 × 512 CT image that uses Hounsfield units for tissue density that ranges from −1000 to +1000 requires at the minimum ____ bits of memory for each pixel.

   (a) 11
   (b) 16
   (c) 2001
   (d) 262144

   Correct answer is (a) 11. There are 2001 possible Hounsfield unit values. Therefore, the pixel depth required to cover this range of values is 11 bits since:
   $(2^{10} = 1024) < 2001 < (2^{11} = 2048)$

6. Of the memory selections available, which of the following is the fastest type of memory?

   (a) magnetic disk
   (b) magnetic tape
   (c) solid-state
   (d) CD-ROM

   Correct answer is (c) solid-state. Solid-state memory is constructed of high-speed transistors fabricated onto silicon microchips. Memory type ordered from the fastest to the slowest is as follows: solid-state, magnetic disk, magneto-optical disk, CD-ROM, and magnetic tape.

7. A computer that uses RISC architecture has

   (a) a reduced instruction set optimized for execution speed.
   (b) programs of reduced size optimized for efficient memory usage.
   (c) arrays of redundant inexpensive microprocessors to enhance execution speed.
   (d) reduced size, inexpensive, high-speed components.

   Correct answer is (a) a reduced instruction set optimized for execution speed. RISC is an acronym defined for Reduced Instruction Set Computer. The RISC architecture utilizes a small instruction set of microprocessor functions that are optimized for execution speed. Complex Instruction Set Computers (CISC) utilize a larger instruction that implements more complex microprocessor functions, but at the expense of execution speed. In RISC computers, complex functions must be handled in the program code, resulting in larger program sizes.

8. The main memory of a general purpose computer system consists of

   (a) magnetic disks.
   (b) solid-state memory.
   (c) CD-ROM disks.
   (d) magneto-optical disks.
   (e) magnetic tapes.

   Correct answer is (b) solid-state memory. The fastest memory available is used for the main memory of a computer system.

9. Images from a gamma camera whose useful field-of-view is 380 mm in diameter are digitized to 128 × 128 matrices. The pixel cross-sectional area is ____ mm^2.

   (a) 0.023
   (b) 0.11
   (c) .34
   (d) 9

Correct answer is (d) 9 mm². The 380 mm diameter of the gamma camera field-of-view is subdivided into 128 pixels horizontally and vertically. Assuming that the pixels are square, the cross-sectional area of each pixel is then:

cross-sectional area = (horizontal pixel width) × (vertical pixel width)
= (380 mm/128 pixels) × (380 mm/128 pixels)
= (3 mm) × (3 mm)
= 9 mm²

10. The image resolution in an X-ray CT image is found to be 0.625 line pairs per mm. What is the upper limit of the pixel width in this image?

(a) 0.318 mm
(b) 0.625 mm
(c) 0.800 mm
(d) 1.250 mm

Correct answer is (c) 0.800 mm. Critical sampling requires that there be at least two pixels per line pair. Thus, for a resolution of 0.625 line pairs per mm, 1.25 (2 × 0.625) pixels per mm are necessary, the upper limit in the pixel width is (1 mm/1.25 pixels) 0.8 mm.

Part 11: Principles of Scintillation Detection

1. The intrinsic spatial resolution of current generation Anger cameras is _____ mm FWHM.

(a) 1–2
(b) 2–3
(c) 3–4
(d) 4–5

Correct answer is (c) 3–4 mm FWHM. This range has been consistently achieved in commercial Anger camera systems since the early 1980s. It is not expected that the spatial resolution will improve much beyond 3 mm FWHM.

2. The system spatial resolution of current generation gamma cameras fitted with a high-resolution low-energy collimator for imaging of Tc-99m at a distance of 10 cm is approximately ____ mm FWHM.

(a) 4
(b) 7
(c) 10
(d) 12

Correct answer is (b) 7 mm FWHM. This is the typical imaging distance for planar gamma camera imaging.

3. The following combination of quality control tests are to be carried out on gamma cameras daily:

(a) field uniformity, linearity, spatial resolution
(b) field uniformity, photopeaking
(c) photopeaking, linearity, spatial resolution
(d) photopeaking, field uniformity, sensitivity
(e) photopeaking, field uniformity, linearity, spatial resolution

Correct answer is (b) field uniformity, photopeaking. The photopeak energy window for the isotope to be imaged must be validated daily, as well as field uniformity. Linearity and spatial resolution are generally evaluated weekly. Sensitivity is evaluated on a far less frequent basis, perhaps yearly.

4. The typical energy window used for imaging Tc-99m on gamma cameras is

   (a) 5%
   (b) 10%
   (c) 20%
   (d) 40%

   Correct answer is (c) 20%. A 20% energy window is standard for imaging Tc-99m. On newer gamma camera systems with energy resolution less than 10% FWHM, a 15% energy window may be selected.

Part 12: SPECT Imaging, PET Imaging

1. The bullseye artifact appears in SPECT images as a result of gamma camera nonuniformities. The bullseye is best observed in the ____________ image slices.

   (a) transaxial
   (b) sagittal
   (c) coronal
   (d) sinogram

   Correct answer is (a) transaxial. The concentric rings of the bullseye artifact only appear in the transverse images. Sagittal and coronal images slice perpendicularly across the concentric ring artifacts.

2. SPECT reconstruction algorithms assume that the gamma camera axis-of-rotation projected onto the computer matrix (referred to as the center-of-rotation) coincides with the matrix center. If misalignment of this axis is not corrected, the resultant SPECT images

   (a) show the same bullseye artifact as for gamma camera nonuniformities.
   (b) exhibit overall loss in image resolution.
   (c) become excessively noisy.
   (d) show a hot rim artifact.

   Correct answer is (b) exhibit overall loss in image resolution. Misalignment in the COR causes the reconstruction program to incorrectly back project the count profiles, which leads to a blurring of the image, and if severe enough point sources appear as donuts.

3. The sinogram of a selected slice in SPECT is used to

   (a) detect patient motion.
   (b) display significant nonuniformities in the gamma camera images.
   (c) assign the appropriate filter for image reconstruction.
   (d) determine the patient contour needed for applying photon attenuation corrections.

   Correct answer is (a) detect patient motion. The sinogram of a selected image slice is extremely useful in detecting patient motion, gantry rotation problems, camera failure, and external contamination.

4. Correction for photon attenuation in SPECT is most problematic in the
   (a) head.
   (b) thorax.
   (c) abdomen.
   (d) equally problematic in all sections of the body.

   Correct answer is (b) thorax. In the thorax, especially for cardiac imaging, the lung, muscle, breast, and bone tissues have very different attenuation coefficients. The attenuation effect becomes very complex, and if uncompensated for, resultant image artifacts and distortions may render the clinical result useless.

5. A flood image used to correct gamma camera nonuniformities should have at least _______ counts per pixel so that the uniformity correction itself does not contribute to uniformity artifacts in the reconstructed SPECT images.
   (a) 100
   (b) 1000
   (c) 10,000
   (d) 1,000,000

   Correct answer is (c) 10,000. Statistical variations of pixel values in the calibration flood image should have a standard deviation of less than 1% of the mean, which requires an average 10,000 counts per pixel.

Part 13: Personnel Monitoring Devices

1. Personnel monitoring for external exposure shall be performed on all occupationally exposed individuals who may receive >___% of the applicable MPD during the normal course of their duties or through accidental exposure
   (a) 5
   (b) 10
   (c) 15
   (d) 20

   Correct answer is (b) Current NRC Regulations require badging of any individual who might receive ≥500 mRem, which is 10% of the maximum permissible dose.

2. Long-term visitors to controlled areas should be regarded as occupationally exposed persons if it is likely that they will receive a dose equivalent exceeding ______ mRem per year.
   (a) 5
   (b) 50
   (c) 100
   (d) 500

   Correct answer is (d) Current NRC Regulations require badging of any individual who might receive ≥10% of the maximum permissible dose.

3. The ideal personnel monitoring device must accurately measure the biological dose in Rems received by parts of the body considered to be most vital from the standpoint of chronic low-level exposure, i.e.,

   (a) bone marrow and gonads.
   (b) breast and thyroid
   (c) spleen and liver
   (d) heart and lungs

   Correct answer is (a) The three most radiosensitive parts of the body are bone marrow, gonads, and lens of the eye.

4. Personnel monitoring records must be preserved

   (a) for 1 year
   (b) for 3 years
   (c) for 5 years
   (d) until the US NRC authorizes disposition.

   Correct answer is (d) While most records must be kept only 3 years, current NRC Regulations require that all personnel monitoring records be kept until the NRC authorizes the disposition

5. Exposure records MUST BE furnished to which ONE of the following:

   (a) any currently employed occupational worker who requests it—on a quarterly basis
   (b) any formerly employed occupational worker who requests it—within 24 h
   (c) The individual and the NRC/State in the case of an overexposure—within 30 days
   (d) A colleague requesting it on behalf of a former employee

   Correct answer is (c) (**a**) is incorrect—should be annual basis; (**b**) is incorrect—should be 30 days; (**d**) is incorrect—only the employee can request his/her record.

6. Film badges can distinguish among different energy photons because

   (a) they contain internal circuitry sensitive to different photon energies
   (b) they contain small metallic filters of different densities, enabling them to distinguish among higher energy photons
   (c) they contain a lithium fluoride chip functioning as a thermoluminescent dosimeter
   (d) they only absorb the lowest energy photons

   Correct answer is (b) internal to a film badge are three separate metal filters, usually Cu, Cd, and Al, that are able to differentiate photons of different energies because the densities of the filters are all different. See the photo below.

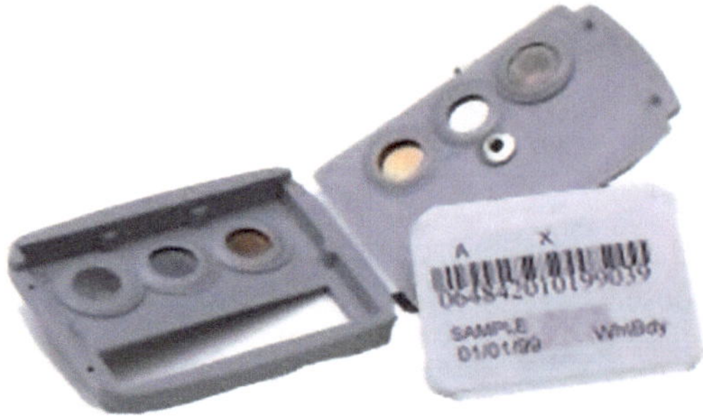

7. To measure the absorbed radiation dose during a 4-h visit to a radiopharmacy, which of the following devices would be the most appropriate one to use?

    (a) GM Counter
    (b) Film badge
    (c) TLD
    (d) Pocket dosimeter

    Correct answer is (d). (**a**) is incorrect as a GM counter can only provide an instantaneous reading; (**b**) and (**c**) would be able to measure the integrated dose, but it would take 2–6 weeks to receive results of this monitoring; (**d**) is correct because a reading is obtained immediately after the end of the 4-h visit.

Part 14: Units of Radioactivity and Radiation Dose

1. Detector efficiency is defined as:

    (a) $\frac{\text{count rate}}{\text{disintegration rate}} \times 100\%$

    (b) $\frac{\text{disintegration rate}}{\text{count rate}} \times 100\%$

    (c) $\frac{\text{disintegration rate} \times \text{count rate}}{100\%}$

    Correct answer is (a) detector efficiency is defined as the count rate of a sample divided by the known disintegration rate of that sample ×100%.

2. A 57Co standard has a disintegration rate of 3.7×10^7 Bq. What is the disintegration rate in disintegrations/s?

    (a) 3.7 × 104
    (b) 3.7 × 107
    (c) 3.7 × 1010
    (d) 3.7×10^{13}

    Correct answer is (b) since by definition, 1 Bq = 1 disintegration/s

3. A sample measures 100 mCi in a dose calibrator. How many GBq does this activity represent?

    (a) 0.37
    (b) 3.7
    (c) 37.0
    (d) 370

    Correct answer is (b) 100 mCi × 37 MBq/mCi = 3700 MBq, which is equal to 3.7 GBq.

4. If a detector is 50% efficient, how many counts/s should be detected for a 10 mCi standard?

    (a) 3.7×10^7
    (b) 3.7×10^8
    (c) 1.85×10^7
    (d) 1.85×10^8

    Correct answer is (d) 10 mCi × 3.7×10^7 dps/mCi × 50 c/100d = 1.85×10^8 cps

5. Occupationally exposed individuals have an annual MPD to the lens of the eyes of 15 Rem. What is this absorbed dose in mSv?

   (a) 1.5
   (b) 15
   (c) 150
   (d) 1500

   Correct answer is (c) 15 Rem × 1 Sv/100 Rem × 1000 mSv/Sv = 150 mSv.

Part 15: Cardiac Radiopharmaceuticals

1. Which of the following is an example of a myocardial viability agent?

   (a) Rb-82 chloride
   (b) I-123 low-density lipoprotein
   (c) F-18 FDG
   (d) Tc-99m RBCs

   Correct answer is (b) best example is F-18 FDG

2. Which of the following are potassium analogs?

   (a) Rb-82 chloride
   (b) F-18 FDG
   (c) Tc-99m Sestamibi
   (d) Tc-99m MDP

   Correct answer is (a). Rb-82 chloride represents monovalent cation like potassium ion; the others are not potassium analogs.

3. A stress/rest study can be completed in less than 45 min with...

   (a) Rb-82 chloride
   (b) F-18 FDG
   (c) Tc-99m Myoview
   (d) Tl-201 chloride

   Correct answer is (a) (b) is not used for stress/rest studies; (c) and (d) take several hours to complete

4. For which one of the following radioisotopes does one form an image using primarily X-rays rather than gamma rays?

   (a) Rb-82
   (b) Tc-99m
   (c) Tl-201
   (d) F-18 FDG

   Correct answer is (c) 71 keV X-rays are routinely used for imaging with Tl-201

5. Which of the following radioisotopes used for cardiac imaging is limited to a total activity for a stress/rest study of 5 mCi because of radiation dosimetry considerations?

   (a) Rb-82 chloride
   (b) F-18 FDG
   (c) Tc-99m Myoview
   (d) Tl-201 chloride

   Correct answer is (d) Testicular dose in males is 3 Rems/mCi and kidney dose in all patients is almost that high. None of the others has doses in that range.

Part 16: Radioisotope Dose Calibrator: Function and Quality Control

1. Dose calibrators detect radiation using

   (a) a NaI(Tl) scintillation crystal
   (b) a liquid scintillation detector
   (c) an ionization chamber
   (d) a solid-state Germanium detector

   Correct answer is (c) none of the others are found in a dose calibrator

2. Dose calibrators can accurately measure radioactivity levels between

   (a) 1 μCi and 1 mCi
   (b) 100 μCi and 5 Ci
   (c) 1 μCi and 1 Ci
   (d) 100 μCi and 100 Ci

   Correct answer is (d) dose calibrators function well across 5–6 orders of magnitude and can reliably calibrate sources between 100 μCi and 100 Ci

3. According to the current NRC Regulations, all administered doses must be within ____% of the prescribed dose

   (a) 5
   (b) 10
   (c) 15
   (d) 20

   Correct answer is (d) Currently within 20%. It used to be 10%

4. For all required QC tests on a dose calibrator, the expected and measured readings must be within

   (a) ±5%
   (b) ±10%
   (c) ±15%
   (d) ±20%

   Correct answer is (b)

Oncology and PET/CT

Medhat Sam Gabriel, Bital Savir-Baruch, Robert H. Wagner, Bruch J. Barron, Davide Bova, Aamna Hassan, Stephanie Rodriguez-Santiago, and Yekaterina Kucerova

Case 1

1. A 30-year-old female smoker complained of an enlarging left neck mass over the past 6 weeks with minimal pain. A physical examination of the mouth and throat did not demonstrate any suspicious area for a primary tumor. An F18 FDG PET/CT was performed as part of her evaluation. A biopsy of the neck mass was demonstrated malignant neoplasm. The most likely finding on the pathology is:

   A. Lymphoma
   B. Adenocarcinoma
   C. Squamous cell carcinoma
   D. Pleomorphic adenoma of the parotid

M. S. Gabriel (✉) · B. Savir-Baruch · R. H. Wagner · D. Bova
Division of Nuclear Medicine, Department of Radiology, Loyola University Medical Center, Maywood, IL, USA
e-mail: sgabriel@lumc.edu; Bital.savir-Baruch@lumc.edu; rwagner@lumc.edu; dbova@lumc.edu

B. J. Barron
Division of Nuclear Medicine and Molecular Imaging, Department of Radiology and Imaging Sciences, Atlanta, GA, USA
e-mail: bjbarro@emory.edu

A. Hassan
Department of Nuclear Medicine, Shaukat Khanum Memorial Cancer Hospital and Research Center, Lahore, Punjab, Pakistan

S. Rodriguez-Santiago
Department of Nuclear Medicine, C.W. Bill Young VA Medical Center, Bay Pines, FL, USA
e-mail: stephanie.rodriguezsantiago@va.gov

Y. Kucerova
Division of Nuclear Medicine, Department of Radiology, Loyola University Medical Center, Maywood, IL, USA

Department of Radiology, Cook County Healthy, Chicago, IL, USA
e-mail: yekaterina.kucerova@cookcountyhealth.org

M. S. Gabriel, B. Savir-Baruch (eds.), *RadTool Nuclear Medicine MCQs*,
https://doi.org/10.1007/978-3-030-69281-0_3

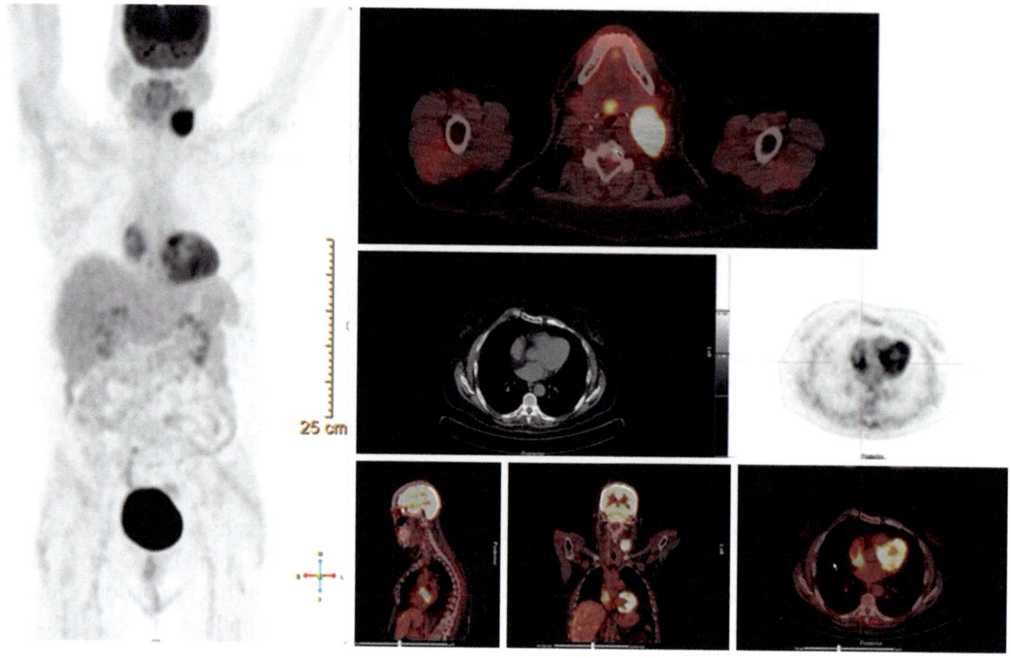

Answer: C: Squamous cell carcinoma

Squamous cell carcinoma is the most common type of head and neck cancer and accounts for approximately 90% of head and neck cancers.

Pleomorphic adenomas are common benign tumors of the salivary glands and may demonstrate significantly increased uptake of F18 FDG. F18 FDG PET/CT imaging is often less helpful in the initial evaluation of salivary gland tumors since uptake may be seen in both benign and malignant tumors. Some malignant tumors (adenoid cystic, mucoepidermoid, or necrotic squamous cell) have little uptake. If distant metastatic disease from a malignant salivary gland tumor is suspected, F18 FDG PET/CT imaging can be useful [1].

While lymphoma commonly involves the head and neck region, it would be uncommon for lymphoma to involve only a single lymph node in the neck as seen in this patient.

2. An unknown primary tumor location can be detected in head and neck cancers using F18 FDG PET/CT:

A. 5–15% of the time
B. 20–40% of the time
C. 50–70% of the time
D. 75–95% of the time

Answer: B

In many cases, the first sign of a head and neck tumor is enlargement of a regional lymph node. The primary focus may not be clinically visible on physical examination, but the point of origin is useful in determining the subsequent clinical care. In the absence of a primary focus, the tonsil is considered a common site. If surgical resection is performed, a tonsillectomy is usually included in the hopes of identifying the primary site. Although PET/CT imaging with F18 FDG is useful only 20–40% of the time, a positive scan may be helpful in identifying the primary location and planning subsequent therapy [2, 3].

3. What is the most likely location for the primary source of tumor in this patient?

   A. A submental salivary adenocarcinoma
   B. A left parotid tumor
   C. The base of the tongue
   D. Metastatic from a right hilar cancer

   Answer: C
   The focus of uptake seen in the submental area is very common and represents physiologic sublingual uptake. In addition, the digastric muscles may often demonstrate uptake further confusing interpretation in this area [1]. The maximum intensity projection image clearly demonstrates that the area of uptake in the left neck is inferior to the parotid. The focal area of uptake in the base of the tongue is suspicious and may represent the primary tumor in this patient. Tonsillar activity can be intense. A tonsillar focus of tumor should be considered if the uptake is asymmetric. It would be uncommon for a lung cancer to metastasize to the head and neck as an initial presentation.

4. The area of uptake seen in the chest likely represents:

   A. A second primary
   B. Metastatic primary head and neck tumor requiring further therapy
   C. An area of myocardium that has ischemia
   D. Benign lipomatous hypertrophy

   Answer: D
   Benign lipomatous hypertrophy is a normal variant that can occasionally be seen on F18 FDG PET imaging. This should not be confused with a metastatic focus to the heart or septum, which is extremely rare. The uptake on fusion corresponds to the interatrial septum [4, 5].
   Cardiac uptake on a F18 FDG study is variable and is dependent on prior diet and length of fasting. Occasionally, a focal area of uptake may be seen in the left ventricle representing a local dependence on glucose metabolism [6].
   Lymphoma involvement of the interatrial septum without additional areas of involvement in the mediastinum would be unlikely. Again, metastasis from the head and neck primary to the interatrial septum would be very uncommon.

5. Head and neck cancers that are P16 (human papilloma virus or HPV) positive:

   A. Tend to have a worse outcome
   B. Account for a minority of head and neck cancers
   C. Are usually seen in older patients
   D. Are staged differently from head and neck cancers that are P16 negative

   Answer: D
   Head and neck cancers with positive HPV are on the rise, now accounting for the majority of head and neck cancer. Tumors that are P16 positive tend to occur in younger patients. However, the outcome is more favorable than P16 negative. The difference in outcome is such that a different staging system has been introduced differentiating cancers that are P16 positive from P16 negative [7].

References

1. Purohit BS, et al., FDG-PET/CT pitfalls in oncological head and neck imaging. Insights Imaging. 2014;**5**(5):585–602.
2. Al-Ibraheem A, et al., Clinical applications of FDG PET and PET/CT in head and neck cancer. J Oncol. 2009;2009:208725.

3. Wong RJ, Current status of FDG-PET for head and neck cancer. J Surg Oncol. 2008;97(8):649–52.
4. Maure, AH, et al., How to differentiate benign versus malignant cardiac and paracardiac 18F FDG uptake at oncologic PET/CT. Radiographics. 2011;31(5):1287–305.
5. Meka M, Depuey EG, Bhargava P, Focal FDG activity in the region of right atrium: coregistered CT identifies three benign etiologies. Radiol Case Rep. 2008;3(1):120.
6. Minamimoto R, et al., Incidental focal FDG uptake in heart is a lighthouse for considering cardiac screening. Ann Nucl Med. 2013;27(6):572–80.
7. Horne ZD, et al., Confirmation of proposed human papillomavirus risk-adapted staging according to AJCC/UICC TNM criteria for positive oropharyngeal carcinomas. Cancer. 2016;122(13): 2021–30.

Case 2

6. A 57-year-old man presents for suspected recurrent Hodgkin lymphoma, 6 years in remission after bone marrow transplant. Recent CT chest performed for upper respiratory infection (URI) symptoms and persistent hoarseness showed bilateral borderline enlarged hilar nodes. Patient has congestive heart failure. Selected images below show:

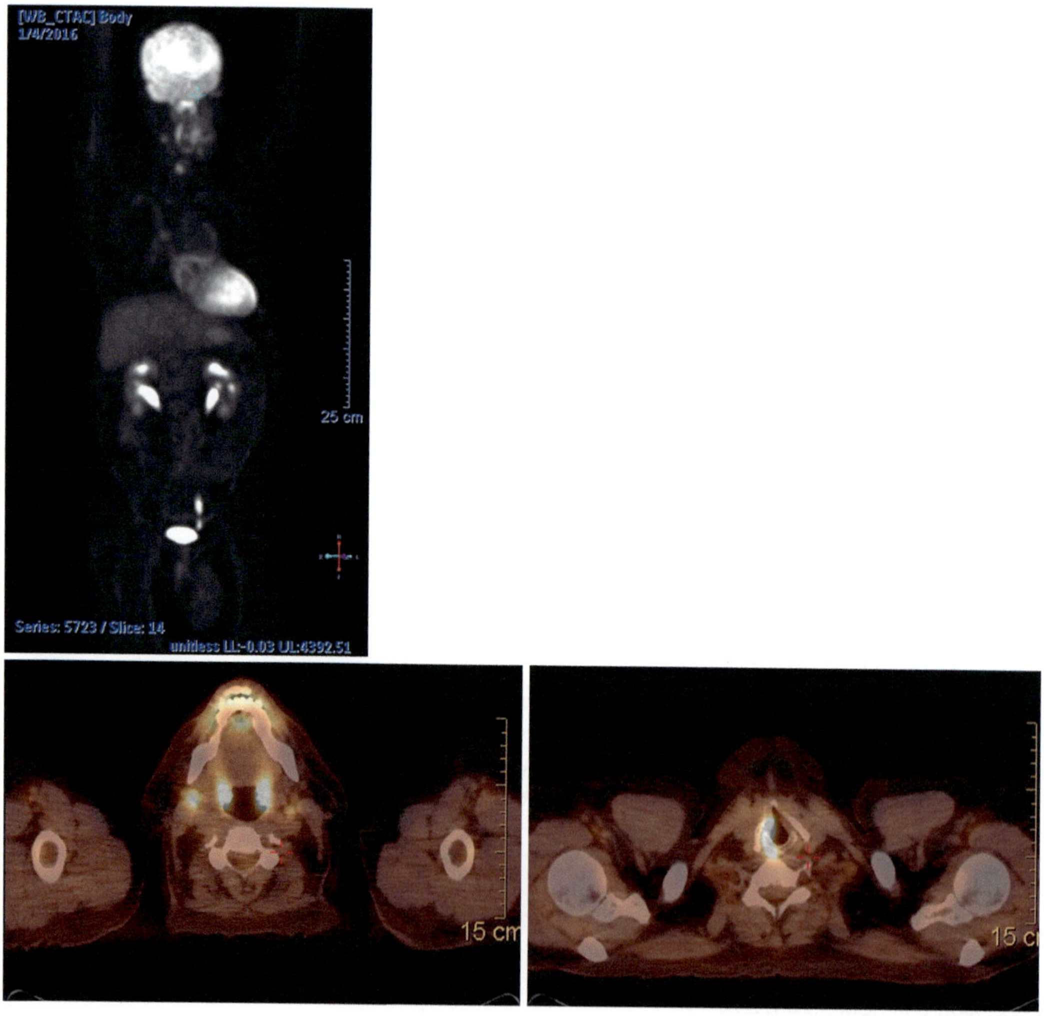

A. Lymphomatous involvement of right cervical nodes, tonsils, and larynx.
B. Right cervical hypermetabolic node and hypermetabolic right vocal cord due to left vocal cord paralysis
C. Findings compatible with an infectious/inflammatory process of the upper respiratory tract
D. New right laryngeal carcinoma with ipsilateral metastatic node.

Answer: B

Right vocal cord is asymmetrically hypermetabolic due to contralateral paralysis, confirmed at laryngoscopy. Right cervical node is moderately hypermetabolic, indeterminate for lymphoma.

7. Additional images of the thorax show:

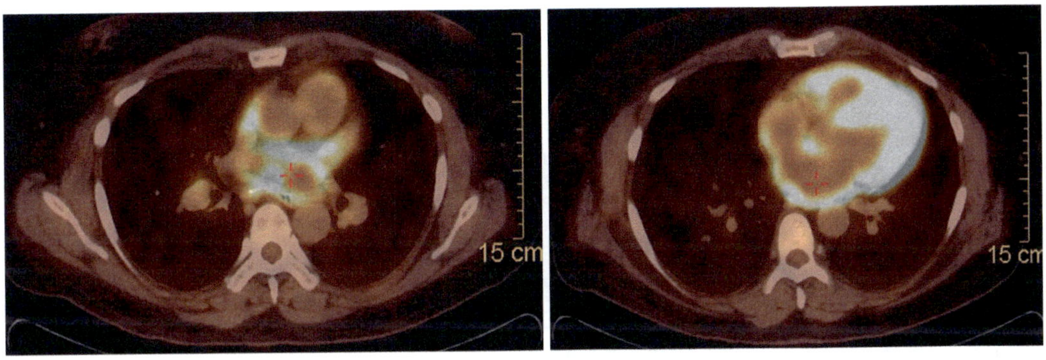

A. Bilateral hilar nodal lymphoma infiltration
B. Lymphomatous infiltration of the atria and ventricles
C. Increased metabolic activity related to congestive heart failure
D. Normal physiologic myocardial activity

Answer: C

Normal physiologic activity can be frequently seen in the left ventricular myocardium, diffusely or focally with preferential postero-basal distribution. Atrial and right ventricular uptake are more commonly associated with a variety of conditions including pulmonary arterial hypertension, congestive heart failure, sarcoidosis, and endocarditis. There is no significant or asymmetric FDG uptake in the hila.

*Maurer AH, Burshteyn M, Adler LP, Steiner RM. How to differentiate benign versus malignant cardiac and paracardiac 18F FDG uptake at oncologic PET/CT. RadioGraphics. 2011; 31:1287–1305.

Case 3

8. 72-year-old male presents for CT follow-up of a vague ground glass opacity described on CT performed 14 months prior. Current (top row) and prior images are provided.

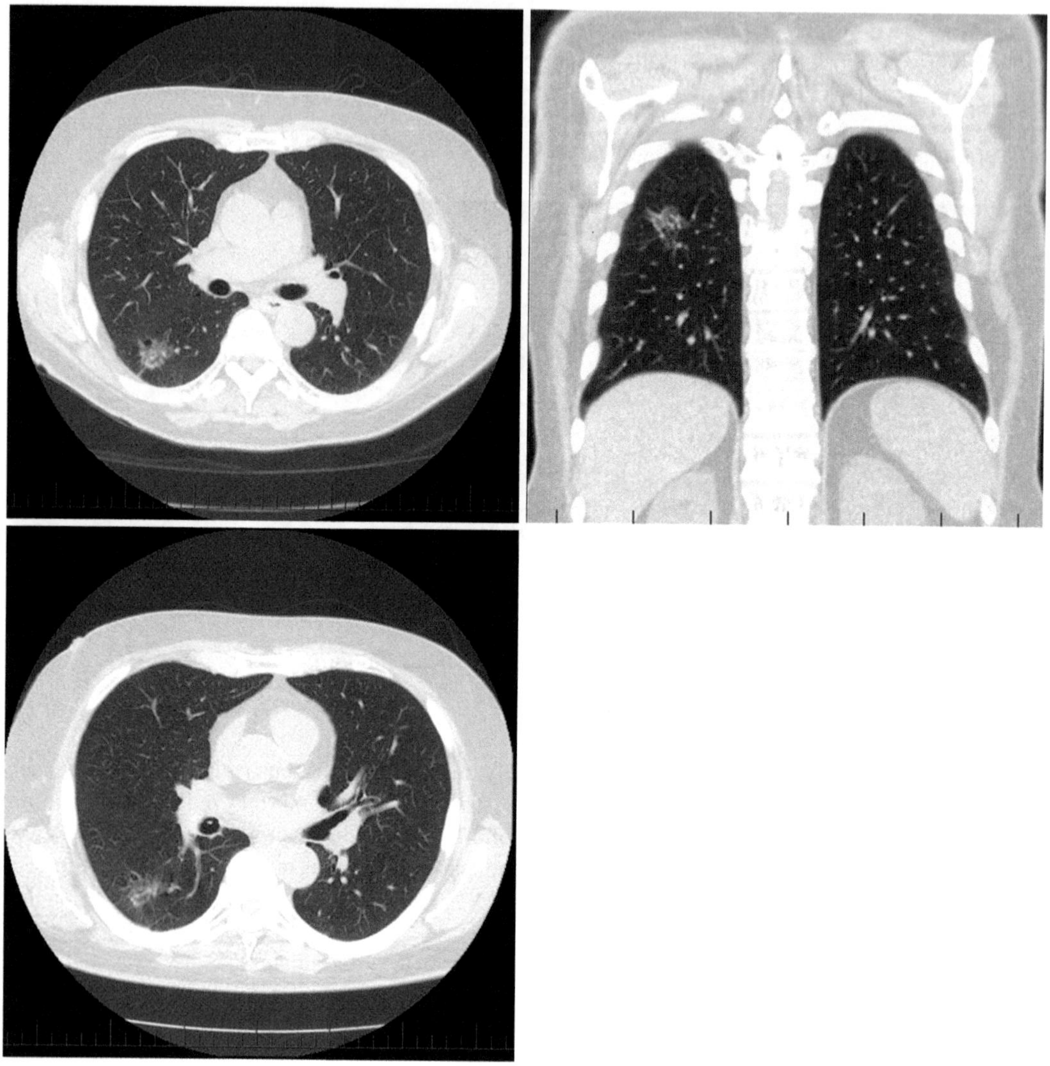

14 months prior.
What is your interpretation?

A. Recommend repeat study in 6–8 weeks after broad-spectrum antibiotic treatment
B. Further evaluation with F18-FDG PET-CT
C. BAL (Bronchoalveolar Lavage) and transbronchial biopsy
D. Stable examination. Continue follow-up in 6–12 months.

Answer: B

The lesion appears denser and slightly larger on current examination, mandating further imaging, rather than follow-up. Antibiotics would probably not be effective, given the long-standing history of this abnormality (14 months at least). The lesion would not be accessible to transbronchial biopsy given its peripheral location. Likewise, BAL would be unlikely to be revealing for the same reason.

9. F18-FDG-PET-CT was performed a few days later. The SUV max of the target lesion was 2.4 (mediastinal background 2.3).

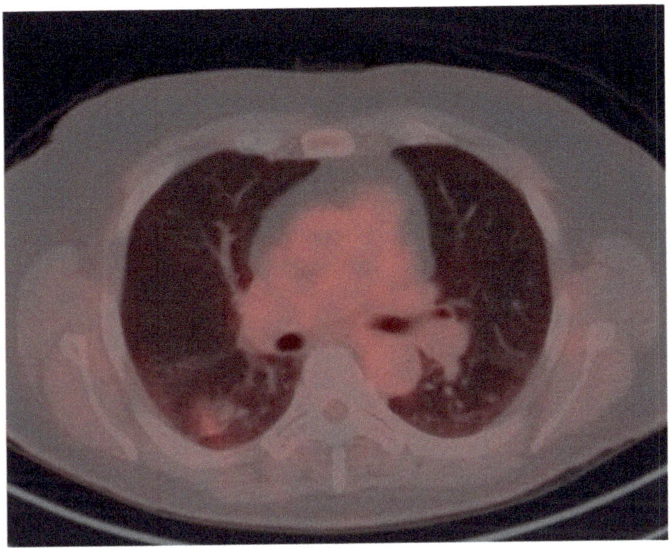

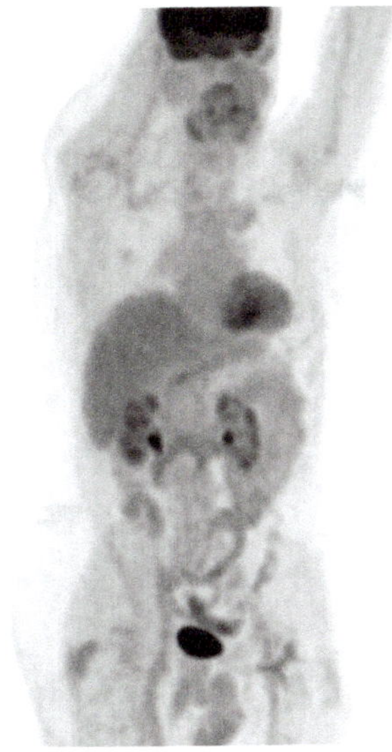

What is your interpretation in light of the new findings?

A. Recommend repeat study in 6–8 weeks after broad-spectrum antibiotic treatment
B. Combined chemoradiation treatment.
C. Further evaluation with CT guided biopsy.
D. PET-CT follow-up in 3–6 months, until 3 years of stability are confirmed.

Answer: C

The lesion appears to have mixed ground glass and solid components, with mildly increased metabolic activity. It is easily accessible through percutaneous approach given its size greater than 10 mm. Degree of metabolic activity has been shown to correlate with the histologic grading as well as the mass of the tumor. Therefore, mixed and purely ground glass lesions must be viewed with a high degree of clinical suspicion for malignancy even if they demonstrate a low metabolic activity.

10. Which of the following statements is true regarding adenocarcinomas of the lung?

A. There is a correlation between the SUV max and survival rate
B. The lepidic growth pattern is the most frequently represented histologic type.
C. They are more frequently central in location.
D. Small cell lung carcinoma (SCLC) and squamous cell carcinomas (SCC) are more frequent in nonsmokers.

Answer: A

Mixed histology pattern is by far the most frequent histologic type, with the acinar subtype frequently more represented. SCC and SCLC are rather typical of smokers and tend to be central in location. Adenocarcinoma are the most frequent histology in never-smokers, especially in younger women. Higher SUV max is frequently associated with poor prognosis and lower survival rate.

References

1. Gardiner N, Jogai S, Wallis A, The revised lung adenocarcinoma classification—an imaging guide. J Thorac Dis. 2014;6(S5):S537–46.
2. Lee HY, Jeong JY, Lee KS, et al., Solitary Pulmonary nodular lung adenocarcinoma: correlation of Histopathologic Scoring and Patient Survival with Imaging Biomarkers. Radiology. 264(3):884–93.

Case 4

11. 55-year-old male presents for staging PET-CT of esophageal carcinoma. Patient also reports radical left neck dissection for squamous cell carcinoma of a branchial cleft cyst 2 years prior to current study; 1/35 nodes was found positive for metastatic disease.

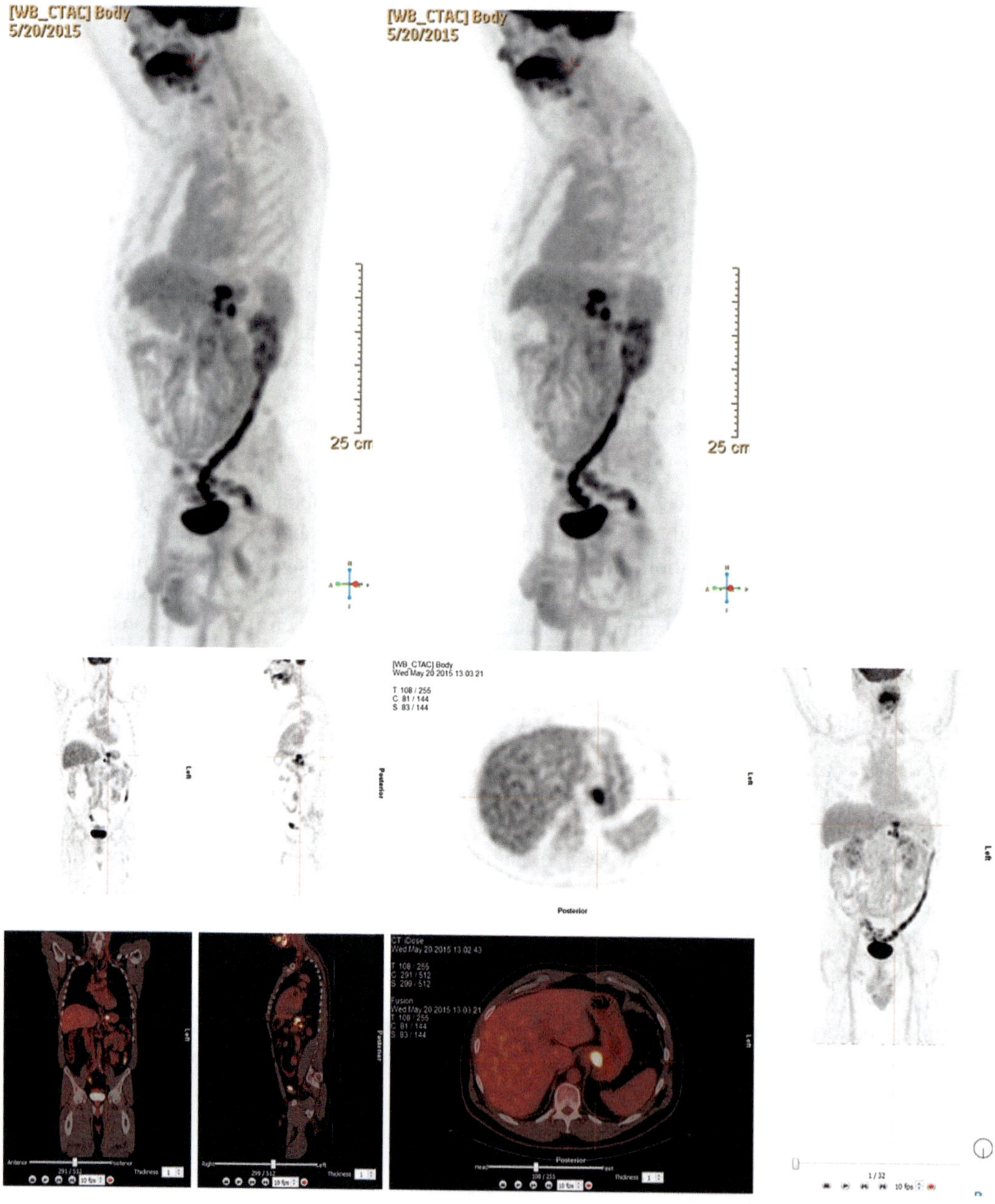

What is your diagnosis?

A. Metastatic head and neck carcinoma to the neck and to the abdomen
B. Metastatic esophageal carcinoma to the neck and to the abdomen
C. Reactive oropharyngeal and cervical uptake from prior surgery and radiation and metastatic lymphadenopathy from GE junction adenocarcinoma
D. GE junction carcinoma and metastatic gastrohepatic lymphadenopathy along with indeterminate cervical and oropharyngeal uptake

Answer: D

The degree of metabolic activity in the head and neck is excessive 18 months after completion of radiation. However, it does not have typical distribution for recurrent malignancy and therefore correlation with direct visualization of the oro-pahryngeal cavity was recommended.

12. Patient presents 3 months later for restaging PET-CT of esophageal carcinoma after neoadjuvant chemotherapy, in preparation to esophagectomy. PET-CT images shown below. ENT evaluation with indirect endoscopy was negative.

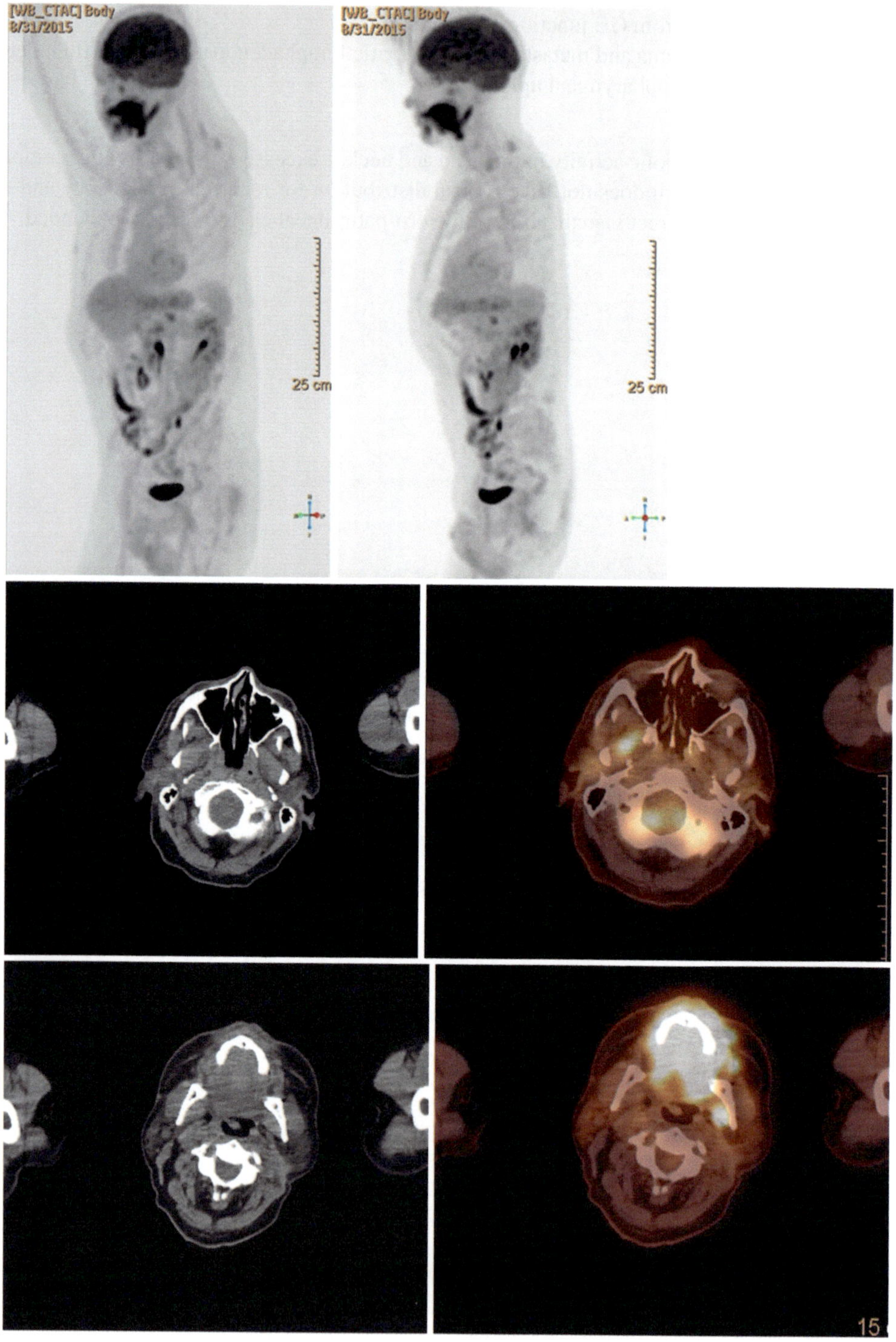

SUV max adjacent to the left mandible is now 20.4

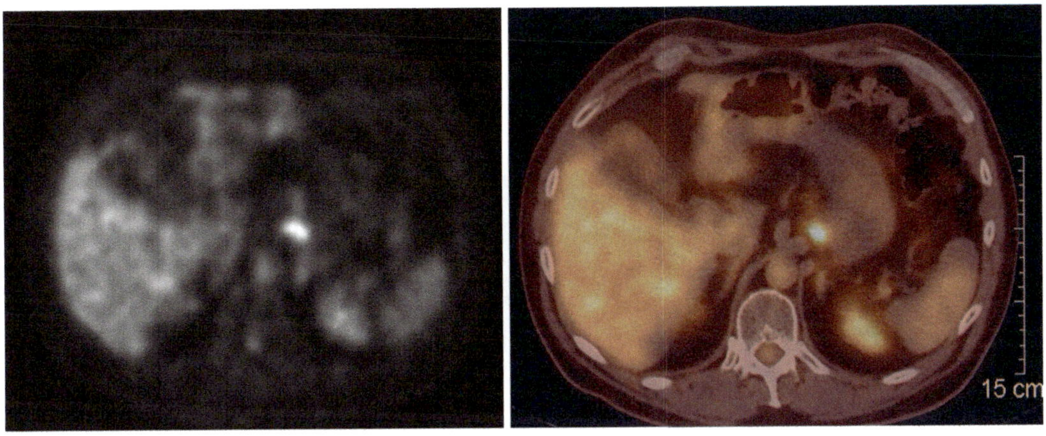

SUV max in the GE junction is now 4.9

What is your diagnosis?

A. Local progression of H and N carcinoma, favorable response of gastro-esophageal malignancy
B. Metastatic esophageal carcinoma to the neck and to the abdomen
C. Reactive lymphadenopathy from prior surgery and radiation in the neck and upper abdomen
D. Favorable response of GE junction carcinoma and gastrohepatic lymphadenopathy, indeterminate or physiologic neck-uptake

Answer: D

There is decreasing metabolic activity in the primary GE junction lesion as well as the gastrohepatic nodes. However, there has been markedly increased metabolic activity in the oral cavity and oropharynx, as well as in the soft tissues medial to the left mandible. The findings are still indeterminate, although could be physiological, particularly the recent ENT evaluation was unremarkable.

13. What is your next step to further evaluate the uptake in the head and neck?

A. CT of the neck with and without contrast
B. Repeat ENT exam with possible random biopsy of the paramandibular structures on the left
C. MRI of the face with IV contrast
D. Repeat patient questionnaire

Answer: D

Although marked, the distribution of metabolic activity does not agree with a neoplastic process; in particular, the uptake medial to the left mandible is linear in distribution. By the time of interpretation, the patient had already left the department. Contacted later in the day, the patient admitted to having been chewing gum the entire prior day, a habit the patient frequently indulged to stimulate salivation after prior cervical surgery and external beam radiation.

Case 5

14. 14-year-old male presents for F18-FDG PET-CT to evaluate enlarging mesenteric nodes and enlarging thymus with chronic abdominal pain. Selected images are shown below.

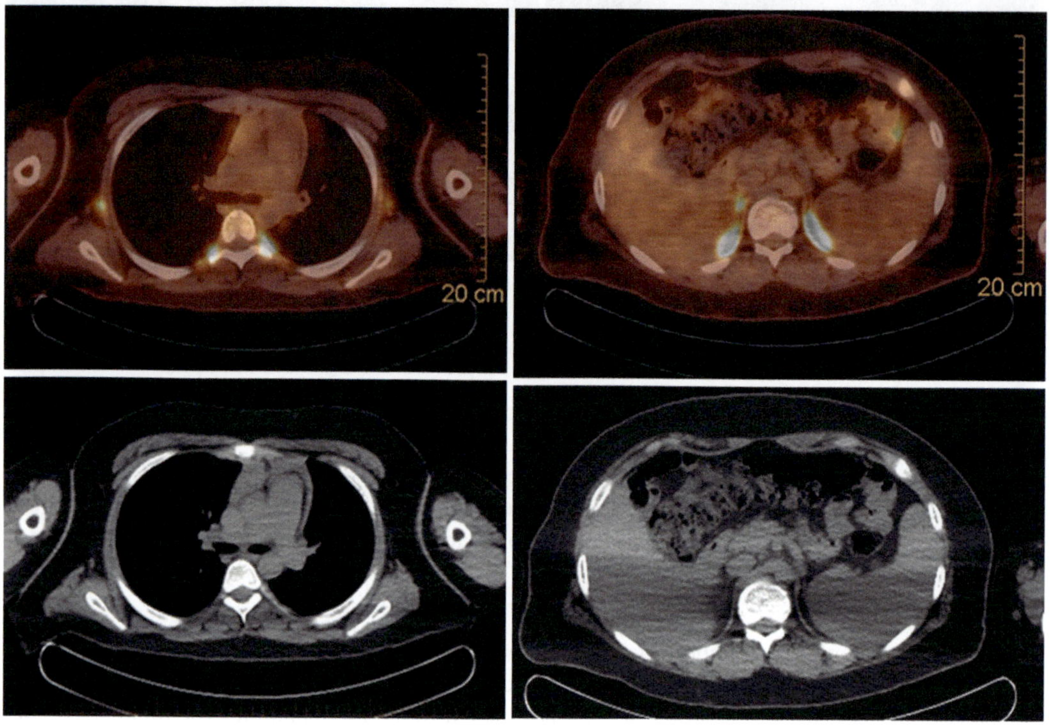

What is your diagnosis?

A. Thymic lymphoma metastatic to the pleural spaces
B. Bilateral adrenal lymphoma
C. Sarcoidosis
D. Normal thymus uptake and physiologic paravertebral uptake

Answer: D

The distribution of metabolic activity does not agree with a neoplastic process. It is symmetric and without associated appreciable soft tissue masses. Sarcoidosis would demonstrate discrete lymphadenopathy.

15. Additional image from the same study provided below.

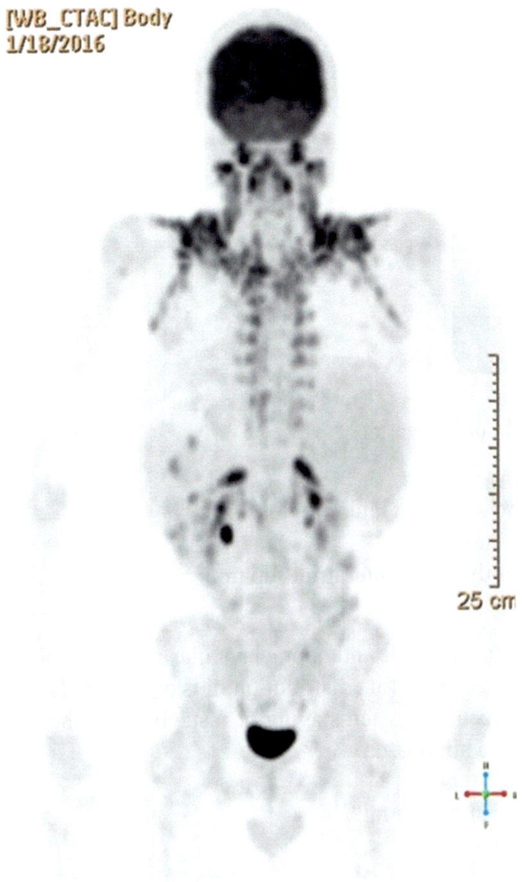

What is your diagnosis?

A. Brown fat activation
B. Bilateral adrenal, pleural, cervical, and supraclavicular lymphoma
C. Sarcoidosis
D. Delayed onset neuroblastoma

Answer: A

The absence of any demonstrable soft tissue masses or lymph nodes at any of the locations of increased metabolic activity, always triangulating to areas of fatty density on associated CT, along with a symmetric distribution and its predominance in the neck and supraclavicular regions, is a clear-cut indication that this patient has brown fat hypermetabolic activity, which is a benign finding. All of the other diagnoses would be associated with some degree of soft tissue abnormality. The contrast CT obtained a few weeks prior demonstrates no abnormality of the adrenal glands (see below).

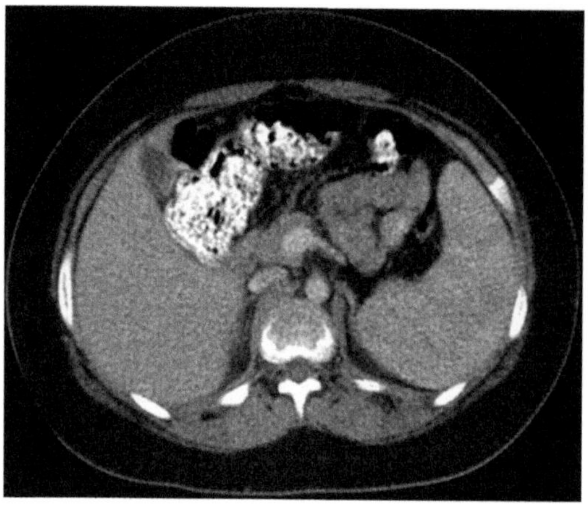

16. Which of the following statements is true about brown fat?
    A. It never occurs in the abdomen
    B. It occurs more frequently in the neck and supraclavicular regions
    C. It occurs more frequently in the mediastinum
    D. Within the thorax, it occurs more frequently in the interatrial septum

    Answer: B

    Distribution limited to, or predominant in, the neck and supraclavicular regions is the most frequent imaging pattern noted on PET-CT. Because of its distribution, it is a potentially dangerous cause of false-positive interpretations, especially when staging or restaging lymphomas and breast carcinomas. Sometimes, it is seen in the abdomen.

17. Which of the following statements is true about brown fat?
    A. It is more frequent in obese patients
    B. It is more frequent in men
    C. It is noted in approximately 5% of patients undergo PET-CT
    D. It is more frequent in summer

    Answer: C

    It is seen in approximately 5% of patients, more frequent in young individuals, lean rather than obese patients, with slight women predominance and more frequently in winter, particularly during cold spells (non-shivering thermogenesis). The day our patient presented the low and high temperatures were respectively −18 and −13 Celsius (−2 and 9 Fahrenheit).

18. Which of the following statements is true about brown fat?
    A. Thermogenin is a protein located in the mitochondria
    B. The fat vacuoles are larger in brown fat than in white fat cells
    C. Barbiturates have been proven to decrease FDG uptake in brown fat
    D. Selective Alfa-1 blocker has been proven to decrease FDG uptake in brown fat

    Answer: A

    Thermogenin, located on the mitochondrial membrane of brown fat cells, generates heat as its name implies. Brown fat cells have more mitochondria and smaller fat vacuoles. Beta-blockers and benzodiazepine have been used with success to decrease brown fat FDG uptake.

*Cronin CG, Prakash P, Daniels GH, Boland GW, et al. Brown fat at PET/CT: correlation with patient characteristics. Radiology. 263(3): 836–43.

Case 6

19. A 19-year-old male presents with 2 day history of a neck mass (large right supraclavicular swelling in supraclavicular fossa, boggy, not fluctuant on physical examination) and an abnormal chest x-ray. Shown below are selected CT images of the neck, chest, abdomen, and pelvis obtained during the admission.

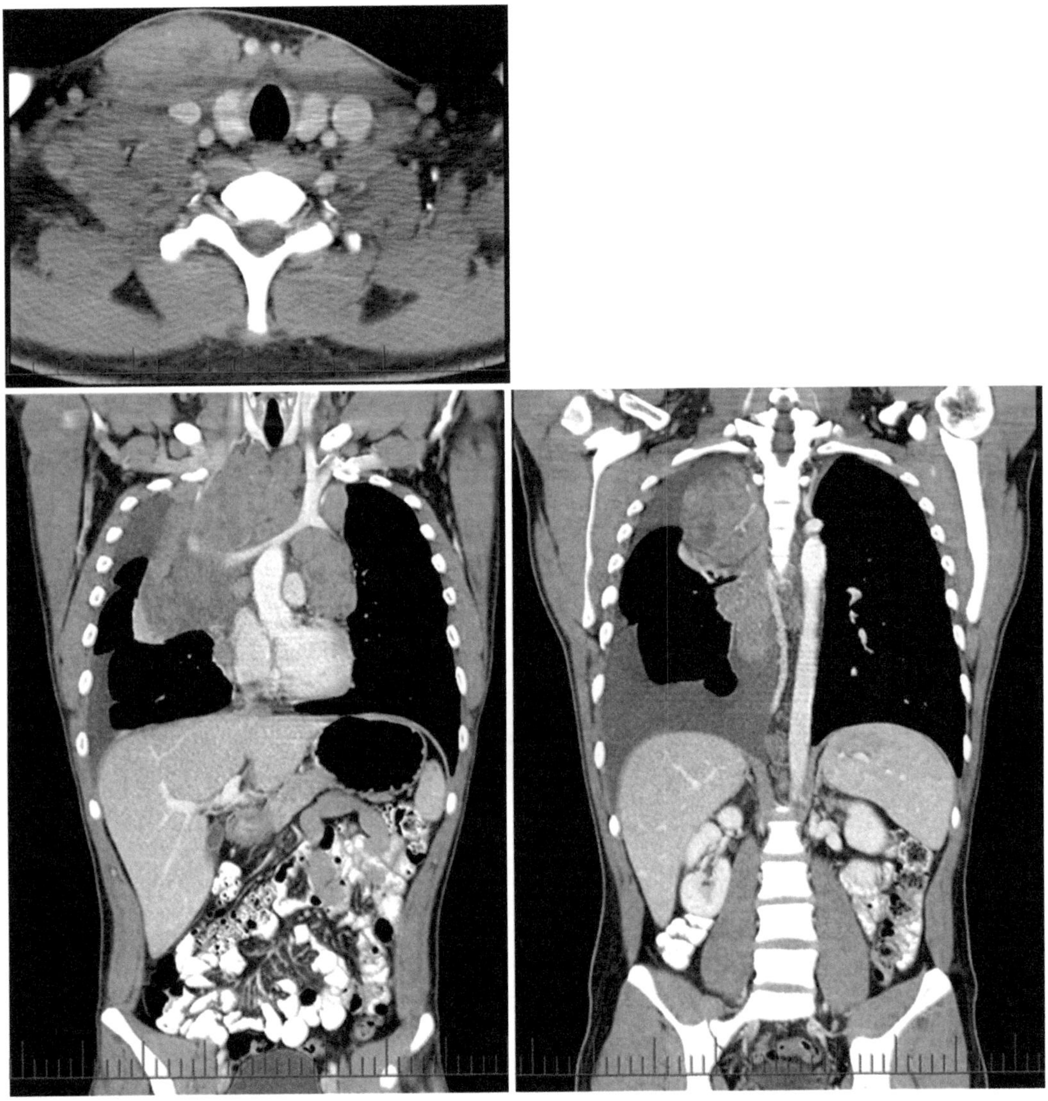

What is your recommendation?

A. Sputum cytology to rule out granulomatous infection
B. Core biopsy or excision of neck mass
C. PET-CT
D. US-guided FNA of abnormal neck nodes

Answer: B

In a patient with bulky lymphadenopathy demonstrating rather homogeneous density and enhancement, lymphoma must be high in the differential diagnosis. Lymphomas can be hard to diagnose at cytology and therefore core or excisional biopsies are most indicated. FDG PET-CT is a universally recognized aid in staging and restaging of lymphomas, once diagnosed. The clinical presentation and the CT appearance are not typical of a granulomatous infection.

20. The patient returned a week later for a staging PET-CT, after being diagnosed with "classic Hodgkin lymphoma of a nodular sclerosis type." Selected images are shown below.

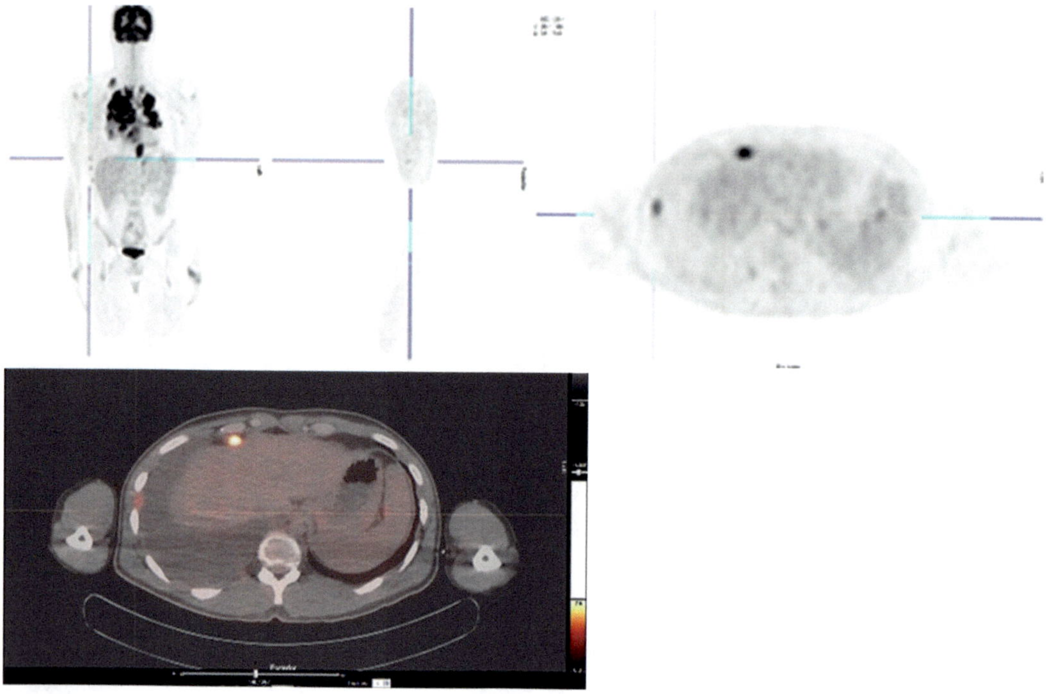

What is this patient's staging?

A. I
B. II
C. III
D. IV

Answer: B

Based on the images provided, there is a metabolic activity in two or more nodal stations located above the diaphragm. No hypermetabolic lymph nodes are seen below the diaphragm.

21. Name the correct location of hypermetabolic activity anterior to the liver.

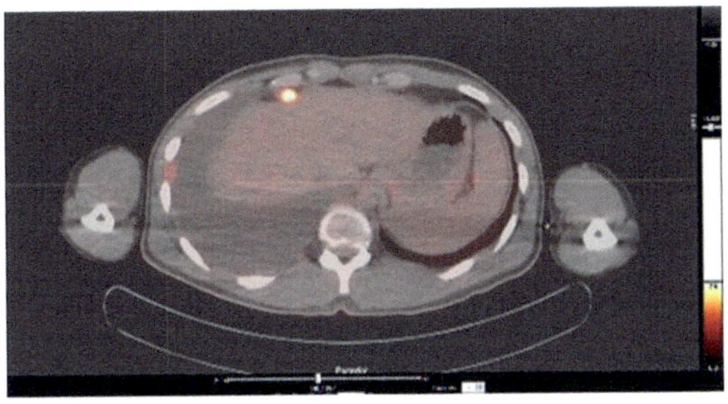

A. Omental
B. Anterior mediastinum
C. Pericardium
D. Cardiophrenic node

Answer: D

Cardiophrenic and retrocrural nodes are intrathoracic, although they may be mistakenly attributed to the abdomen. This can have a negative impact on patient staging, prognosis, and management.

22. Additional images are provided below. SUV max of circled lesions is respectively 2.7 and 2.3.

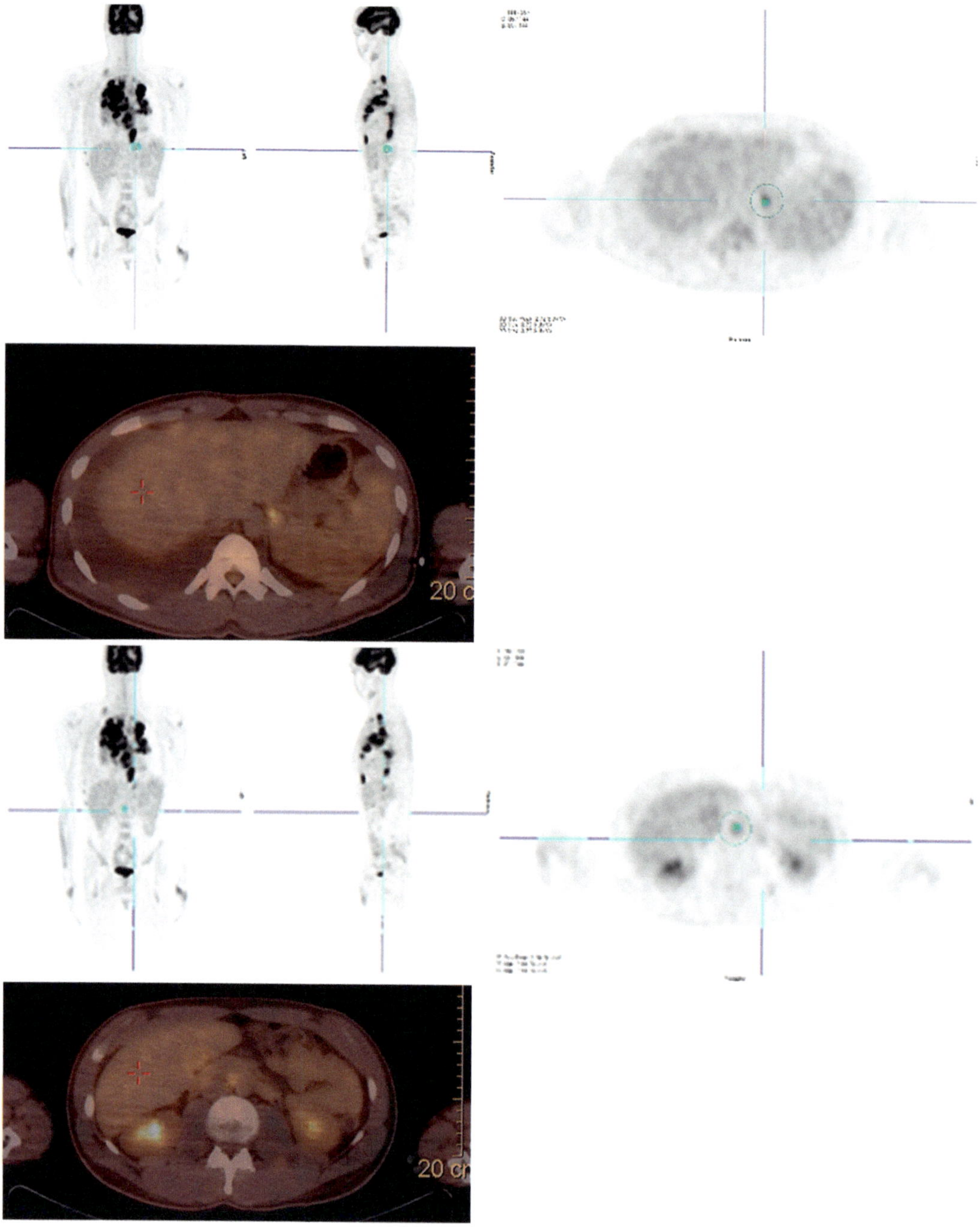

Based on these additional images, what is your interpretation?

A. Reactive non-enlarged subdiaphragmatic nodes
B. Small hypermetabolic retrocrural lymphnodes
C. Subdiaphragmatic Hodgkin Disease, Stage III
D. Inflammatory gastrointestinal uptake

Answer: C

The degree of metabolic activity is worrisome, particularly in light of the subcentimeter size of the hypermetabolic nodes, as demonstrated on the contrast-enhanced CT images obtained a week prior and shown below.

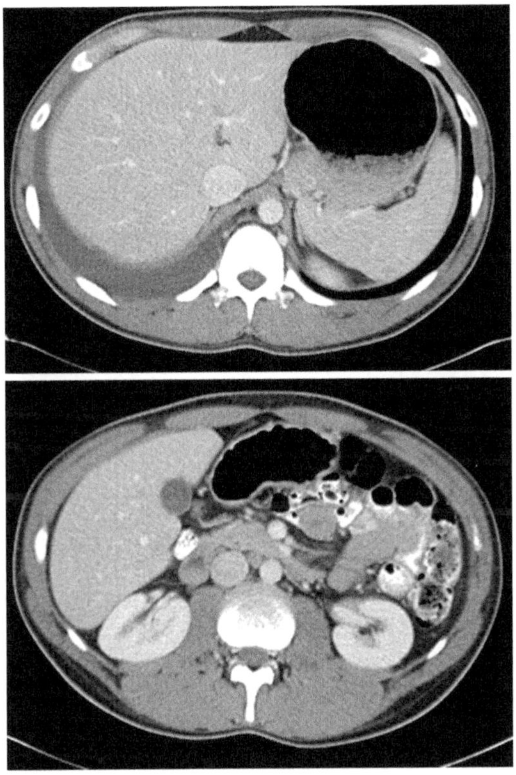

Reference

1. Johnson SA, Kumar A, Matasar MJ, Schöder H, Rademaker J, Imaging for staging and Response assessment in Lymphoma. Radiology. 276(2): 323–338

23. What is the most common site of distant metastasis in head and neck tumors?

    A. Lung
    B. Liver
    C. Bone
    D. Kidney

    Correct answer is A

    By far the most common location is the lung (58%) followed by the liver (29%), bone (22%), and kidney (11%).

24. This 58-year-old male has a history of stage IV rectal adenocarcinoma with metastases to the omentum. He was initially treated surgically and has since received both radiation and chemotherapy. He presents now with some intermittent abdominal pain and fatigue. He has bilateral ureteral stents. The FDG PET/CT study demonstrates suspicious lymph nodes in the pelvis as well as an area of increased metabolic activity in the anterior abdominal wall within the rectus muscle. A focus of new uptake is also seen in his chest. Your thoughts are to:

    A. Ask the technologist about the site of injection
    B. Raise concern of a metastatic focus in the mediastinum
    C. Notify the referring physician of tumor thrombus in the superior vena cava
    D. Repeat the scan the next day after a bowel prep

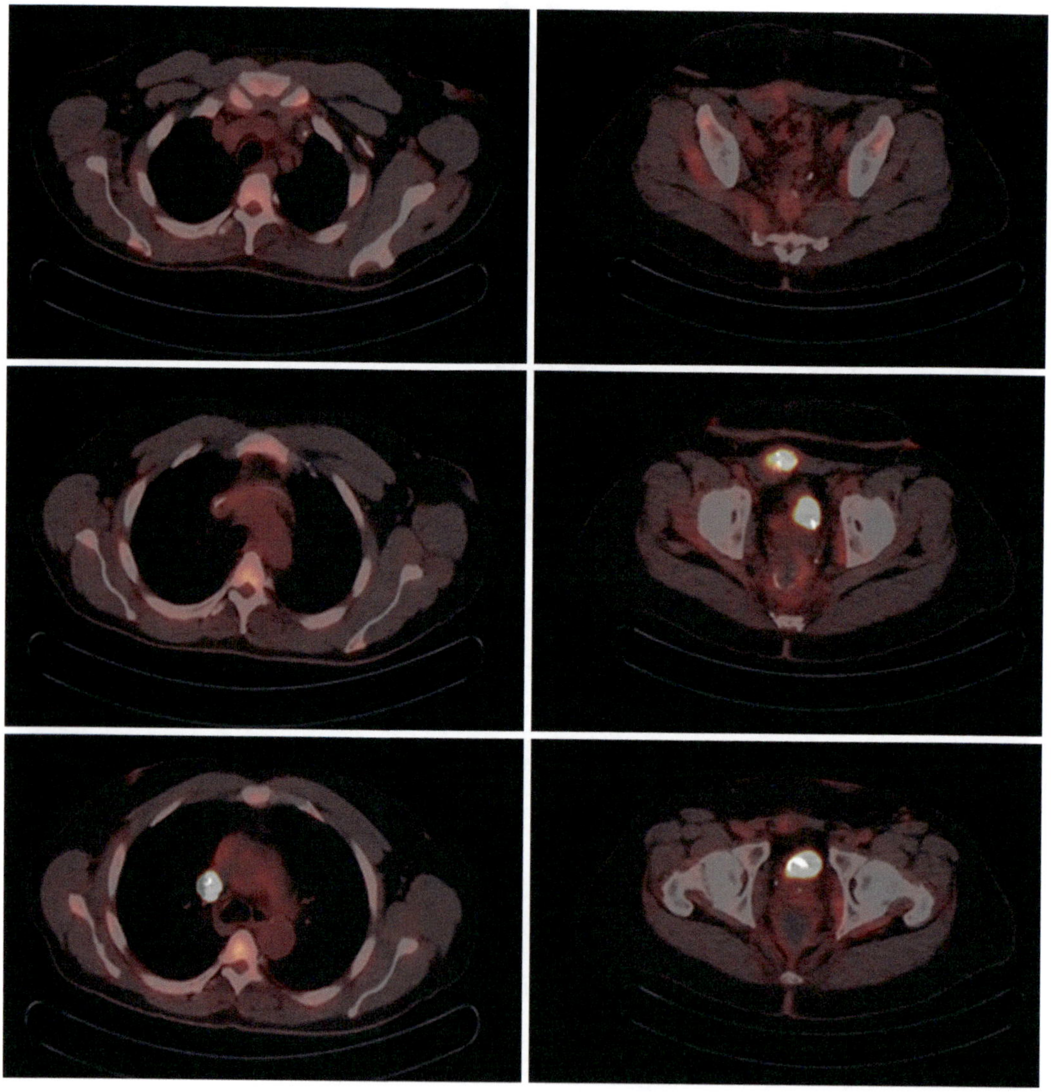

Answer: A

The images demonstrate a focus of intense uptake located in the superior vena cava. The patient has a port-a-catheter in place and insisted on the injection of tracer through that access. Artefactual uptake may occur with tracer injected through a central line, and this should be avoided if possible.

As the focus is within the SVC and at the tip of the port-a-catheter, this is not a likely focus of metastatic disease or tumor thrombus.

Metastatic disease is noted in the abdominal wall. There is no need for the scan to be repeated after a bowel prep.

25. After viewing the MIP image from a FDG/CT study (left) you ask your technologist to:

A. Ask the patient if they exercised prior to the study
B. Ask the patient if they had anything to eat in the past 4 h
C. Confirm that the correct body weight was entered into the FDG/CT study header
D. Confirm that the dose of F18 FDG was entered correctly into the study header

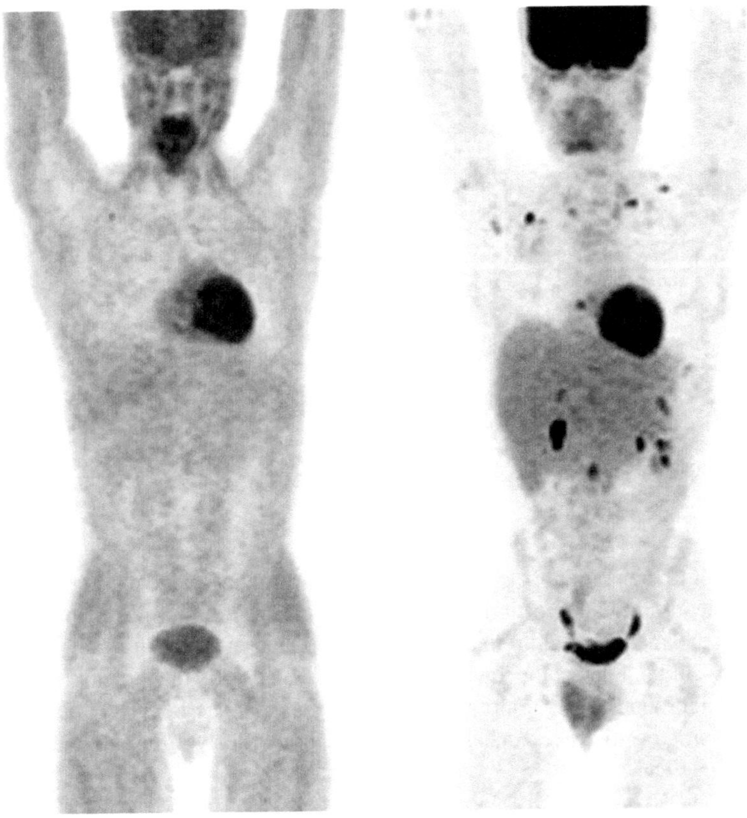

Answer: B

The case illustrates the importance of fasting prior to the injection of tracer. A rising blood sugar from a meal will trigger the release of insulin and the transport of F18 FDG intracellularly. The muscular uptake is very uniform, there is little liver uptake, and even the brain which is usually one of the hottest areas on FDG imaging appears less intense.

In a patient who has recently exercised, the muscle groups used are more intense than others less used.

Entering the incorrect dose of F18 FDG or the body weight will not alter the biodistribution of the tracer, but will result in incorrect SUV values.

Also, the study illustrates the decrease in sensitivity for detection of metastatic disease when the protocol is not properly followed. The patient initially said that he had no breakfast prior to the injection. After the technologist asked again, he said that "Maybe he had a little." The repeat study identifies areas of metastatic disease that were not seen on the initial scan.

26. This 70-year-old male smoker presents for evaluation of a solitary pulmonary nodule identified on a recent CT. The nodule in the lung demonstrates a SUV max value of 5.7 and measures 1.2 × 1.8 cm in size. You note additional findings on the scan. You suspect:

A. The patient had a recent Ga67 study and this is residual tracer
B. The source of the lung nodule is from a head and neck primary
C. The lung nodule is a well-differentiated neuroendocrine tumor primary
D. The patient was noncompliant with instructions

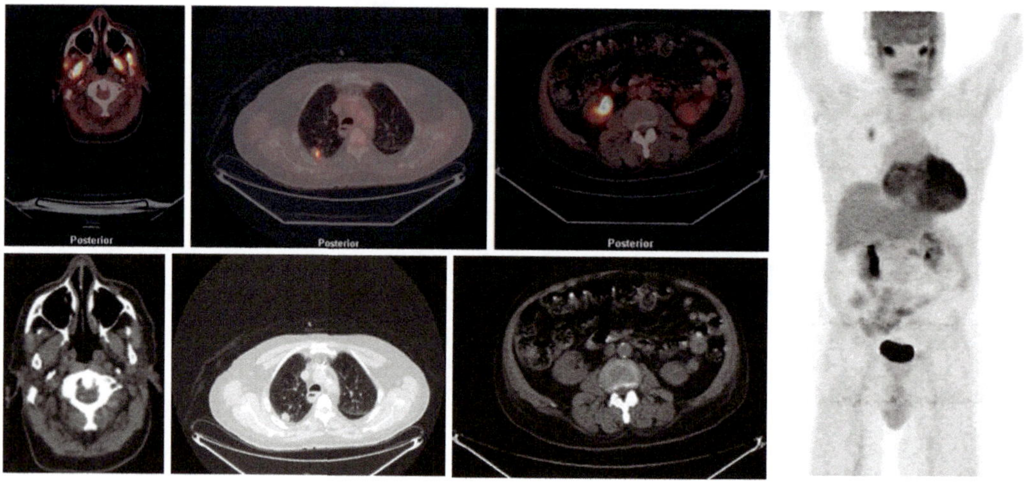

Answer: D

Despite being told not to, the patient began chewing gum during the uptake phase of the study. The areas seen in the head and neck portion of the study represent the muscles of mastication. During the uptake phase, patients should remain quiet, relaxed, and in a dim-lit room.

The PET/CT study would not detect the lower energies of a Ga67 study that are not coincident.

Well-differentiated neuroendocrine tumors have lower uptake on F18 FDG imaging, but express high levels of somatostatin receptors. Those that are more poorly differentiated are more likely to concentrate F18 FDG.

27. This 36-year-old female presented with a history of a parotid tumor (resected) that was identified as non-Hodgkin lymphoma. This FDG/CT study was performed as part of the initial evaluation. After review of the images, the patient should be questioned regarding:

A. Prior family history of breast cancer
B. Failing to comply with fasting prior to imaging
C. Pregnancy or lactation
D. Recent MRI breast imaging

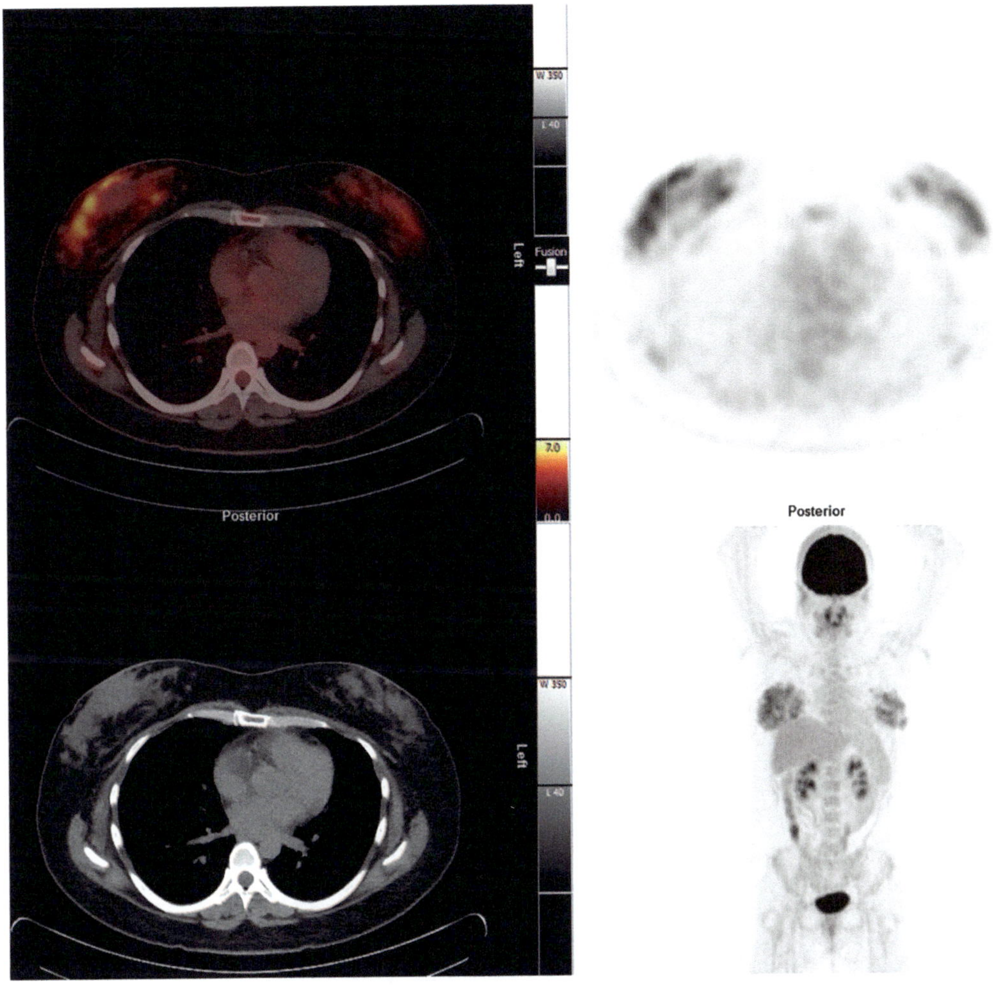

Answer: C: Pregnancy or lactation.

Pregnancy and lactation should be considered whenever increased uptake in the breasts is noted. In this case, the patient had been breastfeeding for the prior 6 months. There appears to be some association with suckling since greater uptake is noted in the right breast that is used more frequently. Little FDG is secreted in breast milk and decreasing radiation exposure to the infant is primarily a function of decreasing close contact. With a half-life of less than 2 h, precautions after F-18 PET study are reasonable for 24 h post-injection.

While breast cancer is an important history to be aware of, a bilateral pattern of diffuse uptake would be unusual. Failure to fast appropriately often forces FDG into the muscular tissues due to insulin release. Breast uptake in this situation would be unlikely. MRI imaging of the breast would have no effect on FDG uptake.

Excellent JNM paper—JNM—2001; 42:1238–1242

28. This 82-year-old male patient presented for an FDG/CT study to evaluate the squamous cell carcinoma of his tongue. Review of the images of the head below indicates decreased cerebral cortical uptake.
Consideration should be given to:

A. The possibility of early frontotemporal dementia
B. Vascular dementia involving the watershed area of the brain
C. Incorrect attenuation correction
D. Time of flight failure

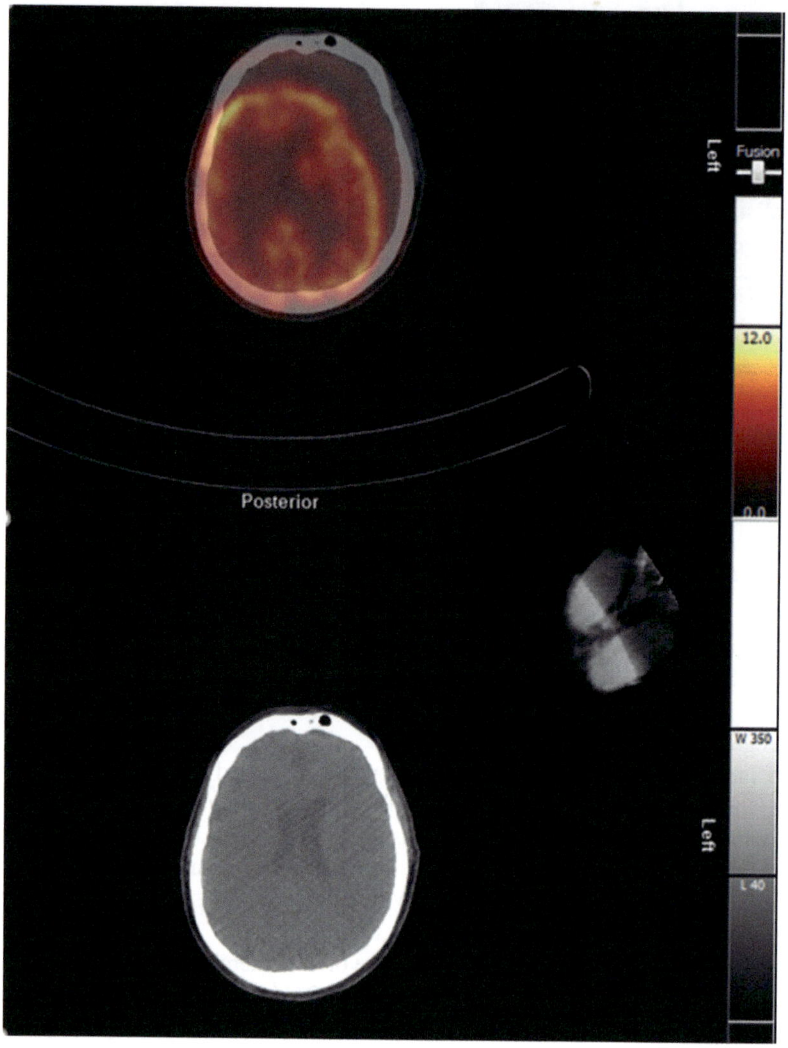

Answer: C: Incorrect attenuation correction.

As can be seen in the fusion image, the brain is clearly outside of the cranial vault. The patient moved their head between the PET and the CT portion of the study. A repeat dedicated study of the head (below) demonstrates symmetric uptake bilaterally.

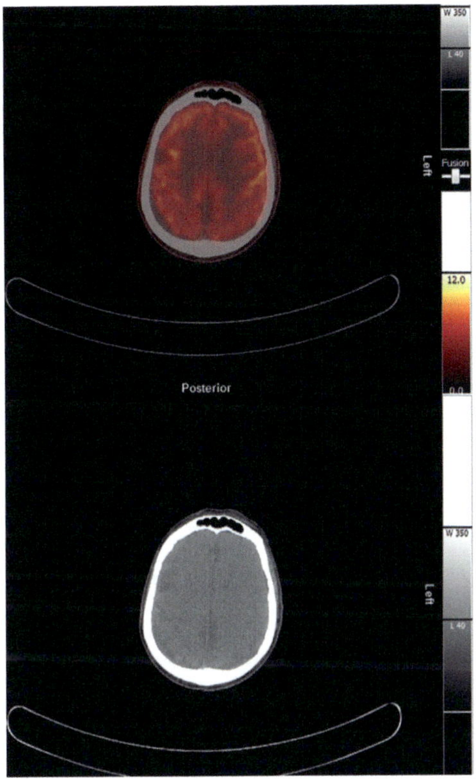

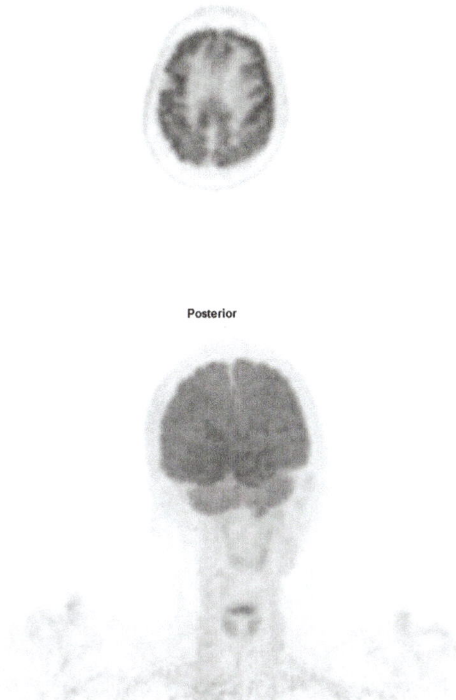

The area of decreased activity seen is not typical for frontotemporal dementia. Although a vascular etiology may be suspected, the poorly fused images are most consistent with a technical artifact that must be corrected prior to establishing a pathologic diagnosis.

Absence of time of flight would not cause this appearance.

Case 7

29. What is the mechanism of tracer distribution on the image shown below?

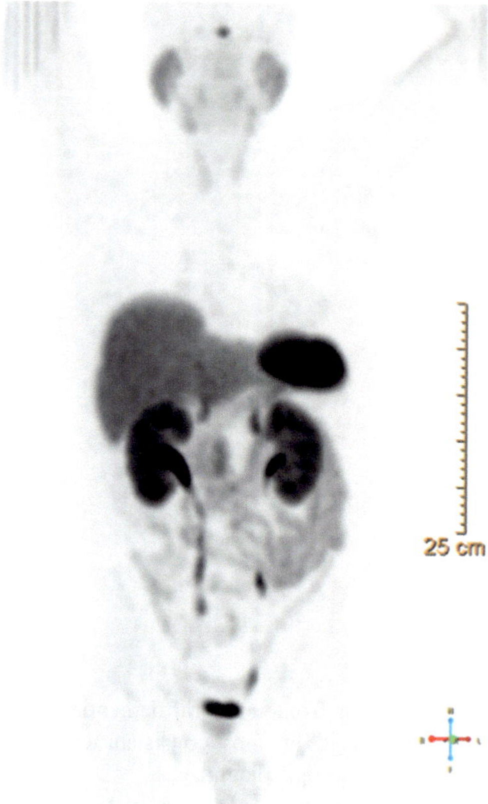

A. Amino acid transport
B. Receptor binding
C. Metabolic trapping by phosphorylation
D. Accumulation in adrenergic tissues

Answer: B

The above image represent a **^{68}Ga DOTATATE scan.**

^{68}Ga DOTATATE is used for PET imaging of neuroendocrine tumors. It is attached to the somatostatin receptors, particularly subtype 3 (SSTR2). These receptors are overexpressed in well-differentiated neuroendocrine tumors. When the tumor becomes less differentiated, it will demonstrate more uptake on F-18 FDG imaging.

Fluciclovine (^{18}F), also known as Axumin, is taken up by amino acid transporters. It is approved by the FDA for prostate cancer imaging, although using it for other types of cancer such as gynecological malignancies is under research.

F-18 FDG, fluorine-18 fluorodeoxyglucose is a glucose analog that is widely used in the initial diagnosis and follow-up of several types of solid tumors. After phosphorylation, F-18 FDG is trapped in the cells with higher concentration in the tumor cells that typically utilize glucose. This "trapping" enables us to detect the site of the primary tumor and the possible areas spread.

MIBG (metaiodobenzylguanidine), labeled either to iodine-131 or relatively recently to iodine-123, is a norepinephrine analog. It accumulates in adrenergic tissues. Although it may be used in a variety of neuroectodermal tumors, it is commonly used in the evaluation of pheochromocytomas, followed by neuroblastomas. It is to be noted that the normal adrenal glands may be seen, particularly with I-123 MIBG but the uptake is usually mild and often symmetrical.

30. What is the physical half-life of the radiopharmaceutical used in the prior study and how is it produced?

    A. 78 h, cyclotron produced
    B. 78 min, generator produced
    C. 68 h, cyclotron produced
    D. 68 min, generator produced

    Answer: D

    In medical practice, the most commercial radioisotopes used are gallium-67 and gallium-68.

    Gallium-68 is a positron-emitting isotope generated from germanium-68 in germanium-68/gallium-68 generator. It has a half-life of 68 min and it decays by positron emission. It is mainly used for the detection of well-differentiated neuroendocrine tumor.

    Gallium-67 citrate is produced in a cyclotron with a much longer half-life of 78 h. It decays by electron capture to Zn-67.

31. Which of the following tumor will likely demonstrate uptake on Ga-68-DOTATATE PET/CT?

    A. prostate cancer
    B. carcinoid tumor
    C. breast cancer
    D. osteosarcoma

    Answer: B

    Carcinoid tumor and other neuroendocrine tumors with higher somatostatin receptors (SSTRs) are best evaluated by Ga-68-DOTATATE PET/CT. As mentioned before, only when these tumors become less differentiated, they might be seen better on F-18 FDG studies.

    Prostate cancer, breast cancer, and osteosarcoma mainly utilize glucose, and therefore are seen on F-18 FDG scans.

32. Ga-68 DOTATATE has the highest affinity to which of the following receptors?

    A. SSTR 1 (somatostatin receptor subtype 1)
    B. SSTR 2
    C. SSTR 3
    D. SSTR 4

    Answer: B

    Somatostatin receptors (SSTRs) are present on the cell surface of neuroendocrine cells. Ga-68 DOTATATE has the highest affinity for SSTR subtype 2 (SSTR 2), which is commonly overexpressed in neuroendocrine tumors.

*Hofman MS, Lau WF, Hicks RJ. Somatostatin receptor imaging with 68Ga DOTATATE PET/CT: clinical utility, normal patterns, pearls, and pitfalls in interpretation. Radiographics. 2015;35(2):500–16.

*Panagiotidis E, Alshammari A, Michopoulou S, Skoura E, Naik K, Maragkoudakis E, Mohmaduvesh M, Al-Harbi M, Belda M, Caplin ME, Toumpanakis C, Bomanji J, Comparison of the impact of 68Ga-DOTATATE and 18F-FDG PET/CT on clinical management in patients with neuroendocrine tumors. Nucl Med. 2017; 58:91–6.

33. A 63-year-old male with suspected lung cancer is referred for PET/CT to evaluate the underlying primary and metastatic tumor burden. On CT imaging he was found to have additional lesions in the right hepatic lobe and left adrenal gland. When attempting to differentiate benign from malignant adrenal lesions on F-18 FDG PET imaging, you will:

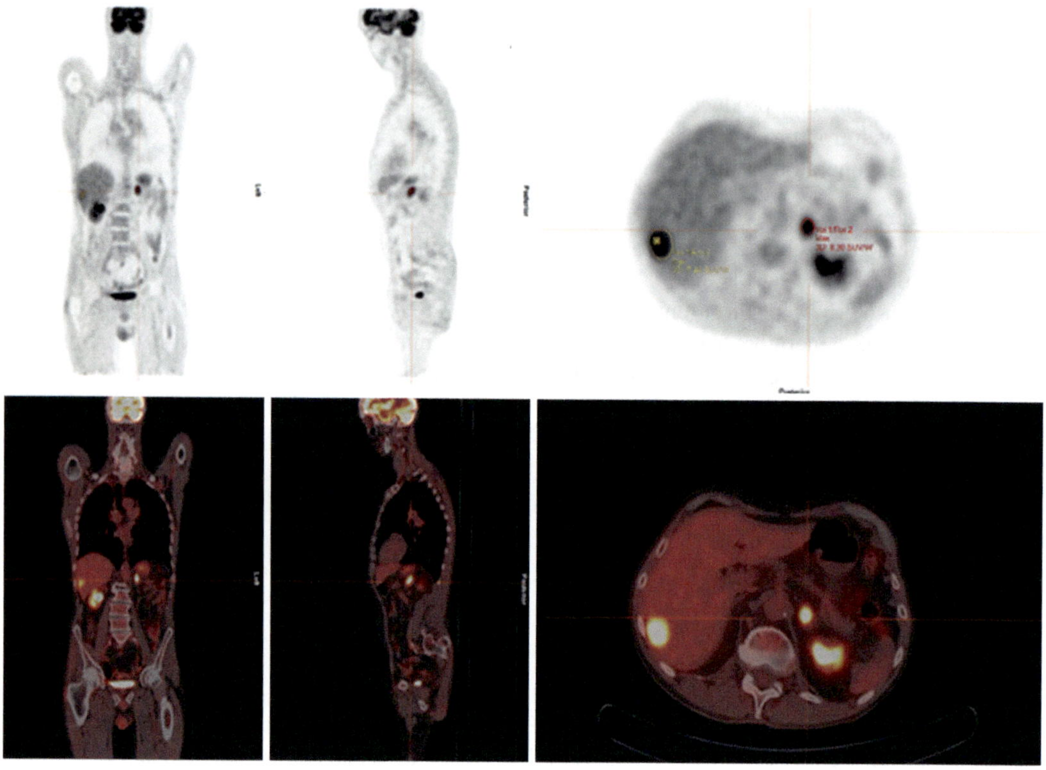

A. Adrenal lesion with SUV similar to the lung parenchyma
B. Adrenal lesion with SUV similar to the mediastinum
C. Adrenal lesion with SUV higher than the mediastinum but less than the liver
D. Adrenal lesion with SUV similar to the liver

Answer: D

Up to 50% of adrenal masses in patients with a known extraadrenal cancer are malignant. Characterization of these masses in this subset of patients is therefore critical to stage the primary disease, direct therapy, and predict prognosis. Several F-18 FDG PET techniques have been described to differentiate benign from malignant adrenal disease. However, Standardized uptake ratio (SUR) technique uses adrenal-to-liver activity to characterize adrenal masses in patients with cancer. It is calculated by dividing adrenal SUV by liver SUV values. It was suggested to use a certain ratio number to differentiate adrenal diseases (with the ratio of at least more than one favoring malignant etiology).

Qualitative analysis of adrenal FDG uptake is performed visually by comparing the signal intensity of the adrenal gland with that of the liver. Activity can be scored as 0 if less than the liver, 1 if equal to the liver, 2 if moderately increased compared with the liver, or 3 if markedly increased compared with the liver. It is suggested, that score of 2 or 3 is considered positive for malignancy. In this case, the signal intensity of the adrenal gland markedly increased compared with the liver, which is supporting malignant adrenal disease.

Case 8 This is a 25-year-old Asian male with a history of Ewing sarcoma. He presented with back pain and a PET/CT was performed.

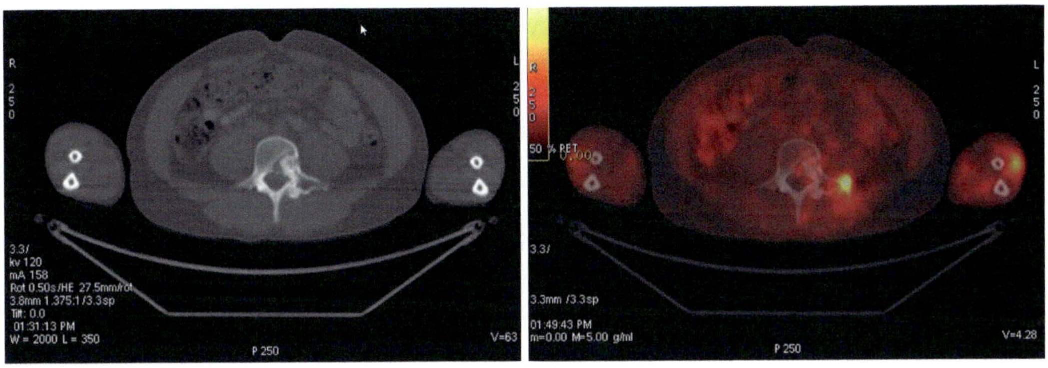

34. The highest incidence of Ewing sarcoma usually occurs in which decade?

    A. First
    B. Second
    C. Third
    D. Fourth

    Answer: B
    Most cases are diagnosed between the ages of 10 and 20.

35. Which statement is true about metastatic lesions?

    A. They are typically diaphyseal in location
    B. They tend to spare the pelvic bones
    C. Lung metastases are uncommon
    D. Pathological fracture is rare

    Answer: A
    Metastatic lesions favor the diaphysis and are more commonly seen in the pelvis and long bones of the lower extremities. Pulmonary metastases are fairly common and in some series actually have better survival prognostication than extrapulmonary metastases.

    Pathologic fractures may be the presenting factor and may also be seen after radiation therapy. Radiation therapy is reserved for inoperable tumors because it may affect the natural growth or the length of the affected limb.

    Bellan DB, et al. Ewing's Sarcoma: epidemiology and prognosis for patients treated at the pediatric oncology institute, IOP-GRAACC-UNFESP. Rev. Bras Ortop. 2012; 47(4):446–50.

36. This gentleman has a squamous cell carcinoma of his tongue. This PET/CT was done as part of his initial workup for tumor staging. Abnormalities were found in his left lung. What is the likely etiology of the abnormality?

    A. Fungal infection
    B. Acute histoplasmosis
    C. Tuberculosis
    D. Aspiration pneumonia

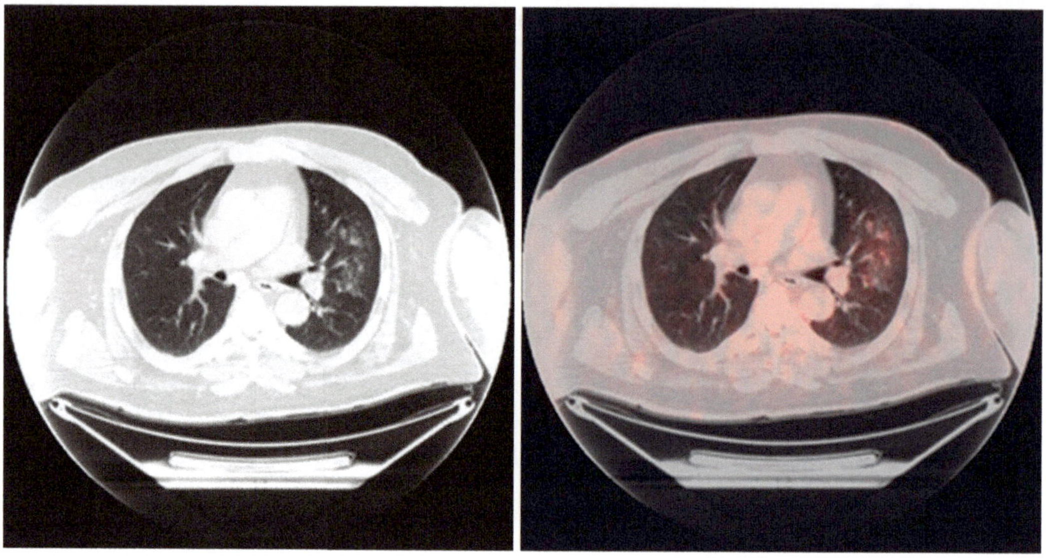

Answer: D

Swallowing difficulties are fairly commonly seen in patients with head and neck cancers. The mild uptake seen in multiple ground glass densities and some areas of tree-in-bud densities make aspiration pneumonitis a likely culprit. This was confirmed on the swallow study and resolved on subsequent PET/CT studies.

Case 9 60-year-old male patient originally referred for evaluation of a newly diagnosed left lung nodule on a chest CT (Fig. 1, Scan b) performed for complaints of shortness of breath. The patient returned 2 years following the first scan for further evaluation of possible progression of disease (Fig. 1, Scan a).

37. Based on the CT characteristics of the left lung nodule shown above, in combination with the different levels of metabolic activity shown in both scans, what is the most likely histopathology of this nodule?

A. Calcified granuloma
B. Squamous Cell carcinoma
C. Carcinoid tumor
D. Adenocarcinoma

Answer: C

The nodule demonstrates soft tissue density (HU) without internal calcifications, well-defined borders with very low level of FDG uptake (bottom scan, Scan b). Although squamous cell carcinoma can be well-circumscribed, these nodules typically demonstrate a higher level of FDG uptake. Adenocarcinoma most commonly demonstrates low to very low levels of FDG uptake however irregular borders and mixed densities are not uncommon. The CT characteristics of this nodule in combination with the low level of FDG uptake and intense level of uptake on the more specific radiotracer (see next question) is most consistent with the typical presentation of a carcinoid tumor.

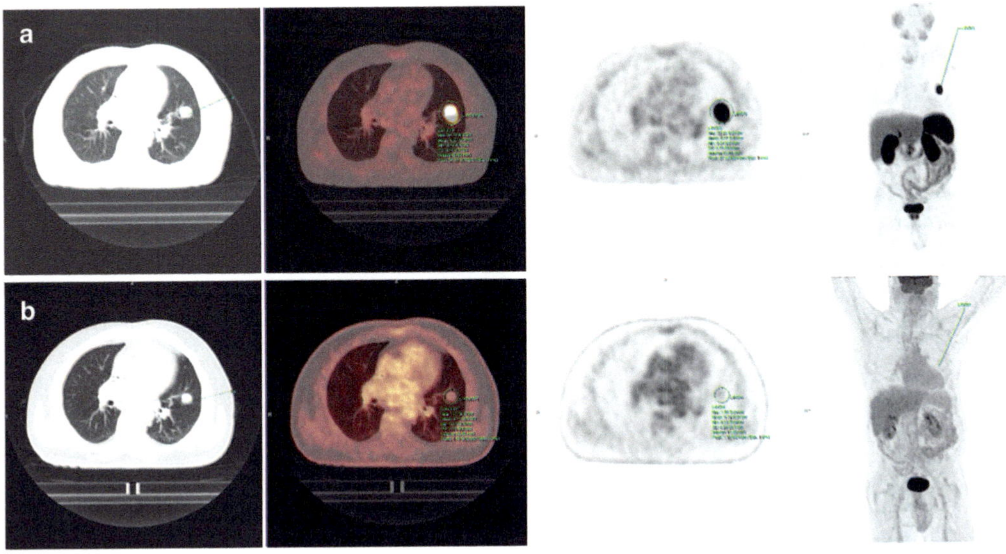

Fig. 1 Current scan (**a**). Old scan (**b**)

38. What radiotracer was used for the scan shown in the Top Scan (a) of Fig. 1?

    A. In-111 Pentetrotide
    B. Tc-99 m MIBG
    C. Ga-68 DOTATATE
    D. F-18 Fluciclovine

    Answer: C

    The excellent spatial resolution and image quality are consistent with the use of a PET tracer. The absence of brain activity helps determine that ^{18}F-FDG was not used. The presence of intense physiologic splenic, genitourinary tract, and adrenal gland uptake is consistent with the distribution of the somatostatin imaging agent Ga-68 DOTATATE [3]. In fact, the CT characteristics of this nodule in combination with the low level of FDG uptake and intense level of Ga-68 DOTATE uptake is most consistent with the typical presentation of a carcinoid tumor [1, 2].

    Fluciclovine shows intense hepatic and pancreatic activity.

39. What is the most likely explanation for the intense liver uptake shown in scan a (Top scan)?

    A. Lymphomatous involvement
    B. Anemia
    C. Physiologic uptake
    D. Chemotherapy-induced changes

    Answer: C

    Somatostatin receptors present in the membranes of neuroendocrine cells are a target for the somatostatin analog Ga-68 DOTATATE. For this reason expected physiologic uptake can be seen in the pituitary gland, salivary glands, spleen, liver, and adrenal glands in addition to any tissue containing neuroendocrine cells [3].

References

1. Carter BW et al. Small cell lung carcinoma: staging, imaging, and treatment considerations. RadioGraphics. 2014;34:1707–21.
2. Kligerman S et al. A radiologic review of the new TNM classification for lung cancer. AJR. 2010;194:562–73.
3. Thomas AH et al. Appropriate use criteria for somatostatin receptor PET imaging in neuroendocrine tumors. Soc Nucl Med Mol Imaging. 2017.

Cardiovascular System

Medhat Sam Gabriel, Bital Savir-Baruch, Robert H. Wagner, Bruch J. Barron, Davide Bova, Aamna Hassan, Stephanie Rodriguez-Santiago, and Yekaterina Kucerova

1. A 50-year-old male with no past medical history of coronary artery disease (CAD) presented to the emergency department with severe retrosternal chest pain. The patient was immediately injected with 25 mCi of Tc-99 m Sestamibi and imaging was performed 3 h later. The gated SPECT demonstrated normal wall motion and thickening of the myocardial walls with estimated left ventricular ejection fraction is 57% (normal value is above 50%). The most likely diagnosis is:

   (a) Normal myocardial scan with some artifact
   (b) Ischemia in one of the territories of LAD artery
   (c) The study is unreliable because scanning was late
   (d) The study is incomplete and stress imaging is needed

M. S. Gabriel (✉) · B. Savir-Baruch · R. H. Wagner · D. Bova
Division of Nuclear Medicine, Department of Radiology, Loyola University Medical Center, Maywood, IL, USA
e-mail: sgabriel@lumc.edu; Bital.savir-Baruch@lumc.edu; rwagner@lumc.edu; dbova@lumc.edu

B. J. Barron
Division of Nuclear Medicine and Molecular Imaging, Department of Radiology and Imaging Sciences, Atlanta, GA, USA
e-mail: bjbarro@emory.edu

A. Hassan
Department of Nuclear Medicine, Shaukat Khanum Memorial Cancer Hospital and Research Center, Lahore, Punjab, Pakistan

S. Rodriguez-Santiago
Department of Nuclear Medicine, C.W. Bill Young VA Medical Center, Bay Pines, FL, USA
e-mail: stephanie.rodriguezsantiago@va.gov

Y. Kucerova
Division of Nuclear Medicine, Department of Radiology, Loyola University Medical Center, Maywood, IL, USA

Department of Radiology, Cook County Healthy, Chicago, IL, USA
e-mail: yekaterina.kucerova@cookcountyhealth.org

M. S. Gabriel, B. Savir-Baruch (eds.), *RadTool Nuclear Medicine MCQs*,
https://doi.org/10.1007/978-3-030-69281-0_4

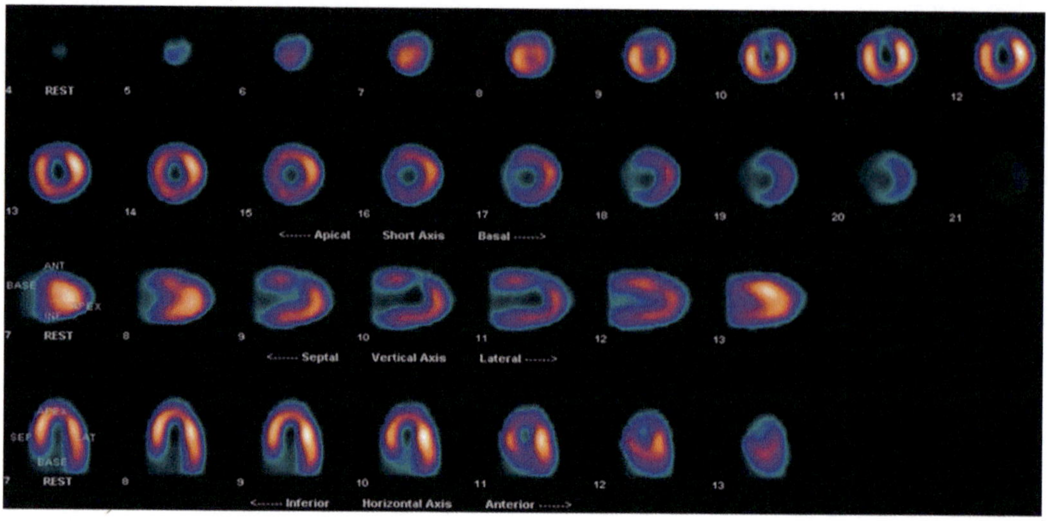

Patients with chest pain are seen routinely in the emergency department. In this patient with typical chest pain, the goal is to rule out serious etiology such as myocardial ischemia or infarction. Some medical centers have cardiac catheterization teams available within a short period of time to evaluate these patients and perform stents if needed. Other hospitals depend on three negative sets of cardiac enzymes, which take about 24 h. This frequently requires admitting the patient to the hospital. A large percentage of these patients will be discharged the next day after negative enzyme sets. These "unnecessary" admissions impose a high cost on the health system. For this reason, some hospitals employ a protocol that evaluates these patients with myocardial imaging, the so-called chest pain protocol. This protocol is based on the fact that the abnormalities seen on the perfusion scan following ischemia usually persists for a few hours after the attack.

Careful evaluation of these scans is crucial. If the scan is completely normal, it is safe to discharge the patient. It is to be understood that normal scan does not exclude the existence of coronary artery disease. It simply indicates that there is no immediate need for hospital admission and further evaluation can be done later.

The quality of the above scan is good. There is a perfusion defect in the apex and distal anterior wall up to the mid ventricle, in the distribution of the distal left anterior descending (LAD) artery. There are no obvious artifacts. In females, breast artifact may result in some decreased activity in the anterior wall but this kind of artifact is unlikely in males. In addition, the defect is transmural and likely to be real.

The images, as shown above, cannot differentiate myocardial infarction from ischemia. However, the additional information from gated SPECT can help make the possible diagnosis. Unless the myocardial infarction is quite small, the wall motion is usually abnormal with focal hypokinesis and likely decrease in the global ejection fraction (EF). Ischemia, on the other hand, may demonstrate normal wall motion and EF. Based on these facts, the likely diagnosis in this patient is ischemia.

Tc-99 m Sestamibi fixes to the mitochondria of the myocardial calls at the time of injection. Scanning at a later time will still reflect the perfusion defect or defects present at the time of injection. This is different from Thallium-201, which, over time moves extra and intracellularly constantly, and therefore the images will reflect the status of perfusion at the time of scanning.

Additional information might be concluded if this patient undergoes stress imaging, specifically if he has other areas of ischemia that might become evident with stress. However, the management of this patient with chest pain and evidence of ischemia is an urgent cardiac catheterization and revascularization, if needed. In fact, the cardiac angiogram showed a single vessel (LAD), high-grade ischemia, and a stent was placed. In conclusion, this test directed the ER physician to admit the patient for further evaluation. If the test is negative, it is appropriate to discharge the patient because the origin of chest pain is unlikely to be cardiac. Still, the patient will need further investigations but not urgent.

The correct answer is b.

2. A 65-year-old male with known critical stenosis (>80%) of the proximal left anterior descending (LAD) artery underwent a treadmill stress myocardial perfusion scan. The patient exercised for 7 min on the Bruce protocol achieving a heart rate of 110 beats/min and blood pressure of 170/95. The myocardial perfusion scan is shown below. The patient stated that he did not take any medication on the day of the study. The most accurate statement explaining the findings on the scan is:

   (a) The patient is probably on a beta-blocker.
   (b) The patient is probably on sublingual nitroglycerin
   (c) The patient should be evaluated with a pharmacological agent, rather than physical exercise because he is older than 60 years.
   (d) The findings of the scan are expected in this patient with LAD artery stenosis

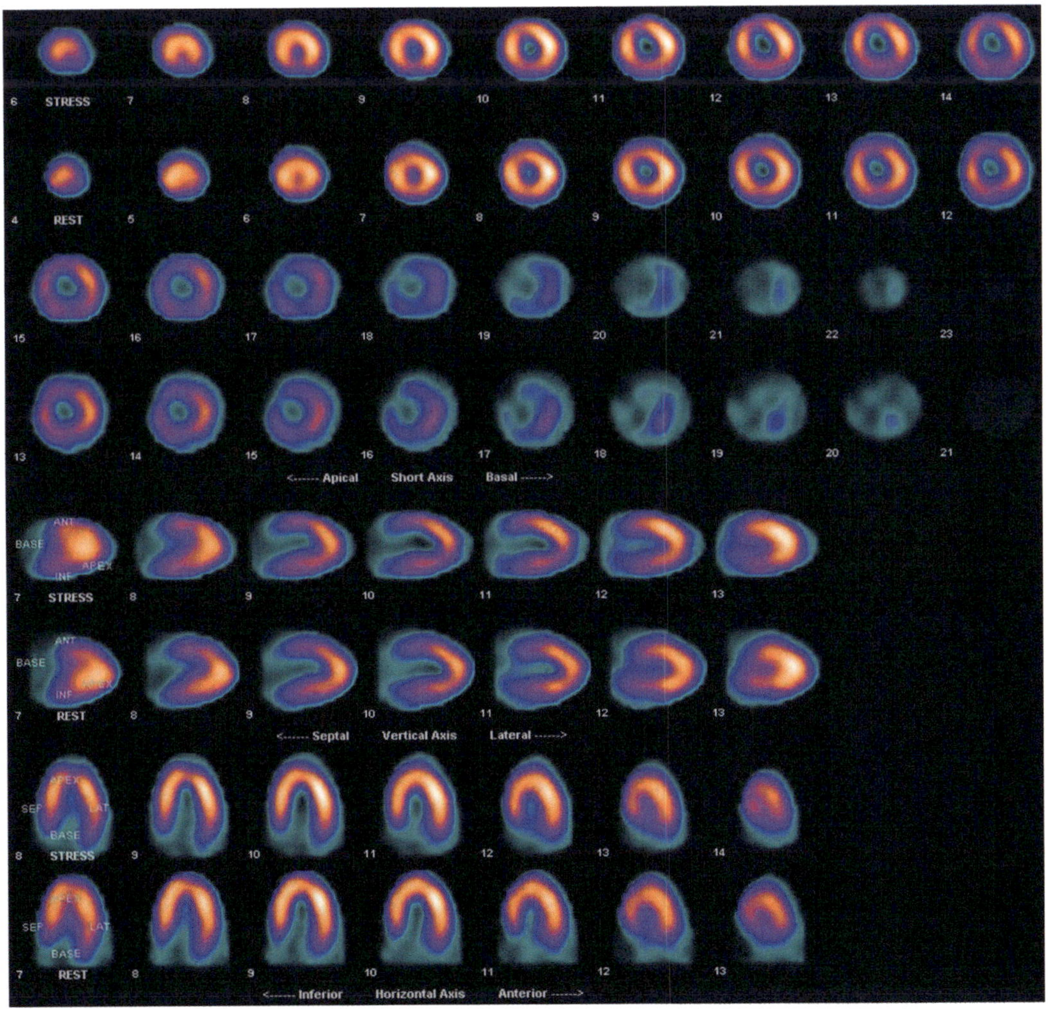

A large reversible perfusion defect in the apex, anterior wall, and septum is expected to be seen on myocardial perfusion scan in this patient. However, the scan appears normal with no reversibility, i.e., false-negative.

Inadequate or suboptimal exercise is a common cause of false-positive scans. In some cases, elderly patients with frequent medical problems or arthritis may not be able to walk on the treadmill long enough to reach an adequate level of exertion. In these cases, a pharmacological stress agent is to be considered. However, this is not the rule and there is no age limit beyond which physical stressing is prohibited.

Adequate exercise level can be achieved if the patient reaches at least 85% of the age-predicted maximum heart rate or has a double product of at least 25,000. The age-predicted maximum heart rate is calculated by the simple formula: 220 minus the patient's age in years. For instant, if the patient is 65-year-old, it is 220–65 = 155. Therefore, adequate exercise is equivalent to 85% of the 155, i.e., 132 beats/min. The double product is calculated by multiplying the maximum heart rate achieved during the exercise by the highest systolic blood pressure. For this patient, it is $170 \times 110 = 18{,}700$, which is below the desired value of 25,000.

Beta-blockers are commonly prescribed in patients with coronary artery diseases. They have been proven to decrease mortality and help to prevent subsequent myocardial infarction and damage. However, they blunt the response to physical exercise by delaying and diminishing the natural increase in heart rate. Therefore, they frequently lead to false-negative results. It is preferable to discontinue these medications for at least 48 h prior to the stress test. Sometimes, for medical reasons, the patient cannot stop taking these medications. In other situations, the cardiologist may require the study while the patient is on a beta-blocker agent to evaluate his or her risk for significant adverse effects while on these medications. In these cases, pharmacological stress should be considered. Alternatively, the study can be performed putting in mind that the patient may not reach the desirable level of exercise.

Short-acting nitrates, such as sublingual nitroglycerine, avoid first pass hepatic metabolism and they converted rapidly into inactive metabolites. They usually do not interfere with myocardial perfusion imaging and there is no need to discontinue them before physical stressing. However, long-acting nitrates circulate in the body for a longer period of time. These vasodilator agents might mask the effect of exercise. In general, long-acting nitrates should be discontinued for about 12 h before stress myocardial perfusion imaging.

In brief, although this patient might be on multiple medications, most likely he was on a beta-blocker, which blunted the increase in heart rate and led to a false-negative study.

The correct answer is a.

3. Thallium-201 has been used for many years as a myocardial perfusion imaging. Correct interpretation of the scan depends mainly on interpreting the tomographic images and looking for direct and indirect evidence of coronary artery disease (CAD). Of the following statements about thallium stress imaging, the most accurate one is:

   (a) Slight left ventricular dilatation during stress indicates severe coronary artery disease.
   (b) Normal wall motion along with normal global left ventricular ejection fraction on gated SPECT excludes multivessel disease
   (c) Increased lung uptake is an indirect indication for severe coronary artery disease in nonsmokers with no history of pulmonary disease
   (d) Left ventricular dilatation, global hypokinesis, and low ejection fraction indicate severe CAD, even if the perfusion is normal

Although Tc-99 m myocardial perfusion agents such as Tc-99 m Sestamibi (Cardiolite) and Tc-99 m tetrofosmin (Myoview) are widely used nowadays, imaging with thallium is still performed, particularly when there is a shortage in Technetium production. Sometimes, thallium-201 imaging gives additional information. This kind of information usually comes from the fact that early imaging with thallium (within 10 min after exercise) is required and is feasible. These early images reflect the condition of the heart under stress. Tc-99 agents, on the other hand, usually require a waiting period, typically 30–60 min, until the clearance of the blood pool and the hepatic activity takes place. It is to be noted that Tc-99 m agents are fixed to the mitochondria of the myocardial cells very shortly after injection and reflect the status of myocardial perfusion at the time of injection. During this "waiting" period, the heart rate, the left ventricular dilatation, and the lung uptake frequently return to normal.

Mild left ventricular dilatation is typically seen as a physiological response to stress. However, marked dilatation during stress and normal cavity at rest (the so-called stress-induced dilatation or transient ischemic dilatation) may indicate severe or multivessel CAD.

In normal individuals, the lung uptake is far less than the myocardial uptake, usually less than 50%. This value is typically obtained from the anterior planar image of the chest before tomographic imaging. Increased lung uptake usually indicates increased pulmonary wedge pressure as a result of increased left atrium pressure secondary to impaired systolic function. Although it is not specific, the increased lung uptake is an indirect indication of severe CAD. However, the lung uptake may also be seen in heavy smokers and in patients with pulmonary diseases but normal coronaries. Therefore, interpretation of this finding should be done with caution and in conjunction with other findings such as perfusion defects and abnormal wall motion.

The literature describes the so-called balanced triple-vessel disease where the degree of stenosis in the three main coronary arteries is comparable. Because the myocardial perfusion scan essentially reflects the difference in regional perfusion, this balanced decreased perfusion may appear homogeneous or look "normal." Currently, the interpretation of a myocardial perfusion study does not depend on the images only but on the wall motion and the ejection fraction from the gated SPECT analysis as well. Patients with multivessel CAD often demonstrate wall motion abnormalities and decreased global ejection fraction. However, sometimes the perfusion, wall motion, and ejection fraction are completely normal in patients with triple-vessel disease.

Coronary artery disease is the most common cause of cardiomyopathy (ischemic cardiomyopathy). Nonischemic cardiomyopathy may result from excessive alcohol intake or myocardial toxicity secondary to chemotherapeutic agents such as Adriamycin. Patients with nonischemic cardiomyopathy may demonstrate diffuse hypokinesis and low global ejection fraction. These patients may, in fact, have normal perfusion on myocardial scan and "clean" coronary arteries on angiography. Besides, severe, untreated, long-standing hypertension may result in decreased ejection fraction and impaired wall motion without ischemia.

The correct answer is c.

4. A 78-year-old male with chest pain underwent adenosine stress test. The myocardial perfusion scan is shown below. The perfusion defect seen on the scan is most likely due to stenosis of which of the following:

   (a) Proximal left anterior descending (LAD) artery
   (b) An obtuse marginal (OM) branch
   (c) A diagonal branch
   (d) Left circumflex artery (LCX)

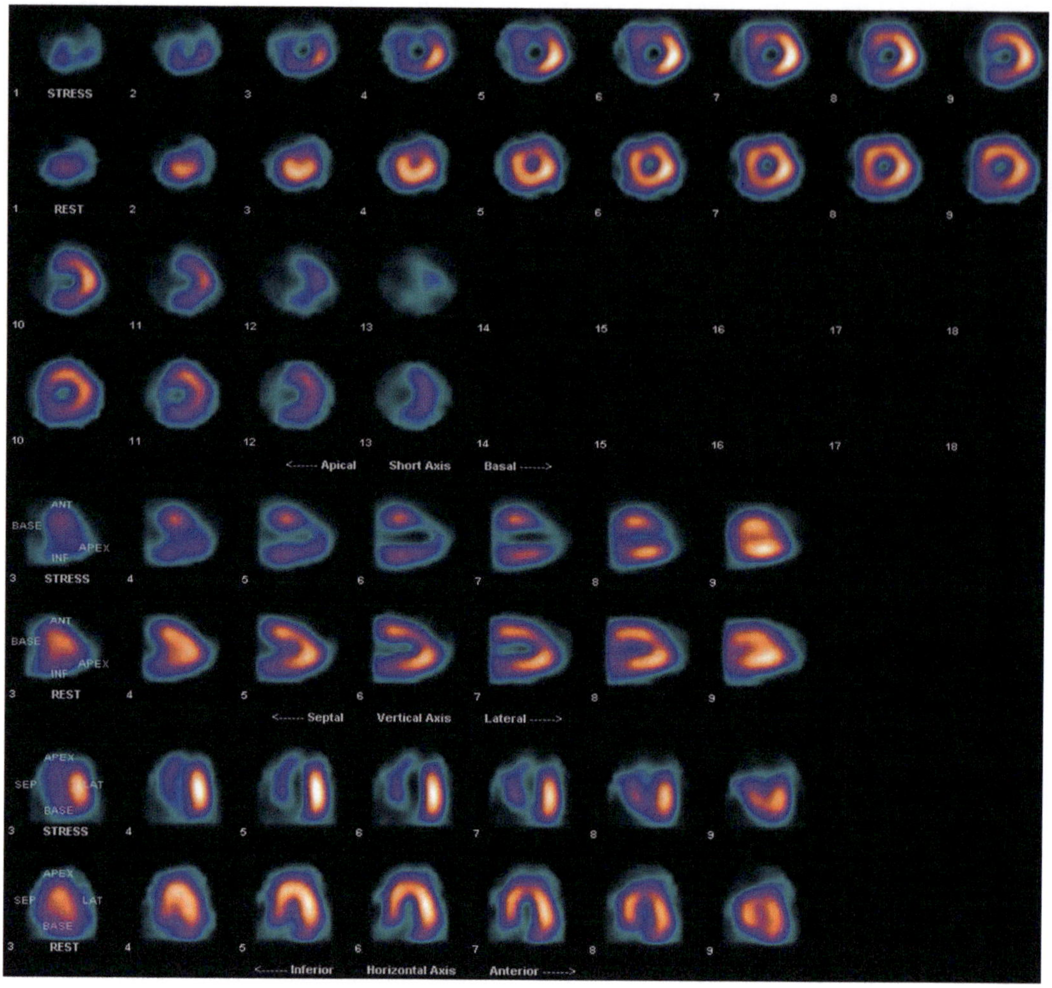

The left main artery is usually a short artery (one cm in length or less) that bifurcates to the left anterior descending (LAD) and left circumflex (LCX) arteries. Occasionally, the left main is absent. The LAD artery supplies the apex, the anterior wall, and the anterior two-thirds of the interventricular septum. Occlusion of the distal part of the LAD usually affects the perfusion of the anterior wall and the apex but spares the septum. The anterolateral wall is supplied by the diagonal branch(s) of the LAD. The number of these branches may vary from one to five. The perforating branches that supply the septum are ranging between one and six branches. These branches also supply the anterior wall of the right ventricle that is close to the septum. In 80% of cases, the LAD artery reaches the apex or wraps around it.

The scan demonstrates a large reversible perfusion defect involving the apex, anterior wall, and the septum, in the typical distribution of the proximal LAD.

The left circumflex (LCX) artery and its branches supply the lateral wall. Most individuals have three obtuse marginal (OM) branches; the second is the largest. It also supplies the left atrium.

Occlusion of the left main artery should lead to perfusion defects in the distribution of LAD and LCX. However, LCX artery is more prone to perfusion defects than the LAD artery.

The right coronary artery supplies the right atrium and ventricle as well as the posterior one-third of the interventricular septum. Then it usually continues as the posterior descending artery (PDA) to supply the inferior wall (in 80% of individuals). In addition, the RCA supplies the atrioventricular (AV) node in about 90% and sinoatrial (SA) in 55% of individuals. The SA branch arises from the LAD in the remaindered 45% of people.

The ramus intermedius artery, usually a large vessel, arises between the LAD and LCX arteries when the left main artery trifurcates. Occlusion of this vessel usually leads to a posterolateral wall perfusion defect in the distribution of the high OM branch or anterolateral defect if it is anterior in position.

The word "dominant" is a misleading one. It does not refer to which artery is the largest but simply refers to which artery gives rise to the posterior descending artery (PDA) and supplies the posterolateral branches. In about 80% of individuals, it is the RCA and in 10% it is the LCX. In the remaining 10%, the PDA arises from the RCA and the posterolateral branches of the LCX artery (balanced or codominant circulation). The dominant artery usually gives rise to the AV nodal artery.

The correct answer is c.

5. Regarding pharmacological stress imaging:

   (a) The physical half-life of adenosine is shorter than dipyridamole or dobutamine
   (b) Dobutamine is contraindicated in patients with bronchial asthma
   (c) Chest pain is a valid indication to terminate adenosine infusion
   (d) Sublingual nitroglycerine is an antidote to adenosine

Physical stress is widely used as a means for the evaluation of ischemia. However, some patients cannot exercise on the treadmill, such as those with peripheral vascular disease, severe osteoarthritis, or other morbidities. In these cases, pharmacological stress is usually performed with adenosine, dipyridamole (persantine), or dobutamine.

The administration of Adenosine or dipyridamole usually results in coronary artery dilatation. The normal coronaries will dilate adequately, up to 4 or 5 folds while diseased arteries will not dilate as much. Because myocardial perfusion imaging measures the difference in regional blood perfusion, the regions of the heart that are supplied by the diseased vessels will appear less perfused relative to the regions supplied by normal vessels. Dobutamine, in contrast, increases the heart rate and blood pressure resembling physical stress.

Adenosine is a direct coronary vasodilator. It is usually administered by infusion at a rate of 140 μg/kg/min for 4–6 min. The radiopharmaceutical agent is administered at half time of infusion. In addition to the vasodilator effect, adenosine causes atrioventricular (AV) conduction delay, which is the basis for using this drug to treat supraventricular tachycardia (SVT). Nonetheless, this delay in conduction might result in heart block (HB). If third degree (complete) HB occurs, the infusion has to be discontinued.

Dipyridamole acts by increasing the levels of endogenous adenosine by inactivating the deaminase enzyme that metabolizes circulating adenosine. In other words, it is an indirect vasodilator. It is infused at a rate of 140 μg/kg/min for 4 min. The radiopharmaceutical is administered 8 min after the beginning of the infusion. The drug might induce bronchospasm and aminophylline at a dose of 125–250 mg given intravenously is the antidote. Although aminophylline is also an antidote to adenosine, owing to the very short half-life of adenosine, the need for its use is infrequent.

Dobutamine, a synthetic catecholamine, is an inotropic agent. It acts through the stimulation of beta-1 and beta-2 receptors. It has a fast onset of action, within 2 min. Unlike adenosine or dipyridamole, dobutamine is given in incremental doses starting at 10 μg/kg/min and increased up to 40 μg/kg/min. Adjuvant atropine may be given to maximize the effect of dobutamine. The criteria for appropriate level of exercise are the same as physical stress, i.e., the patient should reach at least 85% of the age-predicted maximum heart rate or a double product of at least 25,000. The current indication for dobutamine administration is reserved for the patients who cannot walk on the treadmill and cannot tolerate adenosine or dipyridamole as in cases of bronchospasm or asthma.

Chest pressure is common with adenosine infusion. It is nonspecific and does not require intervention. However, the steal phenomenon may occur, i.e., the normal coronary vessels will dilate and directing the blood away from the stenotic or diseased vessels. Anginal pain may occur which is usually accompanied by ECG changes. In such a case, the infusion should be terminated and sublingual nitroglycerine be administered.

The pharmacological stress drugs described above are not radioactive agents, and therefore do not have physical half-lives. Being drugs, however, they have biological half-lives. Adenosine has a very short plasma half-life of less than 2 s, persantine of 20–30 min and dobutamine of 2 min.

The correct answer is d.

6. Myocardial stunning and hibernating myocardium share some similarities, although they are two different entities. True statements regarding these phenomena include which of the following:

   (a) In either condition, the gated SPECT analysis usually demonstrates severe hypokinesis or akinesia.
   (b) Hibernating myocardium usually occurs after acute myocardial infarction and reperfusion.
   (c) Stunned myocardium usually results from severe, chronic ischemia.
   (d) Hibernating myocardium demonstrates normal uptake on N-13 ammonia PET imaging

   Myocardial stunning is a transient phenomenon that may occur upon reperfusion of an occluded coronary blood vessel. Typically, the vessel is occluded for a short period of time, e.g., 15 min. Permanent damage may occur with longer periods of occlusion. On the gated SPECT, stunned myocardium demonstrates itself as wall motion abnormalities but normal perfusion. It is postulated that the reintroduction of oxygen after a period of deprivation may lead to transient calcium overload that results in decreased contractility of the segment supplied by this vessel. Severe hypokinesis or akinesis may be seen early after reperfusion. Gradually, over several hours or days, the contractility often improves.

   Hibernating myocardium refers to viable myocardial tissue that demonstrates markedly diminished function (contractility). It is a severe form of chronic ischemia. It has been postulated that the myocardium is metabolically adapted to low levels of perfusion. Characteristically, the myocardial perfusion scan demonstrates fixed perfusion defects on both stress and rest imaging with severe hypokinesis or akinesis on gated SPECT. This segment with severe ischemia usually requires more time for reperfusion and can be detected on "resting/delayed resting" thallium-201 study. In this technique, the patient is injected with a single dose of thallium-201 chloride and the myocardium is imaged shortly (at rest) and again 18–24 h later (delayed rest). No booster dose of the radiopharmaceutical should be given before delayed imaging. As previously mentioned, this severe, chronic ischemia requires a longer period of time for reperfusion. Uptake will be seen on the delayed imaging that was not detected on the previous resting imaging. If thallium stress/rest technique is acquired, the patient is injected during stress dose and a booster may be given at rest. Again, if viability is sought, no more additional doses of thallium are given.

   Another way to distinguish hibernating myocardial from scar tissue formation is by comparing the resting perfusion study to F-18 FDG uptake on PET imaging. Regardless of the perfusion agent used (thallium-201, a technetium agent such as sestamibi or a PET agent such as N-13 ammonia), the result is similar, i.e., an area of decreased or no perfusion on resting images.

   Myocardial tissue that demonstrates decreased perfusion and normal glucose uptake on F-18 FDG imaging is still viable (hibernating). The literature clearly confirmed the benefit of revascularization in patients with mismatch defect (decreased perfusion and better FDG uptake). On the other hand, patients with matched decreased perfusion and metabolism show no significant improvement after revascularization.

   The correct answer is a.

7. Regarding thallium-201 and Tc-99 m myocardial perfusion agents which statement of the following is true?

   (a) Modern gamma camera best operate around thallium principal energy
   (b) The critical organ in Tl-201 imaging is the kidney
   (c) Tl-201 has a shorter half- life than Tc-99 m agents
   (d) The uptake of thallium by the myocytes in the hypoperfused (ischemic) myocardium is delayed but the clearance is faster than normal tissue

Thallium-201, a potassium analog, was probably the first agent used to measure regional myocardial perfusion. Its relatively low principal energy (mercury X-rays at 69–83 KeV) results in an attenuation problem within the body. The higher energy photons from the technetium agents (140 KeV) make them more desirable for imaging.

Thallium-201 is transported across the myocyte cell membrane via the Na-K ATPase pump. Both uptake and clearance from the myocytes in the hypoperfused (ischemic) myocardium are slower than normal. Because of its long physical half-life (73 h), the radiation dose is high. It is important to remember that the kidney is the ***critical organ****. The gonads and the lens of the eye also receive high radiation dose.

Tc-99 m sestamibi, an isonitrile, is a lipophilic cationic complex that is taken up by myocytes in proportion to myocardial blood flow. Approximately 90% of intracellular sestamibi is associated with mitochondria as a free cationic complex. Because the myocardial clearance of sestamibi is significantly slower than that of thallium, separate stress and rest injections are generally employed.

Generally speaking, Tc-99 m tetrofosmin, a diphosphine cation, is comparable to Tc-99 m sestamibi. It has good myocardial uptake and slow clearance, but rapid clearance from surrounding structure. Unlike sestamibi, heating is not necessary for its preparation. The physical properties of Tc-99 m agents make them ideal for modern Anger (gamma) camera imaging. It is to be noted that Tc-99 m teboroxime is no longer in the market. Its short time window for imaging makes it impractical in everyday use.

The correct answer is "b".

- Most textbooks use the terms "critical" and "target" organ exchangeably, referring to the organ that receives the highest radiation dose from a radiopharmaceutical. Sometimes the term "target organ" is used to refer to the desired organ to be imaged. This might result in some uncertainty. For this reason, in this book, the authors prefer to use the term "critical organ" to refer to the organ which receives the highest radiation dose because they feel that this organ is in a "critical situation." If the term "target organ" is to be used, it will only refer to the organ to be imaged.

8. All of the following radiopharmaceuticals measure myocardial blood flow ***except***:

   (a) Tl-201
   (b) F-18 FDG
   (c) Rb-82
   (d) N-13 ammonia

F-18 FDG (F-18 Fluoro-2-deoxy-D-glucose) differs from many other agents used to image the heart in its ability to measure the glucose uptake by the cardiac tissue rather than the blood flow. It is transported by the same carrier as glucose but undergoes trapping after the first phosphorylation step via the hexokinase reaction. FDG accumulation reflects regional glucose utilization. The half-life of F-18 FDG is 110 min. So, it is possible to transport it from the manufacturing site to satellite facilities and hospitals lacking cyclotron.

Nitrogen-13 ammonia is another positron-emitter agent. Its short half-life of 10 min makes it essential to have an on-site cyclotron. The agent is used predominantly to measure regional blood flow of the heart. Carbon-11 and oxygen-15 are also positron-emitter agents with half-lives of 2 and 20 min, respectively.

The potassium analog, rubidium-82, has a very short half-life of 76 s. It is derived from a generator as a daughter of strontium-82, which has a longer half-life of 25 days.

Thallium-201 has been used for cardiac imaging for many years. Like rubidium-82, it is potassium analog. Low principal energy is the main disadvantage. Also, its high dosimetry limits the allowable injected dose. Tc-99 m sestamibi and tetrofosmin replace thallium in the majority of nuclear laboratories. Both thallium and technetium agents are taken up by the myocardium in direct relation to the regional blood flow.

The correct answer is b.

9. Artifacts make the interpretation of cardiac imaging a challenging task. Which of the following statements is correct?

   (a) Breast implants usually resemble ischemia in the anterolateral wall
   (b) Cardiac creep usually resembles ischemia in the inferior wall
   (c) Right bundle branch block (RBBB) may result in a reversible pattern of perfusion in the interventricular septum
   (d) Hypertrophy of the pectoralis major muscle may mimic ischemia in the anterior wall

The interpretation of the myocardial perfusion scan is often complicated by the presence of artifacts. Artifacts due to body habitus should be put in consideration at the time of interpretation. The reader should at least be provided with the patient's weight, height, and body composition (truncal versus abdominal obesity) as well as the size of the female breast. Other information such as breast implants or muscle hypertrophy is beneficial.

Breast artifacts depend not only on the size of the breast but also on its position and density. Breast tissue artifacts typically result in fixed defects in the anterior or lateral walls. The fact that the borders of this defect are ill-defined serves as a clue to the diagnosis. As a general rule, the borders of any artifact are not as sharp as true perfusion defects. In some cases, the position of the breast changes from stress to rest resulting in a reversible pattern that mimics ischemia. The interpretation of these scans is quite challenging, especially in symptomatic patients. Breast implants usually result in fixed, rather than reversible, defects. The pattern frequently resembles myocardial infarction. However, the gated SPECT is typically normal in these cases that help to make the correct diagnosis.

Cardiac creep or cephalic creep is a term used to describe the effect of the movement of the diaphragm on the position of the heart during (or shortly after) exercise. During exercise, the patient usually breathes deeply and the diaphragm moves vigorously. This movement results in attenuation in the inferior wall. On the other hand, the diaphragm movement is less vigorous at rest and the attenuation is less prominent. This pattern of artifact, in fact, may resemble ischemia and makes the interpretation of the scan difficult. This artifact is seen more frequently with thallium-201 imaging because of the short period of time between injection and imaging.

The literature reported reversible perfusion pattern in the septum in patients with left (not right) BBB who undergo physical stress or are injected with Dobutamine. The depolarization of the right ventricle precedes the left ventricle. When the heart rate increases, the perfusion of the septum decreases by about 25%, compared to the lateral wall although no detection of true ischemia on coronary angiogram (false result). Septal thickening and F-18 FDG uptake are also decreased by the same ratio. This reversible pattern of perfusion is not seen with adenosine or dipyridamole (persantine) because these agents usually do not significantly affect the heart rate. Therefore, patients with known LBBB are advised to have pharmacological stress.

Artifacts secondary to body habitus such as muscles or fat usually result in a fixed pattern of decreased activity. Abdominal obesity (in males or females) impairs, to some extent, the visualization of the inferior wall. This artifact mimics diaphragmatic attenuation seen in males. On the other hand, pectoralis muscle hypertrophy (usually in males) may result in a fixed decrease activity in the anterior wall, mimicking breast tissue artifact.

Attenuation correction is a software program that studies the body composition and adds proportional counts to areas with increased soft tissue. Although it is expensive, it helps eliminate the effect of many of these artifacts. However, it may induce its own artifact.

The correct answer is b.

10. Which of the following conditions may cause a ***falsely*** high ejection fraction on a MUGA study?

(a) The technologist forgets to subtract the background in the calculation
(b) The systolic region of interest (ROI) is too small
(c) Thyrotoxicosis
(d) Atrial fibrillation

The multigated ventriculography (MUGA) scan was introduced in 1971. The LAO view is usually the best view to separate the right ventricle from the left. The scan provides an accurate calculation of the left ventricular ejection fraction (LVEF). The calculation of the EF is straightforward:

$$\frac{\text{End diastolic volume} - \text{End systolic volume.}}{\text{End diastolic volume.}}$$

In MUGA, the volume of blood inside the ventricles is represented by counts. End-diastole is the frame with the highest counts while end-systole represents the frame with the lowest counts. Background activity should be accounted for. Therefore, the EF equation in a MUGA study would be modified to:

$$\frac{(\text{End diastolic counts} - \text{background}) - (\text{End systolic counts} - \text{background})}{(\text{End diastolic counts} - \text{background})}$$

Or simply:

$$\frac{\text{End diastolic counts} - \text{End systolic counts.}}{(\text{End diastolic counts} - \text{background})}$$

So, incorrect increase in diastolic counts such as overlapping of the left ventricle on the right, decrease in systolic counts (small ROI), or increase in background measurement (putting the background ROI on a blood vessel or more commonly on the spleen) would result in a false increase in EF. Omitting the background from the equation will, on the other hand, result in a falsely decreased EF.

The normal range of LVEF is 55–75%. Most nuclear medicine laboratories consider 50% as the lower limit of normal. True causes of high EF include hyperdynamic circulation states as seen in thyrotoxicosis, pregnancy, and anemia. On the other hand, chemotherapeutic agents such as Adriamycin may result in myocardial toxicity and decrease in the EF.

Arrhythmias, in general, lead to lower, rather than higher, values of EF. Atrial fibrillation (AF) usually poses a problem in acquiring the study because of frame dropping. However, if the study is completed, the ejection fraction is typically lower than the real one.

The correct answer is b.

11. Regarding rubidium-82:

(a) The physical half-life is about 10 min
(b) The strontium-rubidium generator is purchased every week
(c) It is not a potassium analog
(d) Stress images can be done only with pharmaceutical agents

Rubidium-82 is a PET agent that measures myocardial regional blood flow. It has a very short half-life of 76 s. It is eluted from strontium-82/rubidium-82 generator. Strontium-82 has a half-life of 25 days. Therefore, the strontium-82/rubidium-82 generator is typically used for a whole month before it loses its efficacy. Similar to thallium-201, rubidium-82 is a potassium analog. It depends on the sodium-potassium pump for intracellular delivery. Rest images are usually performed before stress.

Using rubidium-82 as a myocardial imaging agent has many advantages. Because of its very short half-life, both rest and stress imaging can be done in a short period of time. In addition, its high energy at 511 KeV results in high-quality images putting in consideration that a dedicated PET scanner has a built-in attenuation correction. However, only pharmacological agents, specifically adenosine and dipyridamole (persantine) can be used for stress imaging. Physical stress or dobutamine are not used. Another disadvantage of rubidium-82 is the high cost of the generator.

The correct answer is d.

12. A 65-year-old female was referred to the nuclear medicine department for a myocardial perfusion scan. She has chest pain and abnormal EKG. The scan was done with thallium-201 chloride and it is shown below. The wall motion was normal and the global ejection fraction was normal at 55%. The most appropriate next step is:

    (a) Send the patient home, the scan is essentially normal with soft tissue attenuation artifact
    (b) Schedule thallium viability test, 24 h later
    (c) Admit the patient for cardiac enzyme profile
    (d) Admit the patient for cardiac angiogram

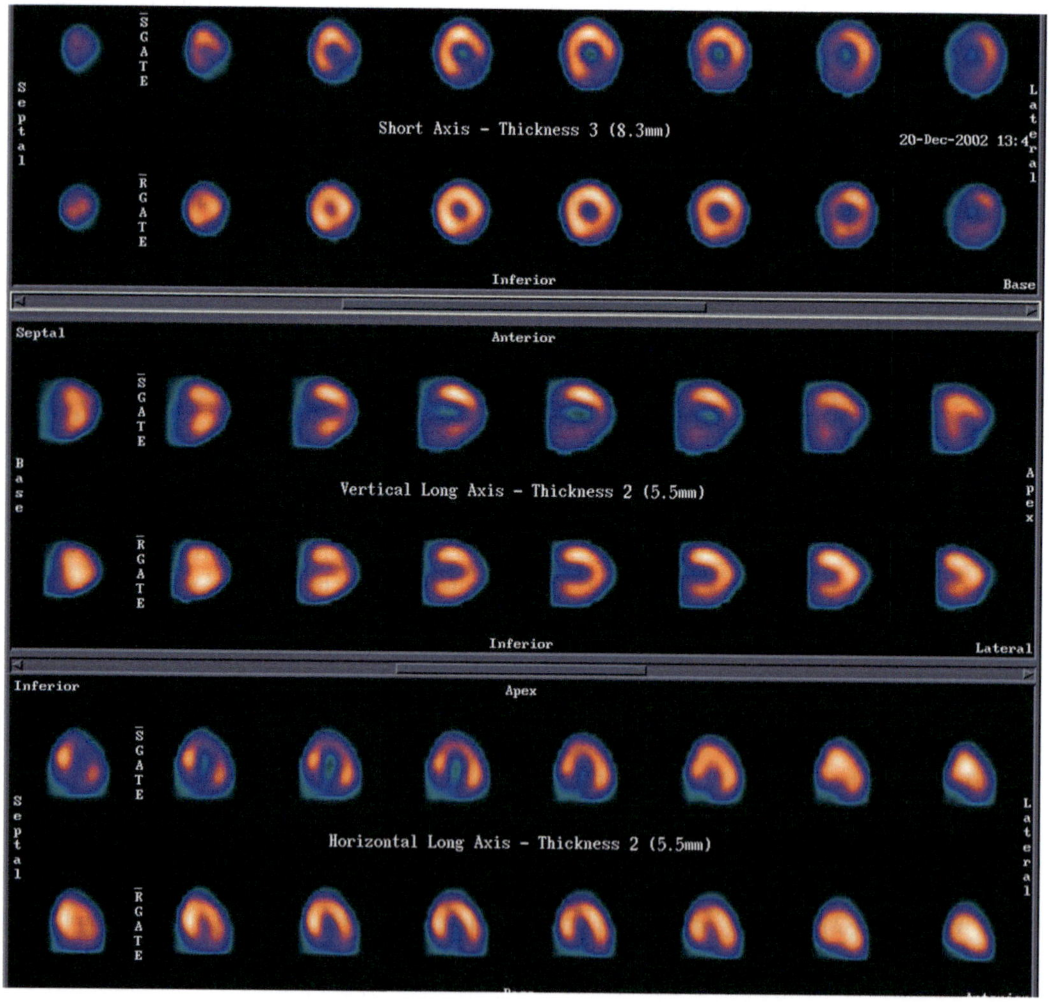

The patient has chest pain and an abnormal electrocardiogram. The scan demonstrates a reversible perfusion defect in the inferolateral region. The defect is transmural and is typical for ischemia. In females, the artifact is typically seen in the anterior wall. Moreover, it is demonstrated on both stress and rest images with no significant change. On the other hand, diaphragmatic attenuation artifact, which might affect the inferior wall, is more seen in males than females. Still, artifacts have ill-defined borders and are not transmural. Therefore, based on the clinical picture, EKG changes, and myocardial scan, this patient has ischemia, probably in the right coronary artery.

Sending this patient with evidence of ischemia home is inappropriate. Cardiac enzyme profile, which requires 24 h to be done, is definitely positive in myocardial damage (infarction). The medical team's ultimate goal is to manage ischemia before infarction develops.

There is no need to confirm the viability of the inferolateral region. Ischemia indicates viable myocardium. If the defect was similar at rest, there might be a need to confirm viability by 24 h thallium or F-18 FDG.

The proper next step is to schedule angiogram which is likely to demonstrate significant stenosis in one (or more) of the coronary arteries. Subsequently, revascularization through stent placement or even coronary bypass surgery may be needed depending on the result of the angiogram.

The correct answer is d.

13. A 76-year-old female is complaining of shortness of breath. She has a history of diabetes mellitus and hypertension but no history of coronary artery disease. Also, she is not complaining of chest pain. Myocardial perfusion scan is shown below. Thallium-201 chloride was the imaging agent. Pharmacological stress was performed with the administration of dipyridamole (persantine) infusion at a dose of 140 μg/kg/min for 4 min. The most appropriate next step would be:

    (a) Repeat the study with treadmill (physical stress); it is more sensitive
    (b) Repeat rest imaging 24 h later
    (c) Start the patient on a beta-blocker
    (d) Send the patient for coronary angiogram

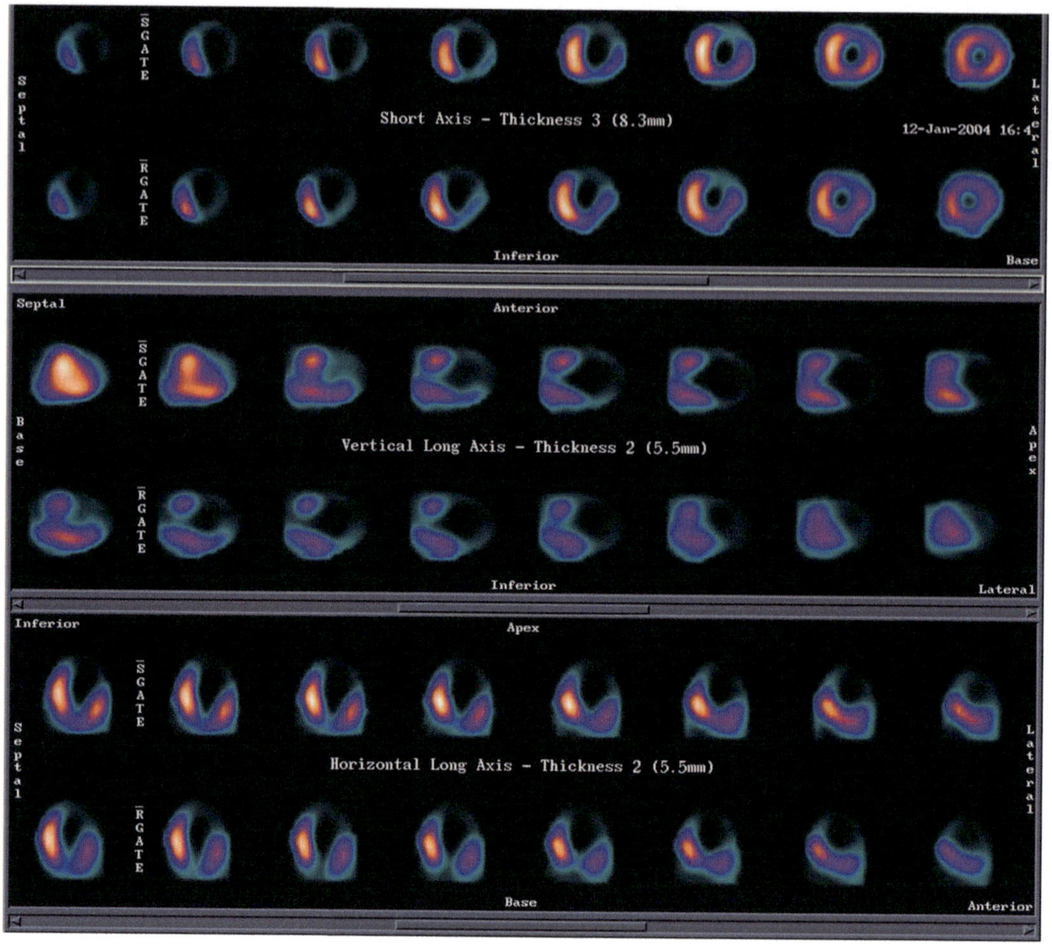

This patient has a large nonreversible perfusion defect in the apex and anterior wall as well as the distal inferior and lateral walls. The findings are likely to represent an old myocardial infarction. In this patient with diabetes mellitus, silent heart attacks are not uncommon. However, the patient has no previous history of coronary artery disease and she has no ongoing chest pain. Accordingly, it is reasonable to further evaluate her condition before rushing into treatment. Obviously, there is no need to repeat the study. Persantine study is valid with essentially similar sensitivity and specificity as physical stress.

Eventually, the patient will need to be on a beta-blocker for life. It has been shown that these agents decrease further cardiac events and mortality in patients with known coronary disease. However, other treatment modalities are usually needed, such as lowering blood glucose levels, controlling hypertension and managing low-density lipoproteins. More important is revascularization or stent placement if appropriate.

Therefore, sending the patient for an angiogram would be rational provided that there is viable myocardium to benefit from increasing blood flow through angioplasty or stent placement. Studies demonstrate that patients with no viable myocardium would not generally benefit from revascularization. Apart from the possible side effects of these procedures, the literature also reveals that patients' condition may actually become worse. For these reasons, the main stream of cardiologists prefers to do a viability study before subjecting patients to angiogram and stent placement. If there is no evidence of significant viability, medical treatment is the preferred option.

Viable myocardium can be assessed by either F-18 FDG study or delayed imaging with thallium-201. FDG is more sensitive than thallium. The theory of delayed thallium imaging is that the severely stenotic segments (hibernating tissue) require more time to show evidence of reperfusion. It is critically important not to inject the patient with another dose of the radiotracer before these delayed rest imaging. Because this patient already had thallium stress/rest scan, it is appropriate to reimage her the following day. If there is significant viability, revascularization is to be considered and if there is no viability, maximizing the medical treatment is preferable.

The correct answer is b.

14. Contraindications to Adenosine infusion as a pharmacological agent in performing myocardial perfusion scan include:

    (a) Hypertension (blood pressure exceeds 200/100)
    (b) Aortic stenosis
    (c) Heart block
    (d) Bronchospastic lung disease

Adenosine infusion usually at a dose of 140 μg/kg/min for 6 min is commonly used as a pharmacological stress for imaging patients with suspected ischemia. Some nuclear medicine departments infuse the medication for only 4 min. Minor side effects are very common with these infusions. Chest pressure, shortness of breath, facial flushing, hypotension, and tachycardia, although bradycardia may occur. Because the half-life of adenosine is very short (usually less than 10 s), all these effects frequently reverse shortly after discontinuing the infusion.

First, second, or third degree heart block may occur with adenosine infusion. Again because of the very short half-life, the test may not need to be terminated. The preexisting first-degree heart block and even second-degree type I are not contraindication to adenosine infusion. Third-degree block is a contraindication. Also, second-degree type-II may progress to third-degree block.

Hypotension is one of the side effects of adenosine. Therefore, the drug is preferred to physical stress in patients with severe hypertension. Besides, the drug is safe in patients with aortic stenosis.

Bronchospasm is a known side effect of both adenosine and dipyridamole. The package insert states clearly that the drug should not be administered in patients with bronchospasm. Dobutamine is the customary alternative in these cases who cannot walk on the treadmill.

The correct answer is d.

15. Regarding ejection fraction (EF):

    (a) Left ventricular EF equals Right ventricular EF
    (b) Left ventricular EF is higher than Right ventricular EF
    (c) Left ventricular EF is lower than Right ventricular EF
    (d) During exercise, LVEF is higher than RVEF, and at rest, RVEF is higher than LVEF

The amount of blood ejected from each ventricle (cardiac output) is the same. If the output from the right ventricle is decreased, the amount of blood going to the lung will also be decreased. As a result, the blood going to the left atrium and subsequently to the left ventricle will be diminished, decreasing the cardiac output. On the other hand, if the left ventricular output is decreased, the venous return to the right ventricle will be decreased. However, the fact that the cardiac output is similar does not necessitate that the ejection fraction from each ventricle is identical. The reason is that cardiac output is the product of the blood volume inside the ventricle multiplied by the ejection fraction. If the blood volume is different, the ejection fraction has to be different to result in the same output. The right ventricular cavity is larger than the left. Therefore, the left ventricular ejection fraction must be higher.

The LVEF equals the stroke volume (difference between end diastolic end systolic volumes) divided by end diastolic volume. Normal range is 55–75%. Many authors consider 50% as the lower limit of normal. LVEF can be reliably measured with MUGA scans. The RVEF is harder to measure and depends on a variety of technical aspects. However, the normal range is considered to be about 40–55%.

The correct answer is b.

16. A 73-year-old patient has 85% stenosis of the proximal left anterior descending (LAD) artery, 30% stenosis of the right coronary artery (RCA), and 40% stenosis of the left circumflex artery (LCX). All of the following are true statement regarding myocardial imaging ***except***:

    (a) Resting thallium-201 is expected to be normal
    (b) PET study utilizing F-18 FDG is expected to be normal
    (c) Adenosine PET study utilizing Ru-82 is expected to be abnormal
    (d) Shortly after successful angioplasty and revascularization, stress Tc-99 m tetrofosmin scan is expected to be essentially normal

Patients with significant stenosis of one or more of the coronary arteries usually demonstrate reversible perfusion defects on stress/rest myocardial perfusion scans. Stenotic lesions less than 50% are usually not detected on myocardial imaging. Significant stenosis of 70% or more is typically identified on myocardial perfusion stress imaging utilizing a technetium agent, thallium-201 or rubidium-82. Rest images are characteristically normal. If myocardial infarction develops, the perfusion defect will be seen on rest imaging as well.

This patient is expected to have a reversible perfusion defect in the distribution of LAD, i.e., the anterior wall and the septum on any of the perfusion imaging agents. Typically, there is no need to do a viability study in patients with ischemia and no fixed perfusion defects. However, if it is done, it is expected to be normal.

Chest pain in the first 2 days after angioplasty and stent should direct the attention to the procedural complications such as MI, thrombosis, or embolization. These patients should not be evaluated with myocardial imaging; rather, they ought to go directly for cardiac catheterization. The incidence of in-stent restenosis increases with time and might reach 20% by one-year post-implant. However, its incidence is decreased with drug eluted stent. Stress imaging is usually performed to evaluate symptomatic patients, although silent restenosis occurs. Following successful angioplasty and revascularization, the stress images should return to normal. However, false-positive scans were reported with early imaging, particularly in the first week. Later imaging usually demonstrates normal perfusion.

The correct answer is d.

17. Contraindications to adenosine use as a pharmacological stress agent include:

    (a) First-degree heart block
    (b) Second-degree heart block
    (c) Recent history of GI bleeding
    (d) Transient ischemic attacks (TIA)

Adenosine is a common pharmacological agent used in patients who undergo myocardial perfusion imaging. These patients usually are unable to exercise on the treadmill. Adenosine is a safe drug with a very short half-life of a few seconds. Its main contraindications are bronchial asthma and second and third-degree heart block. First-degree heart block is not a contraindication.

Adenosine may lead to tachy- or bradycardia. It may cause atrioventricular block and, this is the basis for using it to manage supraventricular tachycardia. However, it may induce atrial fibrillation or junctional rhythm.

Recent history of gastrointestinal bleeding or transient ischemic attacks is not a contraindication to adenosine use.

The correct answer is b.

18. A 73-year-old diabetic male with prostate cancer had a bone scan with Tc-99 m MDP. You noticed increased uptake in the anterior chest but does not seem to be in the ribs. Oblique views confirm your suspicion of myocardial uptake. The most accurate sentence is:

    (a) The patient had myocardial infarction between 12 and 24 h ago
    (b) The patient had myocardial infarction between 24 and 72 h ago
    (c) The patient had myocardial infarction between 3 and 14 days ago
    (d) The patient had myocardial infarction of indeterminate age

    Silent myocardial infarction may occur particularly in diabetics. Incidental findings on bone imaging may reveal them. In the past Tc-99 m pyrophosphate was used to confirm the clinical suspicion of myocardial infarction and to predict the prognosis based on the size of the infarcted area. The sensitivity of the scan was highest between 24 and 72 h after the onset of symptoms. The rationale is that the radiopharmaceutical binds to the calcium and therefore accumulates in the necrotic tissue after acute infarction. Bone imaging agents such as Tc-99 m methylene diphosphonate (MDP) have the same characteristic. Clearly, the current diagnosis of acute myocardial infarction is based on cardiac enzymes and myocardial perfusion scanning.

    Typically the scan is not positive until at least 12 h had passed after the onset of chest pain. In other words, before 12 h, the scan may be falsely negative. Also, the scan may return negative in a few days and up to 2 weeks. However, persistent uptake may be seen in some patients even months after the onset of acute infarction. Some authors believe that these patients have a bad prognosis and they are more prone to other attacks.

    In conclusion, if the patient did not have recent symptoms of myocardial infarction, based on the scan alone, the age of infarction cannot be certain.

    The correct answer is d.

19. Which of the following sentences regarding paradox image on MUGA scan is correct?

    (a) Paradox image is obtained by subtracting end systolic counts from end diastolic counts
    (b) Paradox image assists in evaluating myocardial aneurysm
    (c) Normal individuals should have positive counts on paradox image
    (d) Patient with myocardial ischemia should have positive counts on paradox image

    MUGA scan is essentially designed to calculate the global left ventricular ejection fraction and to evaluate for regional or focal hypokinesia. The study is not intended to assess myocardial ischemia.

    Normal image on MUGA scan is obtained by subtracting the end systolic counts from end diastolic counts. The end-result should be positive. It reflects the stroke volume. Paradox image, as the name implies, is obtained by subtracting the end diastolic volume from the end systolic volume resulting in a negative value in normal individuals. Having a positive value in a specific region on this image indicates that there is more blood in systole than diastole. Therefore, aneurysm is suspected in that region. In addition, aneurysm also demonstrates regional dyskinesia on cine imaging.

On the left corner of the scan below is the paradox image. No counts are seen in the left ventricle indicating no areas of dyskinesia or aneurismal formation.

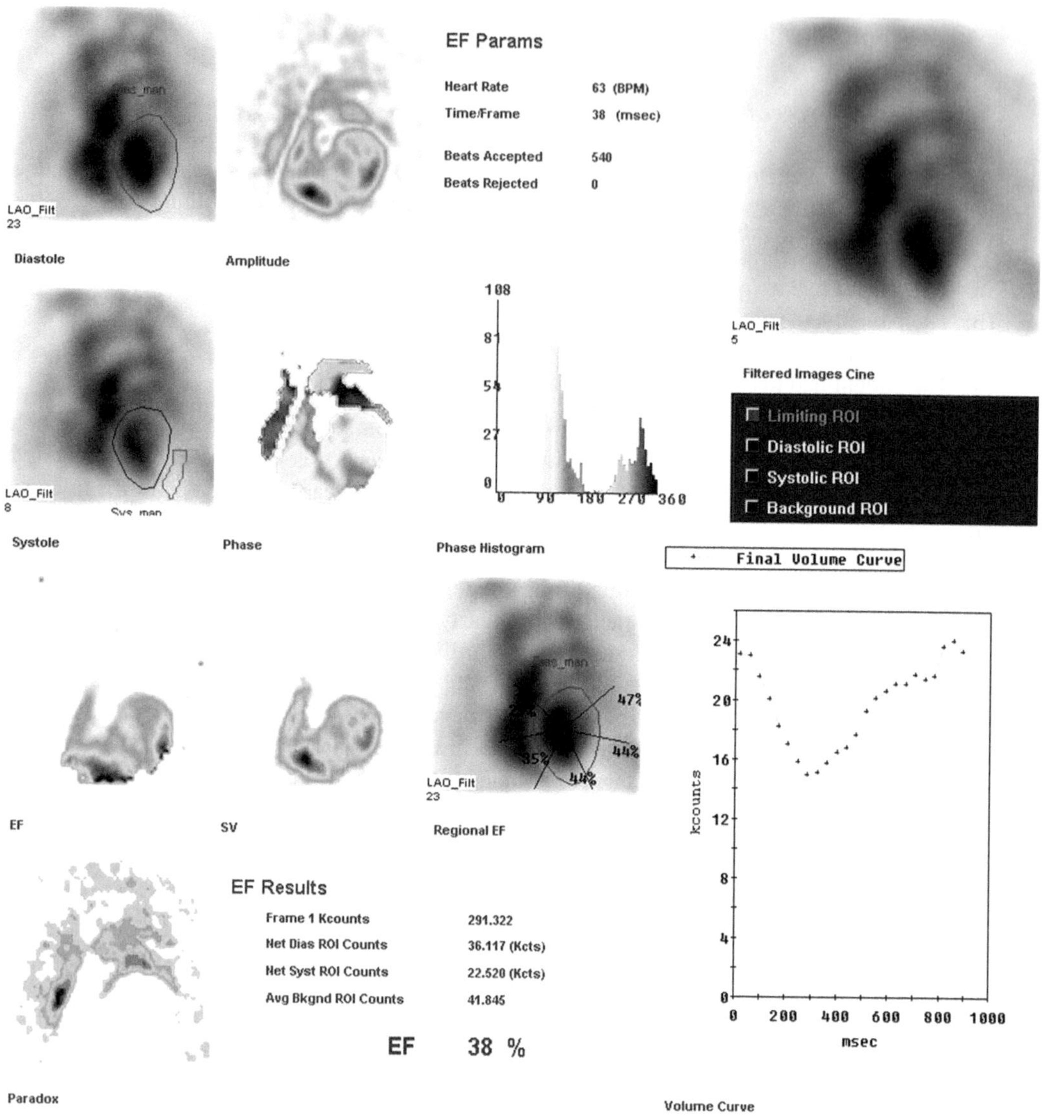

The correct answer is b.

20. A 45-year-old female patient had breast cancer and was given chemotherapy. Baseline MUGA scan done 6 months earlier before starting the chemotherapy was normal and the ejection fraction was 63%. The following scan was done to evaluate her before a second round of the same chemotherapeutic agent. Based on the result of this scan, the most acceptable next step is:

    (a) Administer the chemotherapy with no concern about heart failure
    (b) Administer the chemotherapy with caution
    (c) Discontinue this chemotherapeutic agent
    (d) Order echocardiography for correlation

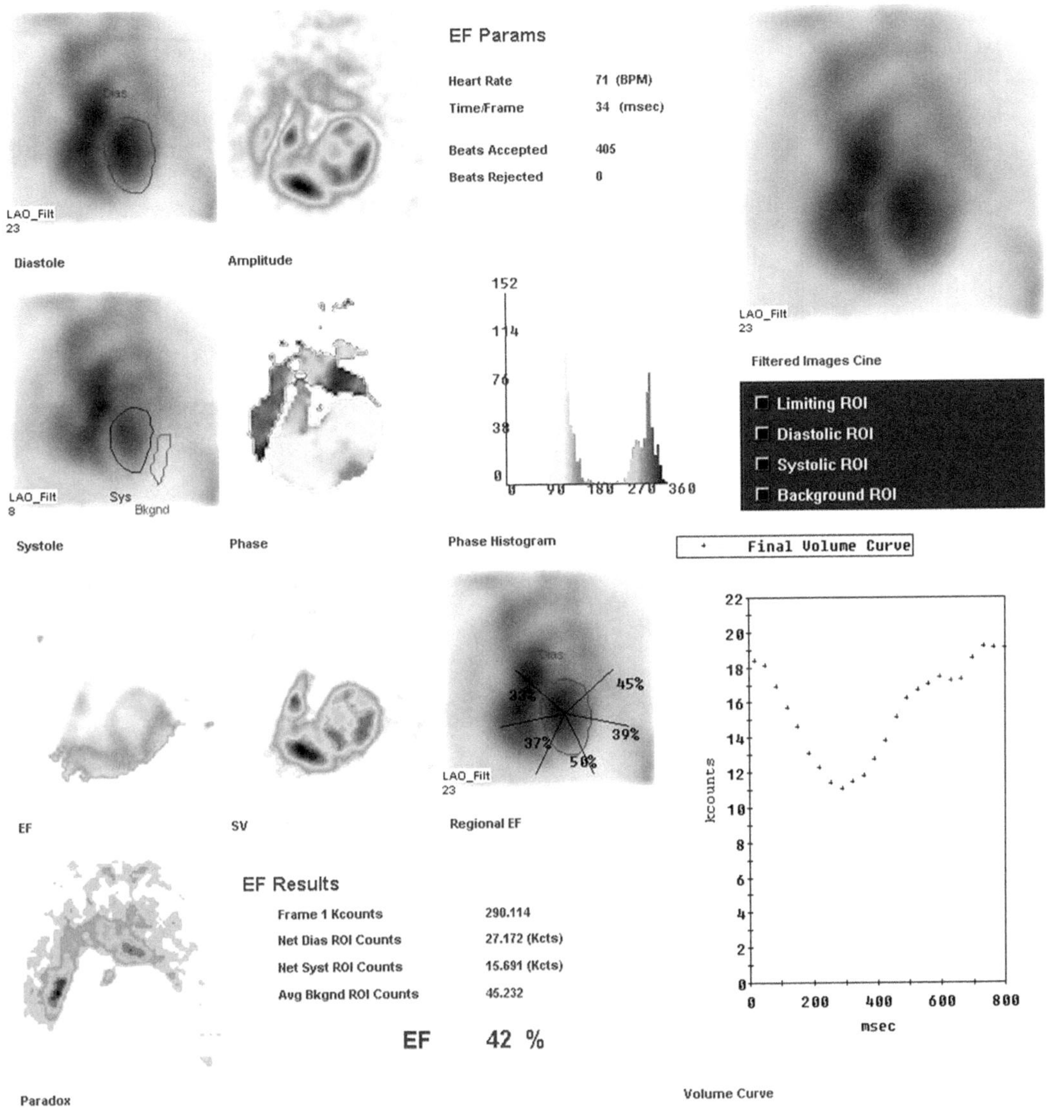

Certain chemotherapeutic agents result in cardiotoxicity such as Adriamycin that was given to this lady. The ejection fraction had significantly decreased from 63% to 42%. Although every patient is a unique case and management should be costumed according to his or her condition, there are some general guidelines for chemotherapeutic administration.

It is safe to continue chemotherapy if there is no significant decrease in the ejection fraction. Care should be taken if there is a significant drop in the ejection fraction. This would be in the range of 10–15 units, provided that the ejection fraction is still above 35% (many scholars believe that the lower end of the normal range is 50%). On the other hand, if the ejection fraction is decreased more than 15 units below the normal limit (35% or less), it is recommended to stop this chemotherapeutic agent. Patients with an ejection fraction of less than 35% are at a greater risk for heart failure.

The scan appears technically satisfactory with good separation of the left ventricle from the right ventricle and no rejected beats. It is widely accepted that the MUGA scan is the gold standard for measuring ejection fraction and it is believed to be superior to ECHO in this aspect.

The correct answer is b.

21. A 51-year-old male presented with intermittent pericordial chest pain for the past 2 weeks. The pain occurs with exertion and on climbing the stairs. Myocardial perfusion scan was ordered. Resting EKG was negative. The patient exercised on the treadmill for 6 min achieving a heart rate of 121 beats/min and blood pressure of 150/90. During peak exercise, he had chest pain similar to the pain he gets with climbing the stairs. The EKG during peak exercise demonstrates 2 mm ST depression in the inferior lead. The myocardium perfusion scan is shown below. The most appropriate next step is:

    (a) Repeat the scan with a pharmacological stress agent
    (b) Send the patient home based on the scan results
    (c) Perform a viability study
    (d) Send the patient for cardiac catheterization

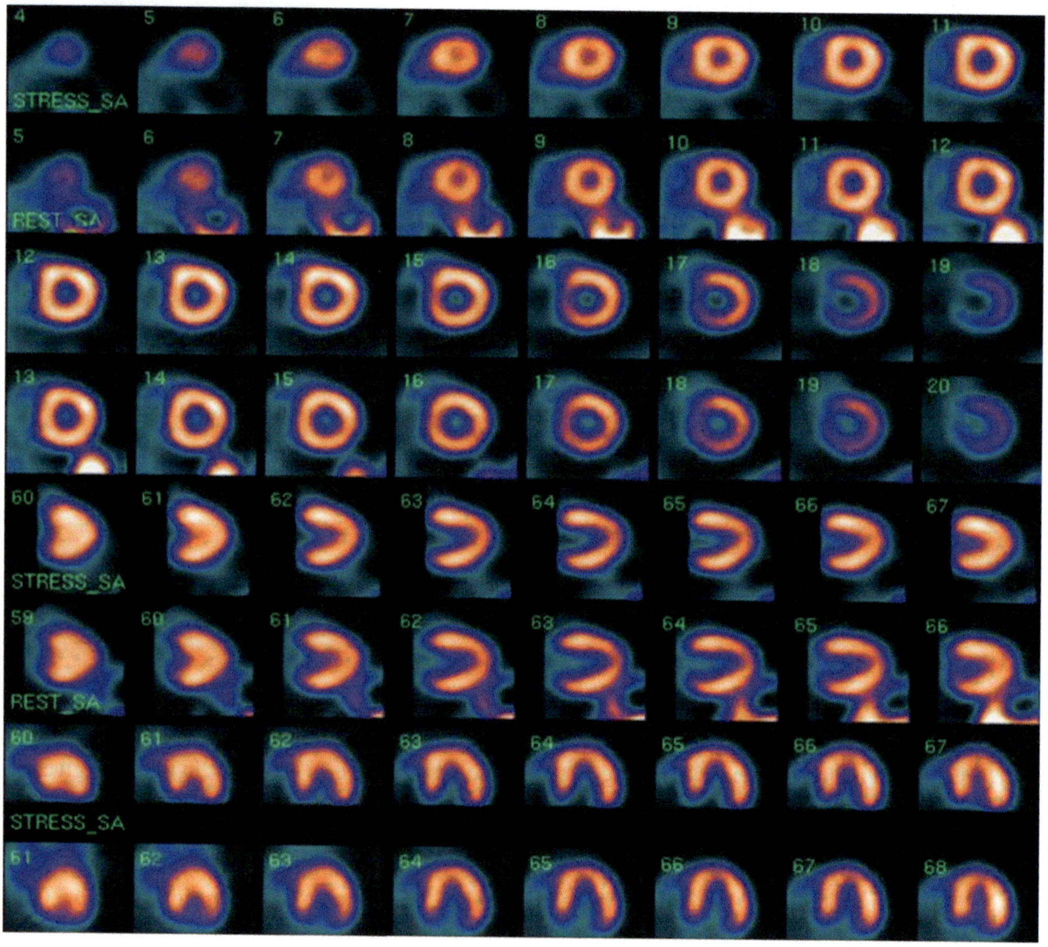

The patient has typical chest pain that is pericardial and associated with exercise. Based on the other risk factors, such patients usually are sent directly to cardiac catheterization. However, his primary care physician elected to do myocardium perfusion scan.

The patient did not reach an adequate level of exercise. His heart rate was below the target (85% of the maximum predicted value for his age). His double product (maximum heart rate multiplied by highest systolic blood pressure) was also below 25,000. Inadequate level of exercise might invalidate the study, and in fact the scan was completely normal. However, there was another piece of information obtained from the treadmill part. The patient has symptoms (chest pain) and EKG changes (2 mm ST depression) suggestive of ischemia. Therefore, the appropriate next step is to send the patient for cardiac catheterization for further evaluation of the patency of his coronary arteries. On the other hand, false-positive EKG changes do occur, particularly in women. However, the specificity increases when they are associated with typical chest pain.

Sending the patient home based on a normal myocardial scan is inappropriate because the scan is technically suboptimal due to a low level of exercise. Also, the sensitivity of myocardial perfusion scan, at best, is about 90%. Repeating the scan with a pharmacological agent is not indicated when the EKG part of the test is clearly abnormal. Also, a viability study is indicated if the scan shows a large fixed perfusion defect with no history of previous myocardium infarction.

The correct answer is d.

22. A 53-year-old male with a history of coronary artery disease (CAD) but no other significant medical or surgical history. He is on atenolol, Lisinopril, Lasix, and potassium supplement. He was referred to the department of nuclear medicine for a viability study because of an abnormal myocardial perfusion scan demonstrating a fixed defect. The following resting Tc-99 m Sestamibi (upper rows) and F-18 FDG (lower rows) images were obtained.

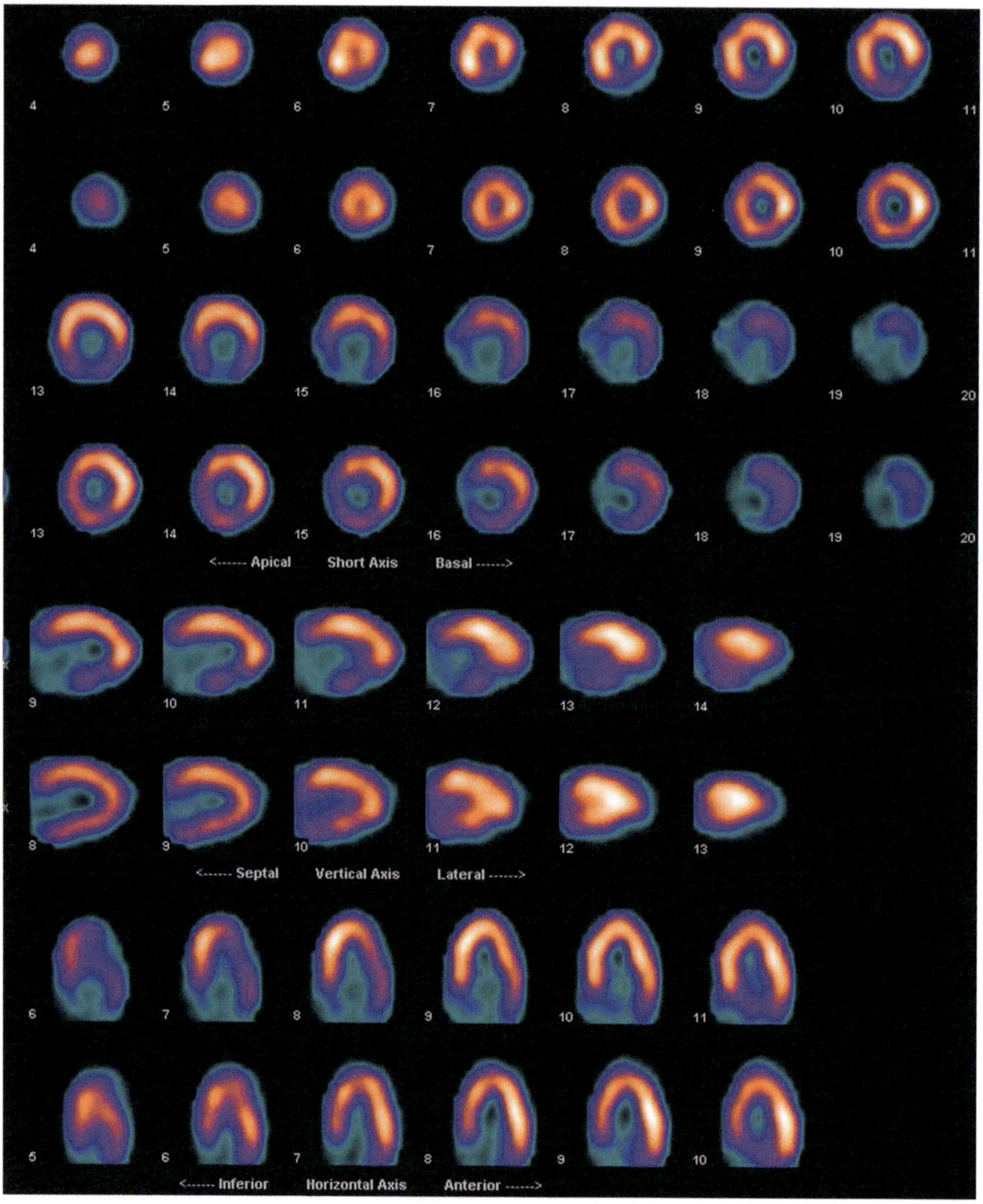

Based on the scan results, your recommendation would be:

(a) Maximizing his medical treatment
(b) Adding carvedilol (Coreg)
(c) Discontinuing lisinopril
(d) Sending him for cardiac catheterization

The patient has a large perfusion defect in the inferior wall extending to the lateral wall as seen on the perfusion scan (upper rows). However, the metabolic study with F-18 FDG is essentially normal, indicating viable myocardium in this region.

The patient is on the recommended lines of treatment for CAD, i.e., a beta-blocker (atenolol), an ACE inhibitor (Lisinopril), and a diuretic (Lasix) along with potassium supplement. Maximizing the medical treatment is always advisable. However, for this patient more can be done.

Coreg is a beta-blocker that is mainly used in patient with heart failure. There is no indication that this patient has such a condition. Also, the left ventricular cavity is not significantly dilated as seen in patient with heart failure. Discontinuing Lisinopril is not recommended as ACE inhibitors are one of the main lines of CAD management.

The literature clearly demonstrates that patient with viable myocardium benefits from revascularization with either stent placement or CABG. From the history, the patient has no contraindication to either option. Therefore, depending on the result of the cardiac catheterization, revascularization (stent or CABG) would be beneficial. On the other hand, if the scan demonstrates no significant viability, maximizing medical treatment would be recommended.

The correct answer is d.

23. A 50-year-old female had an adenosine myocardial perfusion scan. Her weight is 121 kg and height is 174 cm. The study was done for preoperative (knee joint replacement) evaluation. Below are the non-attenuated and the attenuated-corrected images. There was no focal wall motion abnormality and the global ejection fraction was 64%. The most accurate diagnosis is:

(a) Normal scan
(b) Ischemia in the distribution of the left anterior descending artery
(c) Ischemia in the distribution of the left circumflex artery
(d) Ischemia in the distribution of the right coronary artery

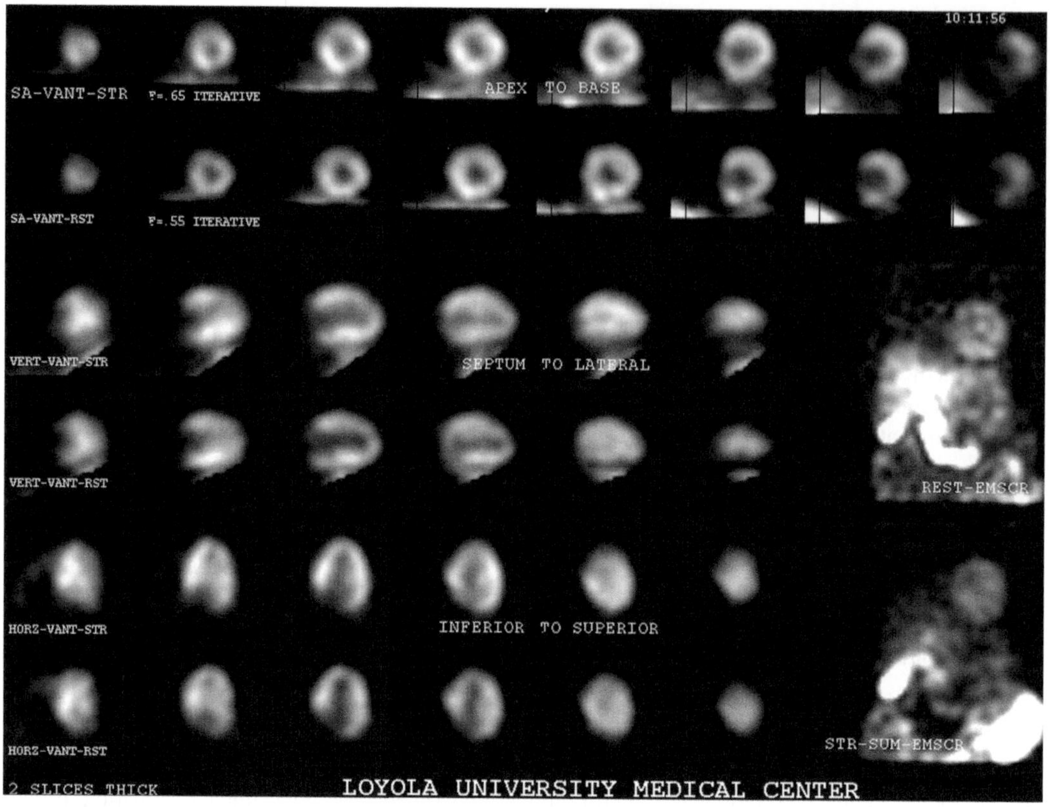

Attenuated-corrected images

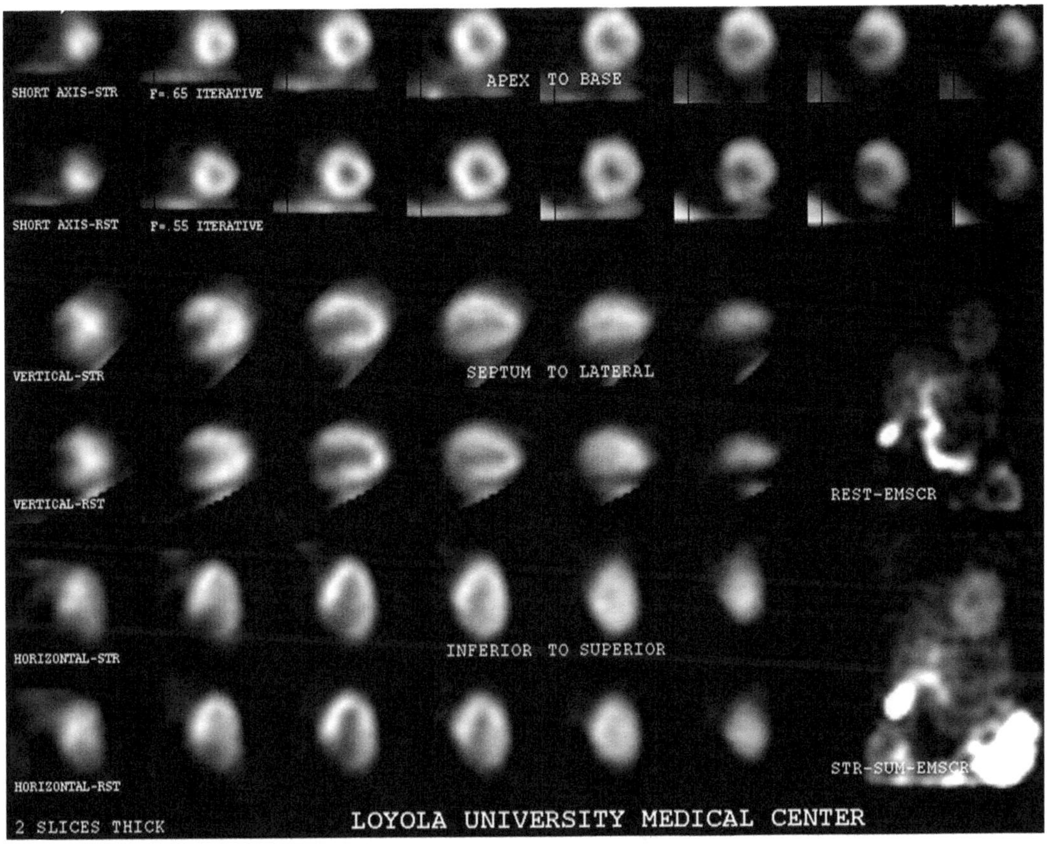

Non-corrected images

The attenuated-corrected images demonstrate subtle decreased activity along the anteroapical region during stress, which is also seen at rest. The other walls demonstrate normal perfusion. On the other hand, the non-attenuated images demonstrate normal perfusion of the apex and the anterior wall but subtle decreased activity along the inferolateral segment.

Attenuation correction is done through transmission rays that identify the body configuration similar to a low-quality CT scan. Then, a software program adds counts to the myocardial segments in proportional to the thickness of the soft tissue overlying them. Attenuation correction is very useful in distinguishing soft tissue artifacts such as female breast attenuation and diaphragmatic artifact from true perfusion defects. However, it is not without problems, particularly in obese patients.

In this patient, what appeared to be some decreased activity in the anteroapical region is most likely an artifact introduced by the attenuation correction program because the non-corrected images are normal in this location. On the other hand, the mild decreased activity in the inferolateral region on the non-corrected images is most likely due to soft tissue attenuation because it is not seen on the corrected images. Therefore, the study should be read as normal.

The correct answer is a.

24. A 67-year-old male was referred to the department of nuclear medicine because of chest pain. A previous myocardial perfusion scan done 10 months earlier was normal with global ejection fraction of 55%. The current scan (attenuated-corrected images) is shown below. The most likely diagnosis is:

    (a) Normal scan
    (b) Ischemia in the LAD distribution
    (c) Ischemia in the RCA distribution
    (d) Ischemia in the LCX distribution

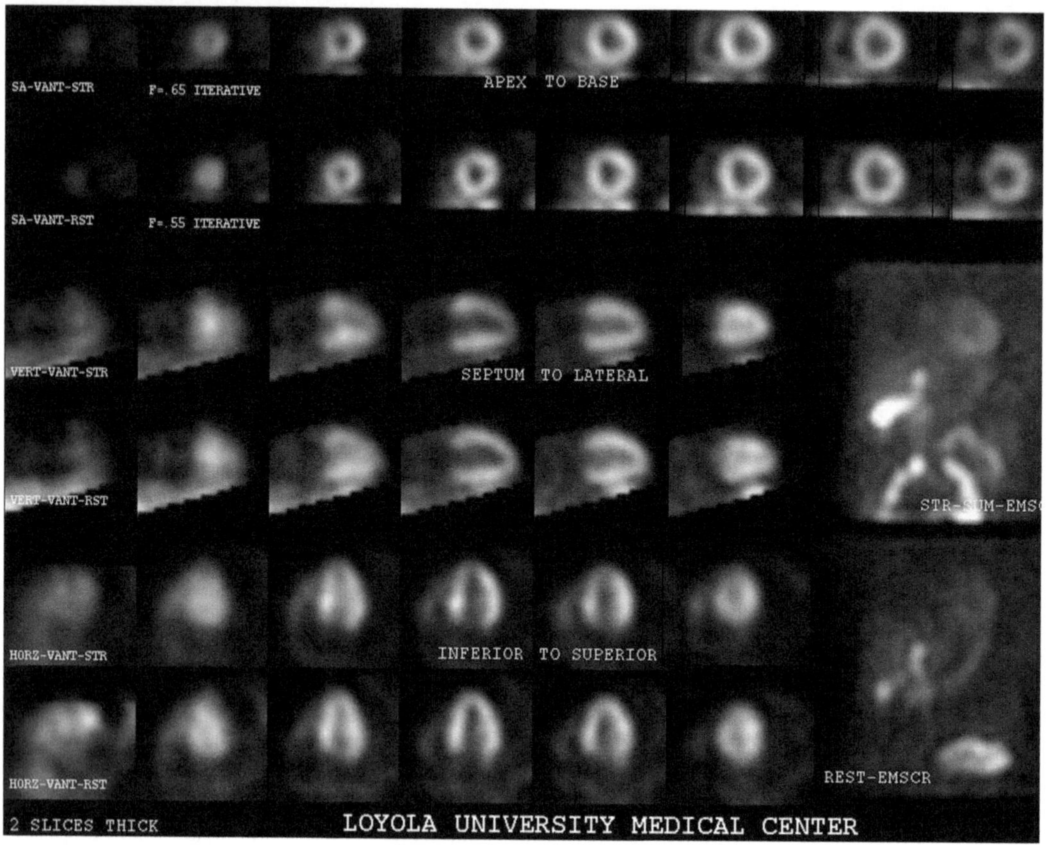

The scan demonstrates decreased perfusion in the anteroapical region during stress with normal perfusion at rest. The fact that these images are corrected for attenuation makes this defect unlikely to be an artifact. Also, anterior wall artifact usually occurs in females due to breast tissue attenuation rather than males. In addition, soft tissue artifact typically results in decreased activity on both stress and rest images with no significant change. In this patient, the decreased perfusion is seen only with stress.

The patient had a previous normal study 10 months earlier. Patients with normal perfusion scans have less than 1% chance of having myocardial event in the following 12 months. This value is low but not zero and should not discourage the primary physician from ordering another scan, particularly if the patient develops new symptoms or EKG changes.

The patient has a reversible perfusion defect in the anteroapical region, likely to represent ischemia in the distribution of the anterior descending artery.

The correct answer is a.

25. A 72-year-old male with type-II diabetes mellitus and kwon coronary artery disease. He was referred for a myocardial perfusion scan utilizing Tc-99 m tetrofosmin (Myoview) because of recent episodes of retrosternal pain. The scan demonstrates a large fixed perfusion defect in the apex, anterior wall, and septum with severe hypokinesia in these regions. The global LVEF was 32% (normal range is 50% and above). Fluorine-18 FDG PET study was then done at an outside facility, and it was read as no areas of decreased metabolic activity in the entire myocardium. Based on these results, patient went for cardiac catheterization, which demonstrated 99% occlusion of the LAD, 80% occlusion of a dominant RCA, and no significant disease in the left circumflex artery. The most appropriate next step is to:

(a) Repeat the viability study with thallium-201
(b) Stent the LAD artery
(c) Stent the RCA
(d) Send the patient for coronary artery bypass surgery (CABG)

The patient has significant stenosis of the LAD and RCA. The perfusion scan demonstrates a fixed defect in the distribution of the LAD but the PET study demonstrates good viability in this region. The perfusion scan also was suggestive of good perfusion in the distribution of the RCA (no perfusion defects on Tc-99 Myoview study). Because the myocardial perfusion scan typically measures the difference in perfusion, sometimes it detects only the most prominent lesions. In this case, it is LAD. However, from the cardiac catheterization, the RCA was also at great risk.

PET scan is more sensitive and more specific than thallium-201 for detecting myocardial viability. Hence, there is no need to repeat the viability study. It is reasonable to stent the LAD artery, particularly it is probably the cause of this patient's chest pain. However, the RCA also has significant stenosis and need to be fixed. On the other hand, stenting the RCA only is not a good choice because it might slightly improve this patient's condition in the presence of significant LAD stenosis.

Complete revascularization with CABG would probably benefit this patient the most. It is likely to improve the perfusion in the distribution of both LAD and RCA. In addition, the hibernating myocardium in the apex, septum, and anterior wall will likely regain function and might return to normal or near-normal motion.

The correct answer is d.

26. A 56-year-old male with insignificant past medical history underwent a treadmill myocardial perfusion scan because of intermittent episodes of chest pain. The scan was normal with an estimated ejection fraction of 63%. The annual risk for myocardial infarction or death secondary to heart disease in this patient is:

(a) Essentially zero, the patient is healthy
(b) Less than 1%
(c) Three percent to 5%, his age is over 55 years
(d) More than 5%, considering his sex, age, and symptom (chest pain)

A normal myocardial perfusion scan has a risk of less than 1% a year for myocardial infarction or death regardless of age, sex, or symptoms. This risk is never zero but it is low enough not to pursue other investigations. However, if the stress part of the study is abnormal, for example, if there is an ST depression or elevation, further examinations such as cardiac angiogram are needed.

Being a male is one of the risk factors and this patient's age is another risk factor. Still, the scan is normal and the ejection fraction is within the normal limits.

The correct answer is b.

Case 1

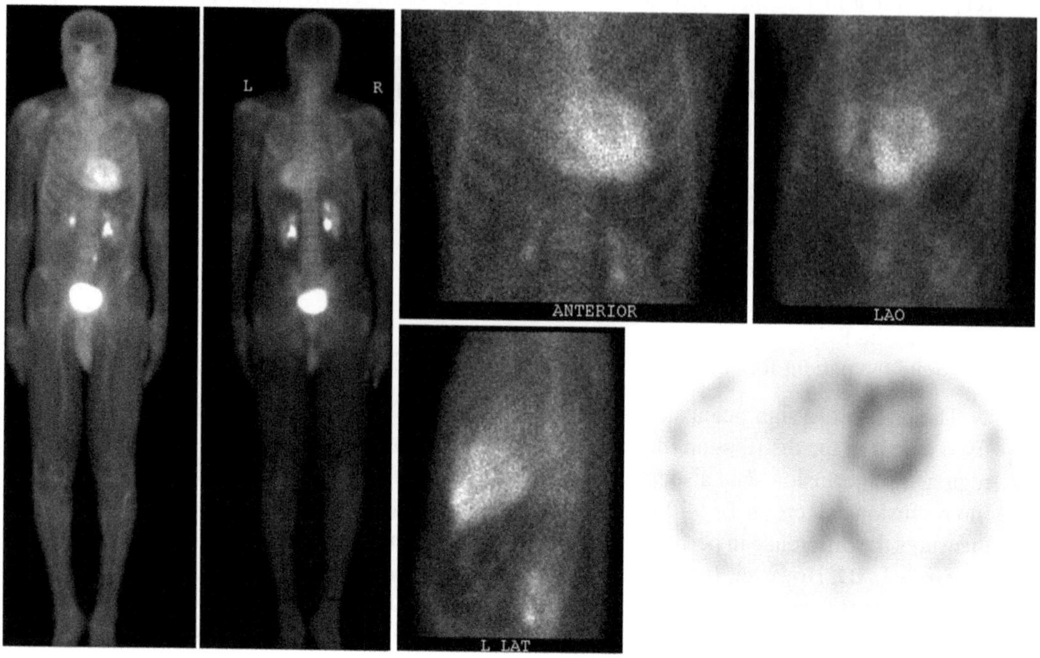

27. The radiopharmaceutical used for this exam was:
    (a) Ga67 citrate
    (b) I-123 MIBG
    (c) In111 capromab pendetide
    (d) Tc99m bone agent

Answer: D

Hepatic uptake is characteristic of Ga67 citrate scans and is usually one of the most intense areas of accumulation. In addition, there is often uptake in the lacrimal glands.

Although I123 MIBG can be used for cardiac imaging, the normal distribution of MIBG includes liver and salivary glands with little uptake in the bones.

In111 capromab pendetide (Prostascint) is a monoclonal antibody that also demonstrates significant hepatic accumulation. Initially introduced for imaging of prostate cancer and prostate-specific membrane antigen, its use has been largely replaced by F18 fluciclovine (Axumin). Like most antibody agents, hepatic uptake is significant.

The old bone-seeking radiopharmaceutical, Tc99m pyrophosphate has been largely replaced by other phosphonate bone imaging agents such as Tc99m HDP and Tc99m MDP. It is commonly used to evaluate cardiac amyloidosis. Bone agents do not demonstrate significant hepatic uptake.

28. The findings on the scan demonstrate:

A. Cardiac uptake consistent with myocardial infarction
B. Amyloid involvement of the myocardium
C. Imaging too early with significant blood pool activity and urinary excretion
D. Nonspecific uptake in a lung mass

Answer: B

Tc99m pyrophosphate was historically used for the detection of and imaging of cardiac infarcts. The test was most sensitive when imaging was performed between 24 and 72 h of the infarct. The test is rarely used anymore since the introduction of cardiac enzyme assays.

The cardiac activity seen on the scan is located in the walls with a more photopenic center. If the blood pool activity were significant, the central area would be more intense and there would be greater uptake seen in other vessels.

Soft tissue masses can accumulate tracers in a nonspecific fashion. The intensity seen here however is quite great and corresponds to the cardiac walls.

Tc99m pyrophosphate accumulates in cardiac amyloidosis to a significant degree and is an inexpensive and noninvasive way of determining myocardial involvement.

29. The critical organ in this study is:

A. The heart
B. The bones
C. The bladder
D. The lungs

Answer: C

The critical organ is the organ that receives the greatest dose of radiation, in this case, the bladder. The target organ is the organ that imaging is directed at and may change with the application of the radiopharmaceutical. When used for bone scanning, the bones are the target organ for Tc99m pyrophosphate. In this case, when cardiac amyloid is suspected, the target organ is the heart. The critical organ remains the same for each radiopharmaceutical while the target organ may change. Another example is the use of Tc99m Sestamibi. When used for cardiac imaging, the heart is the target organ. When the agent is used in a protocol for the detection of parathyroid adenoma, it is the parathyroid that is the target.

Case 2

30. An 80-year-old male with history of coronary artery disease, hypertension, and heart failure with preserved ejection fraction presented with echocardiographic findings of increased LV wall thickness and echogenicity that was consistent with an infiltrative cardiomyopathy. He was further evaluated with the nuclear medicine study that is shown below.

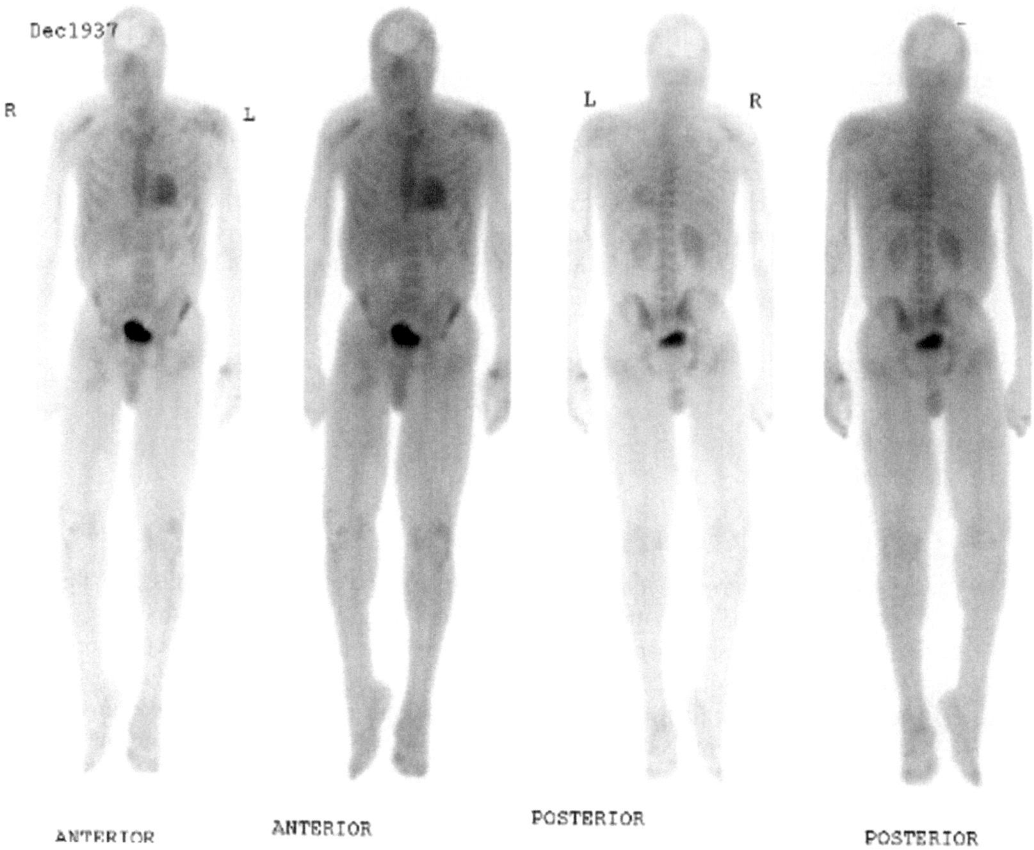

Image 1

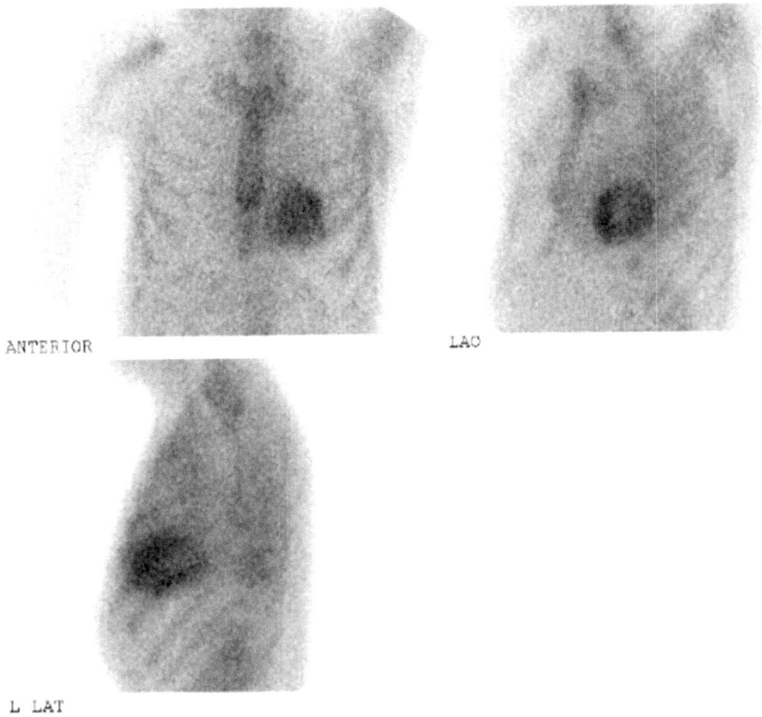

Image 2

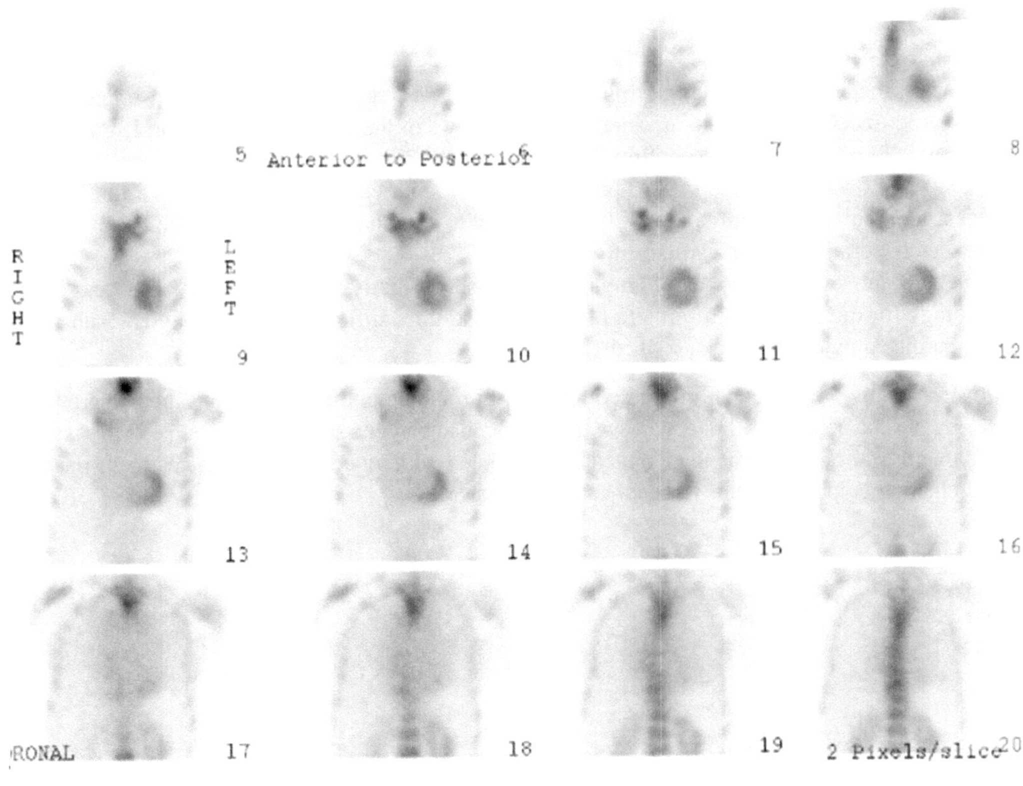

Image 3

Which of the following radiotracer was used in the above whole-body scan to evaluate for infiltrative cardiomyopathy?

A. Tc-99 m methyl diphosphonate (Tc-99 m MDP)
B. Tc-99 m pyrophosphate (Tc-99 m PYP)
C. Tc 99 m-labeled red blood cells (Tc 99 m-labeled RBC)
D. Tc-99 m Sestamibi

Answer: B

As mentioned before, Tc-99 m pyrophosphate (Tc-99 m PYP) is an old nuclear medicine bone agent. Newer agents such as Tc-99 m MDP are more utilized nowadays. Recently, the PYP agent is used as a cardiac imaging agent for evaluation of myocardial injury, primarily when amyloidosis is suspected.

Tc-99 m methyl diphosphonate (Tc-99 m MDP) is primarily used as a bone imaging agent to delineate areas of altered osteogenesis. It is not used for cardiac imaging.

Tc 99 m-labeled red blood cells (Tc 99 m-labeled RBC) are used for blood pool imaging, including cardiac first pass and gated equilibrium imaging, and for detection of sites of gastrointestinal bleeding.

Tc-99 m Sestamibi is a myocardial perfusion agent that is indicated for detecting ischemia and myocardial infarction.

31. Upon evaluation of the above images, you conclude that the study is:

A. Normal
B. Consistent with cardiac sarcoidosis
C. AL (Light-chains) cardiac amyloidosis
D. Transthyretin-related cardiac amyloidosis (ATTR)

Answer: D

Whole-body Tc-99 m pyrophosphate (Tc-99 m PYP) scan (image 1) demonstrates diffuse increased tracer uptake in the myocardium. Spot views of the chest in the anterior, LAO and left lateral projections (image 2) confirm this finding. SPECT images of the chest (image 3, coronal views) demonstrate diffuse tracer uptake in the left ventricular myocardium. The findings are consistent with cardiac amyloidosis. It is important to note that a positive Tc-99 m PYP scan will be seen in certain types of cardiac amyloidosis.

There are several subtypes of amyloidosis based on the type of protein involved and/or the etiology for the formation of protein fibrils. At least, three different pathophysiologic substrates for cardiac amyloidosis are described:

1. In AL amyloidosis, the fibrils are composed of immunoglobulin light-chains that are produced by plasma cells in the bone marrow. It has a rapid and aggressive clinical course with systemic manifestations. Cardiac involvement is associated with a poor prognosis. Treatment involves chemotherapeutic agents targeted at the plasma cell.

2. Mutant transthyretin-related cardiac amyloidosis (ATTRm), which is also referred to as familial amyloid cardiomyopathy, occurs secondary to mutation, leading to misfolding of normal tetrameric transthyretin protein (TTR) that deposits in the myocardium.
3. Wild-type transthyretin-related amyloidosis (ATTRwt) is a slowly progressive disease that tends to affect the heart of elderly men and is commonly referred to as senile systemic amyloidosis.

   Several studies have shown, that Tc-99 m PYP is able to differentiate between AL and ATTR amyloidosis noninvasively. While Tc-99 m PYP is avidly taken up by the heart in ATTR amyloid cardiomyopathy, there is no significant tracer accumulation in cardiac tissue in patients with AL type disease.

   Tc-99 m PYP scintigraphy is not used for evaluation of cardiac sarcoidosis.

32. Which of the following is the likely mechanism of PYP cardiac uptake?

    A. binding to cytoplasmic mitochondria
    B. localization secondary to high calcium levels
    C. binding to the beta chain of the hemoglobin molecule in red blood cells
    D. uptake by sympathetic nerve terminals

    Answer: B

    It is thought that a bone-seeking radiotracer agent, technetium Tc-99 m pyrophosphate (Tc-99 m PYP) is selectively localized to injured muscle tissue where calcium deposition is high, such as that of cardiac amyloidosis.

    Tc-99 m Sestamibi is a lipophilic radiotracer that is extracted by the myocardium with lengthy myocardial retention, primarily through fixed binding to cytoplasmic mitochondria.

    Tc 99 m-labeled red blood cells (Tc 99 m-labeled RBC) binds to the beta chain of the hemoglobin molecule (after being reduced by stannous ion).

    Metaiodobenzylguanidine (MIBG) is a norepinephrine analog that is actively taken up by sympathetic nerve terminals, enabling imaging cardiac sympathetic innervation. I-123 MIBG scintigraphy is used to image cardiac sympathetic denervation (not amyloidosis). Myocardial defect of MIBG activity provides direct evidence of impaired cardiac sympathetic nerve endings due to amyloid deposits. The normal distribution of the radiotracer is in organs with rich adrenergic innervation, including the heart, salivary glands, spleen, liver, kidneys, and thyroid gland. No skeletal uptake will be seen.

Case 3

33. A 70-year-old male with a history of coronary artery disease, s/p recent CABG and heart failure s/p biventricular defibrillator as well as symptomatic bradycardia, was evaluated with Tc-99 m pyrophosphate scintigraphy for suspected cardiac amyloidosis. Images are shown below.

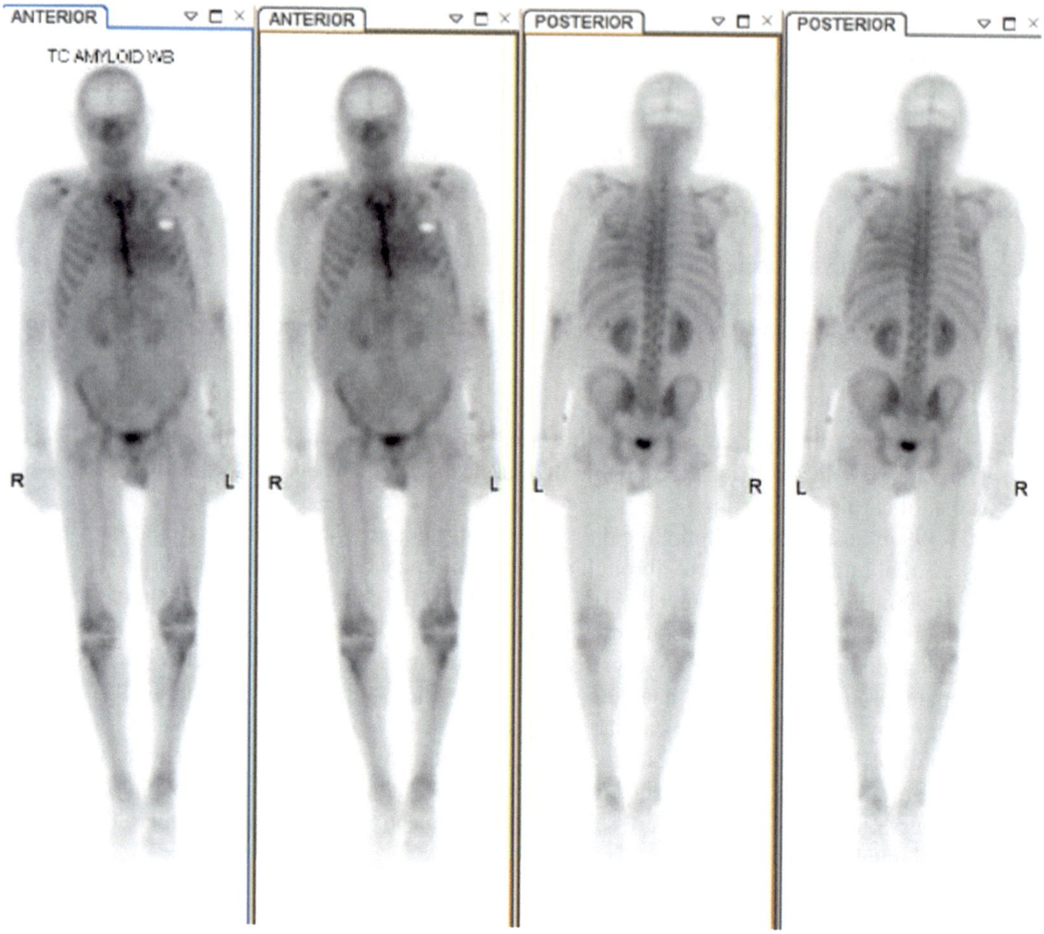

Image 1

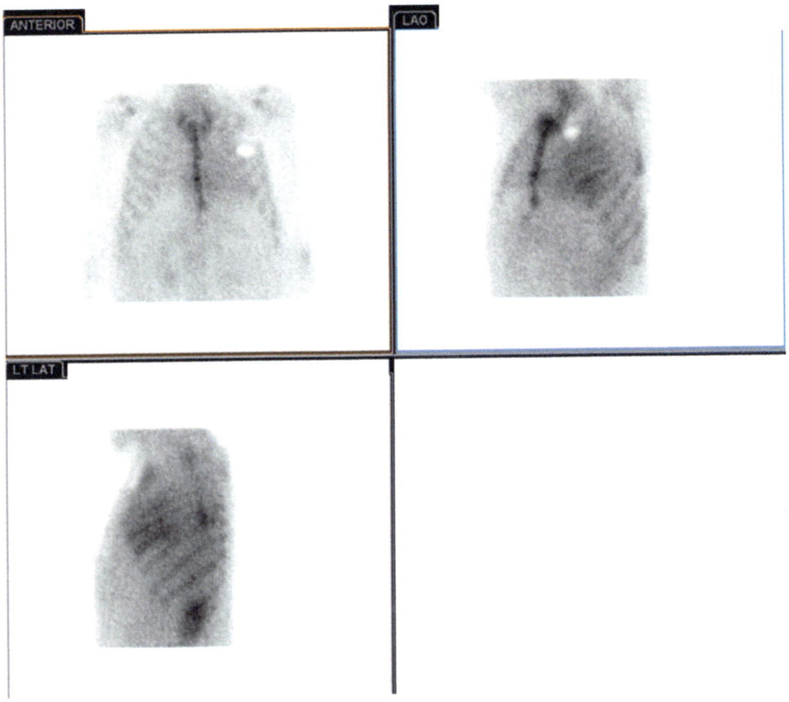

Image 2

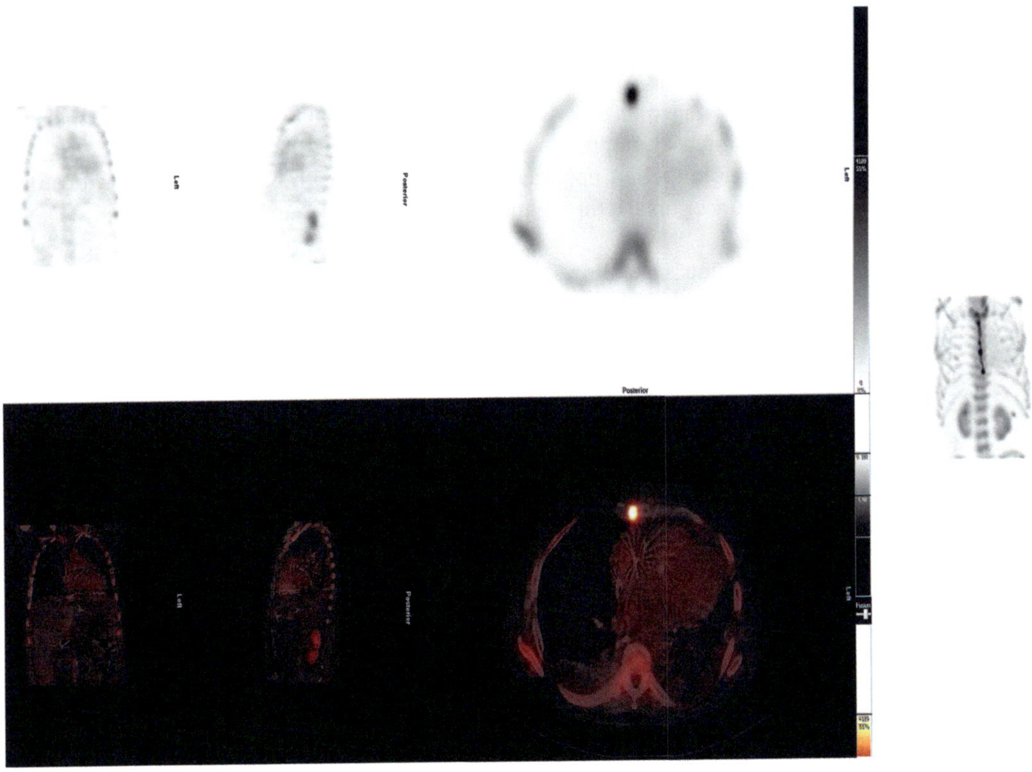

Image 3

Upon evaluation of images, you conclude:

A. The left upper chest photopenic area is consistent with cardiac amyloidosis
B. The patient likely has TTR cardiac amyloidosis
C. Further evaluation with serum protein electrophoresis would be beneficial
D. linear tracer uptake in mid-chest is consistent with esophageal amyloidosis

Answer: C

Tc-99 m pyrophosphate (Tc-99 m PYP) scintigraphy is used to confirm the suspected diagnosis of cardiac amyloidosis. The scan also helps to determine the type of amyloid. The most common types of amyloid fibrils that affect the heart are derived from either light-chains produced by a plasma cell dyscrasia (amyloid light-chain [AL] amyloidosis) or transthyretin (TTR), a protein mainly synthesized by the liver from either wild-type or mutant TTR-related amyloidosis.

Tc-99 m pyrophosphate is thought to be taken up via a calcium-mediated mechanism and seem to be helpful to identify (both mutant and wild-type) ATTR amyloidosis. While a positive Tc-99 m pyrophosphate scan suggests ATTR amyloidosis, a negative scan does not rule out AL cardiac amyloid.

In above case scenario, whole-body Tc-99 m Tc-99 m PYP images (image 1) and spot views of the chest (image 2) demonstrate mild diffuse tracer uptake in the region of the heart. However, the intensity of the uptake is significantly less than the surrounding thoracic bones, which is not characteristic of amyloid deposits. SPECT-CT images of the chest (image 3) demonstrate no significant tracer accumulation in the myocardium. Only physiological minimal blood pool accumulation is noted. Therefore, Tc 99 m PYP study has no scintigraphic evidence of TTR cardiac amyloidosis. Findings, however, do not exclude the presence of the other types of cardiac amyloidosis, such as AL amyloidosis, and additional testing should be performed. To establish the presence of AL amyloidosis several diagnostic criteria have to be met. One of them is the evidence of a monoclonal plasma cell proliferative disorder (e.g., presence of a serum or urine M protein, abnormal serum free light-chain ratio, or clonal plasma cells in the bone marrow). A serum or urine monoclonal (M) protein, for example, can be detected by performing serum protein electrophoresis (SPEP).

The photopenic area in the left upper chest is consistent with the patient's known cardiac device and the linear tracer uptake throughout the sternum is due to recent sternotomy.

34. Which of the following radiotracers allows detecting amyloidosis by assessing myocardial sympathetic nerve activity?

A. Tc-99 m pyrophosphate
B. I-123 MIBG
C. F-18 florbetapir
D. C-11 Pittsburgh B compound

Answer: B

Radionuclide imaging of cardiac amyloidosis can be performed using radiotracers such as direct amyloid imaging agents (I-123–labeled serum amyloid P component), bone imaging agents (Tc-99 m pyrophosphate or Tc-99 m DPD), or imaging cardiac sympathetic innervation (I-123 MIBG).

Metaiodobenzylguanidine (MIBG), a norepinephrine analog, is actively taken up by sympathetic nerve terminals and not further metabolized, enabling imaging cardiac sympathetic innervation. It is to be noted that I-123 MIBG scintigraphy is used to image cardiac sympathetic denervation and not amyloidosis. Myocardial defect of MIBG activity provides direct evidence of impaired cardiac sympathetic nerve endings, which could be due to amyloid deposits.

Tc-99 m pyrophosphate (Tc-99 m PYP) is selectively localized to injured muscle tissue where calcium deposition is high, such as that of cardiac amyloidosis. The accumulation of technetium Tc-99 m pyrophosphate to these areas allows for their specific imaging of cardiac amyloidosis.

C-11 Pittsburgh B compound and F-18 florbetapir are PET tracers used to direct imaging of beta-amyloid plaques in neuronal tissue, assisting in the diagnosis of Alzheimer's disease. The efficacy of these radiotracers in the diagnosis of cardiac amyloidosis seems promising and is currently under investigation.

References

1. Bokhari S, Castaño A, Pozniakoff T, Deslisle S, Latif F, Maurer MS, 1Tc-Pyrophosphate scintigraphy for differentiating light-chain cardiac amyloidosis from the transthyretin-related familial and senile cardiac amyloidosis. Nuclear Cardiology Laboratory, Columbia College of Physicians & Surgeons, New York, NY, USA. Center for Advanced Cardiac Care, Division of Cardiology, Columbia College of Physicians & Surgeons, New York, NY, USA.
2. Falk RH, Quarta CC, Dorbala S, How to image cardiac amyloidosis.
3. Glaudemans AWJM, Slart RHJA, Zeebregts CJ, Veltman NC, Tio RA, Hazenberg BPC, Dierckx RAJO, Nuclear imaging in cardiac amyloidosis. Eur J Nucl Med Mol Imaging. 2009;36:702–14.
4. https://www.drugs.com/

Case 4

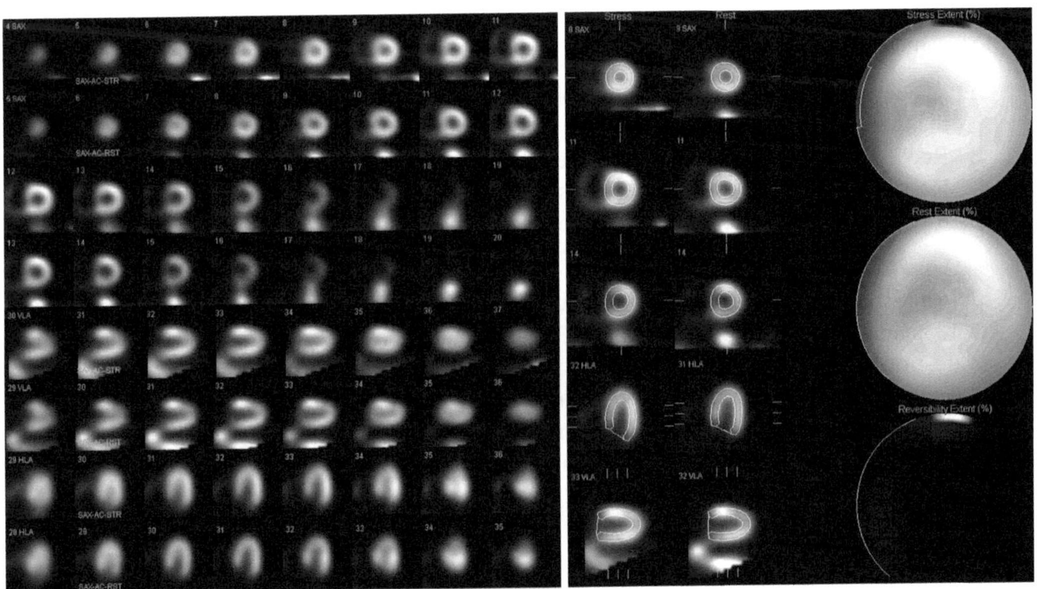

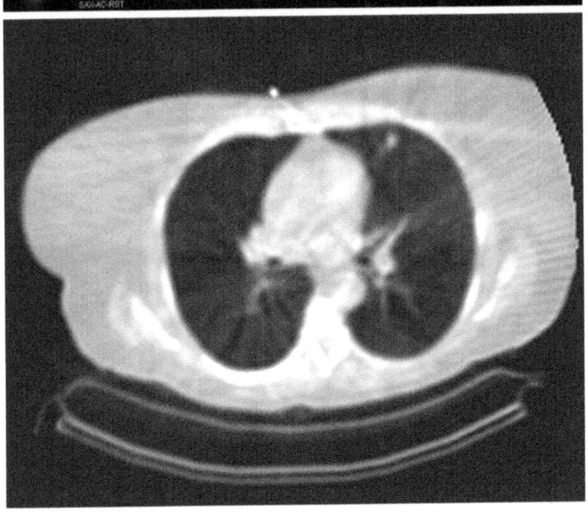

35. This otherwise healthy 67-year-old female smoker presents to the referring physician with symptoms of dyspnea that is clinically suspected to be of cardiac origin. The referring physician requests a pharmacologic stress Tc99m myocardial perfusion study. You perform the study on a gamma camera with low dose CT attenuation correction. The ejection fraction is 70%, the summed stress score is 2 and the TID is 1.05. Your impression:

    A. The study should be repeated in the prone position.
    B. A septal perfusion defect should be evaluated with CT angiography.
    C. The perfusion to the myocardium is normal.
    D. A CT attenuation correction artifact is present.

    Answer: C

    The myocardial perfusion is normal.

    Gastrointestinal activity can sometimes prevent accurate evaluation of the inferior wall. In such cases, one option is to repeat the study with the patient in the prone position. In this patient, the inferior wall can be well seen from apex to base and prone imaging is not necessary.

    A portion of the patient is outside the field of the attenuation correction image. If large, this may create a reconstruction artifact. In this patient, no attenuation correction artifact is seen despite the inability to fit the entire patient into the field of view.

36. After making the determination about the results of the myocardial perfusion study you:

    A. Encourage the technologist to use a binder in the future to ensure that the entire patient is in the field of view.
    B. Recommend additional imaging to the referring physician
    C. Recommend pulmonary function tests to evaluate the source of dyspnea.
    D. Suggest an immediate lung scan be performed to evaluate for pulmonary embolism.

    Answer: B

    While pulmonary function tests may be useful in evaluating a patient with a history of COPD, there is a nodule in the left lung on the CT attenuation correction image which requires further evaluation.

    Binders are not necessary when performing myocardial perfusion studies.

    A perfusion lung study could be performed in 24 h if pulmonary embolism is suspected. Some residual tracer activity may still be present in the heart.

37. A follow-up CT of the chest was performed for evaluation. Selected CT images are shown. The next best step for evaluation should be:

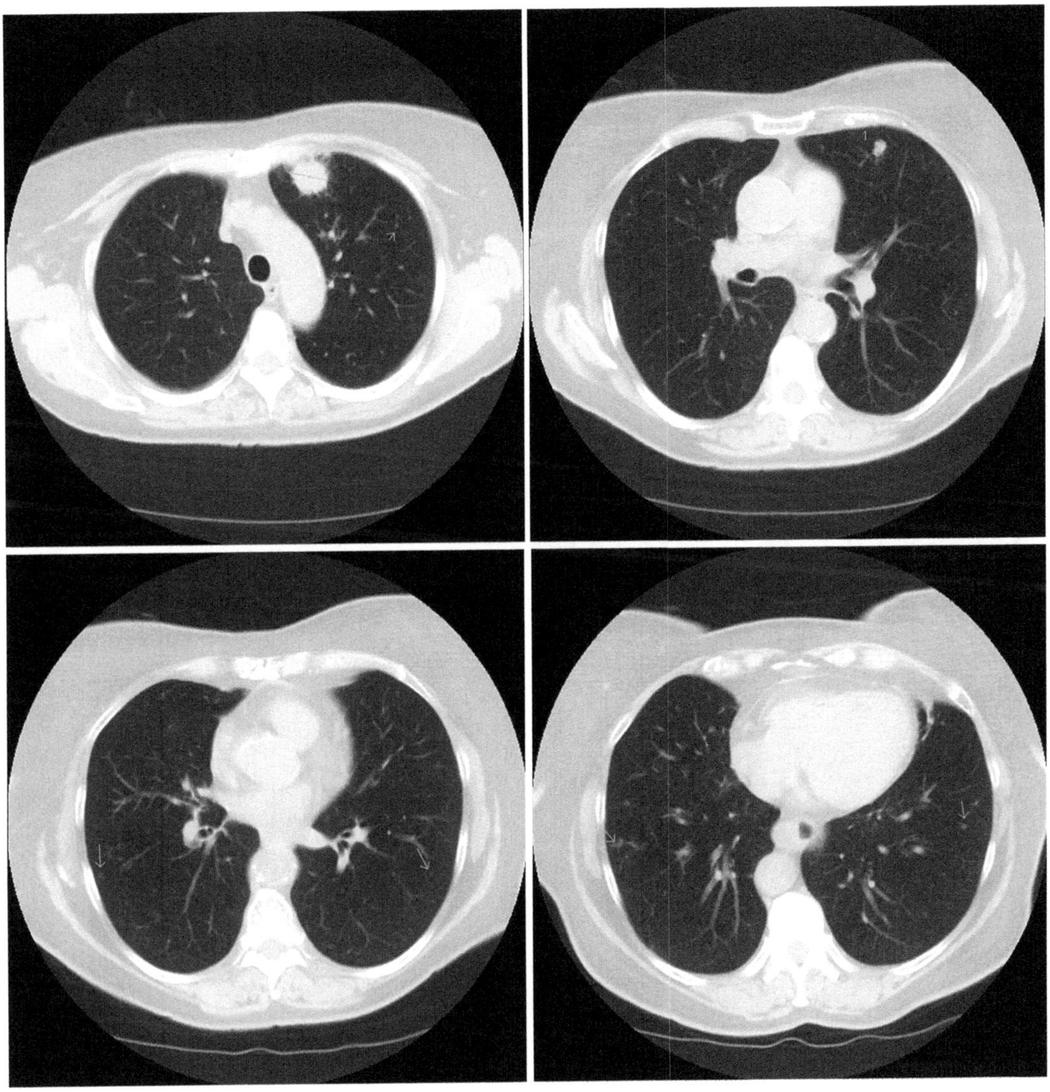

A. A metastatic workup including a FDG/CT study
B. Biopsy of the largest nodule seen.
C. Begin a course of antibiotic therapy
D. Follow up of the nodules in 6 months to determine if they have resolved.

Answer: A

Multiple previously unknown lung nodules are identified in the study. A metastatic workup including a FDG/CT study is appropriate in this patient. It is unknown if the lung nodules are malignant and their origin is unknown. Determination of their site of origin and tissue type will be important in the subsequent management of this patient.

Although infection can produce nodules on CT that look suspicious for malignancy the abnormalities seen must be further evaluated for malignant potential prior to initiating a course of antimicrobial therapy.

38. After reviewing the FDG/CT images, no abnormalities are identified outside of the lung nodules seen bilaterally. The next best step is:

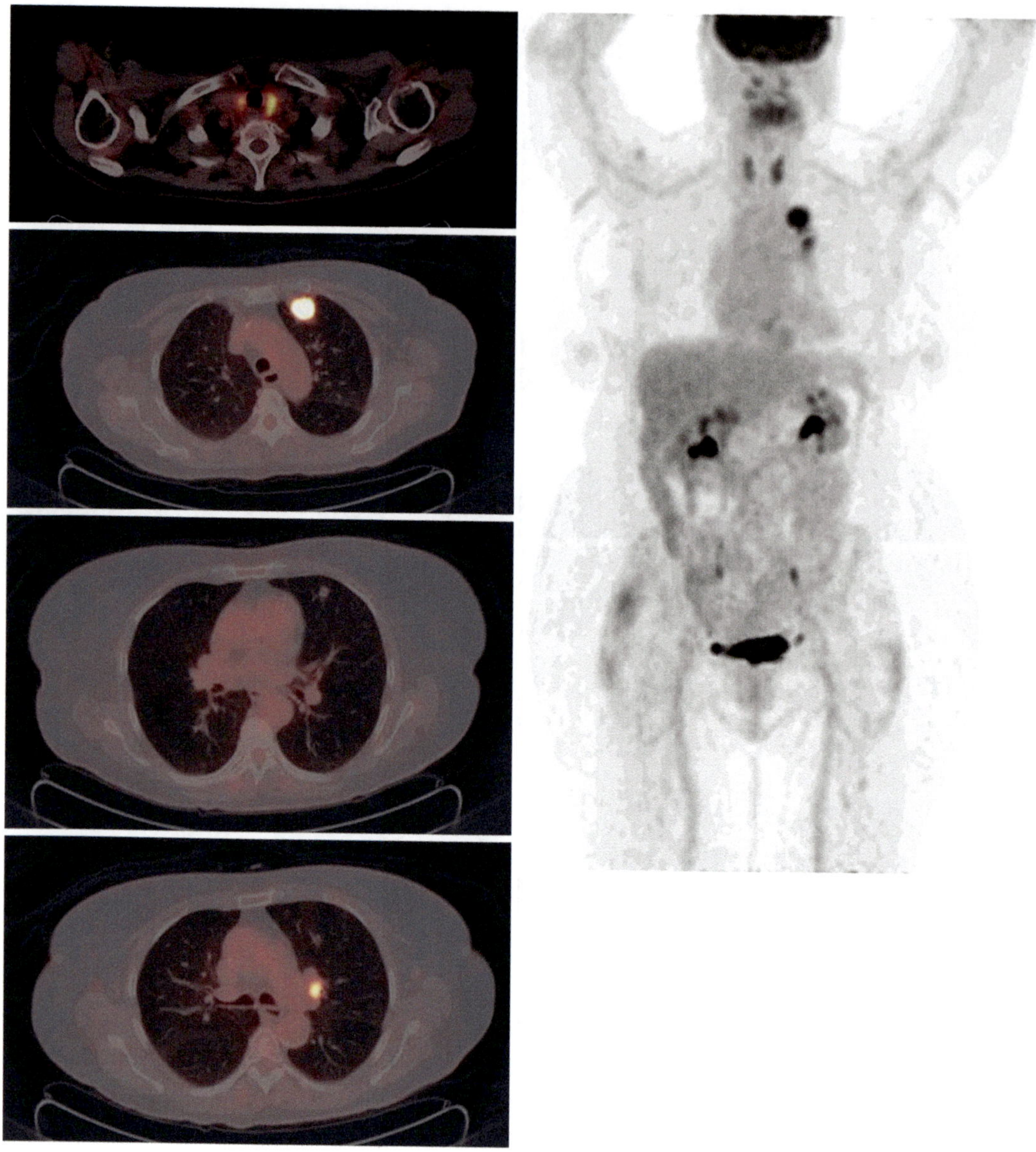

A. Thyroid ultrasound to evaluate the asymmetric uptake seen on FDG imaging.
B. Establish a tissue diagnosis to plan for definitive therapy
C. Consult medical oncology and radiation oncology to provide definitive therapy.
D. Consult thoracic surgery for left upper lobectomy.

Answer: B

Most likely malignant, the largest nodule demonstrates the greatest uptake and is the best choice for obtaining tissue diagnosis. Alternatively, a bronchoscopy may be performed with an attempt to biopsy the left hilar node, though this may be more difficult to sample.

Interestingly, the nodule seen on the CT attenuation correction image of the myocardial perfusion study does not demonstrate significant uptake although it served to prompt further investigation.

Surgery, medical oncology, and radiation therapy may at some point be needed in the management of this patient and further discussion is warranted at a multispecialty conference.

39. This 78-year-old male presented to the emergency room with a history of chest pain that occurred after eating a heavy meal. After reviewing the pharmacologic stress myocardial perfusion study images you suggest that:

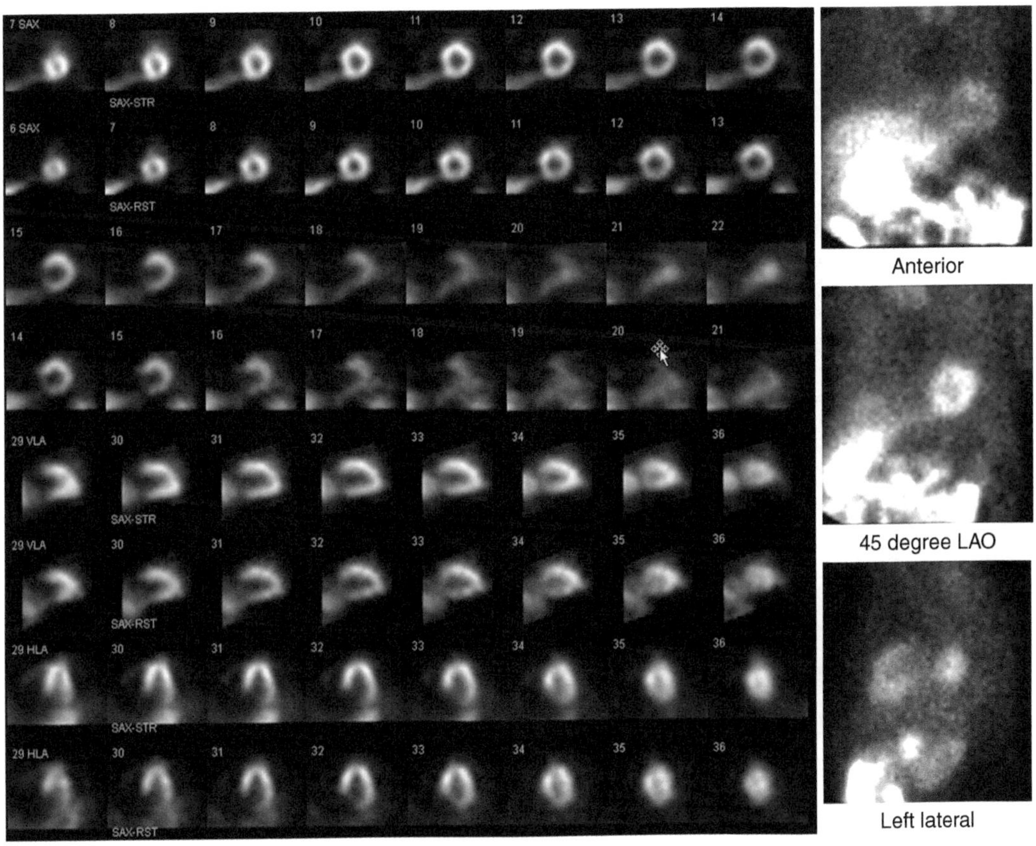

A. The myocardial perfusion study is abnormal.
B. A hiatal hernia may be the cause of the symptoms.
C. Further evaluation with cardiac CT angiography is warranted.
D. Uptake in the thyroid should be further evaluated.

Answer: B

Best seen on the left lateral planar image obtained from the cine portion of the study and visible on the horizontal and vertical long-axis images, a blush of activity can be seen posterior to the heart. Since the myocardial perfusion agents also accumulate within the stomach, this is in the location for a hiatal hernia. A chest radiograph or CT study of the chest would be confirmatory.

No perfusion abnormalities are identified. Uptake in the thyroid is a normal distribution for the Tc99m myocardial perfusion agents.

40. A 72-year-old female presents with a history of intermittent angina and is referred for a pharmacologic myocardial perfusion study with Tc99m tetrofosmin and Regadenoson. You perform the study and review the images. Wall motion appears normal, and the ejection fraction is estimated at 66%. The summed stress score is 25, summed rest score is 24 with a summed difference score of 1. The extent of the perfusion abnormality is estimated at 34% of the myocardium. The TID is calculated at 1.0. After reviewing the images and findings you:

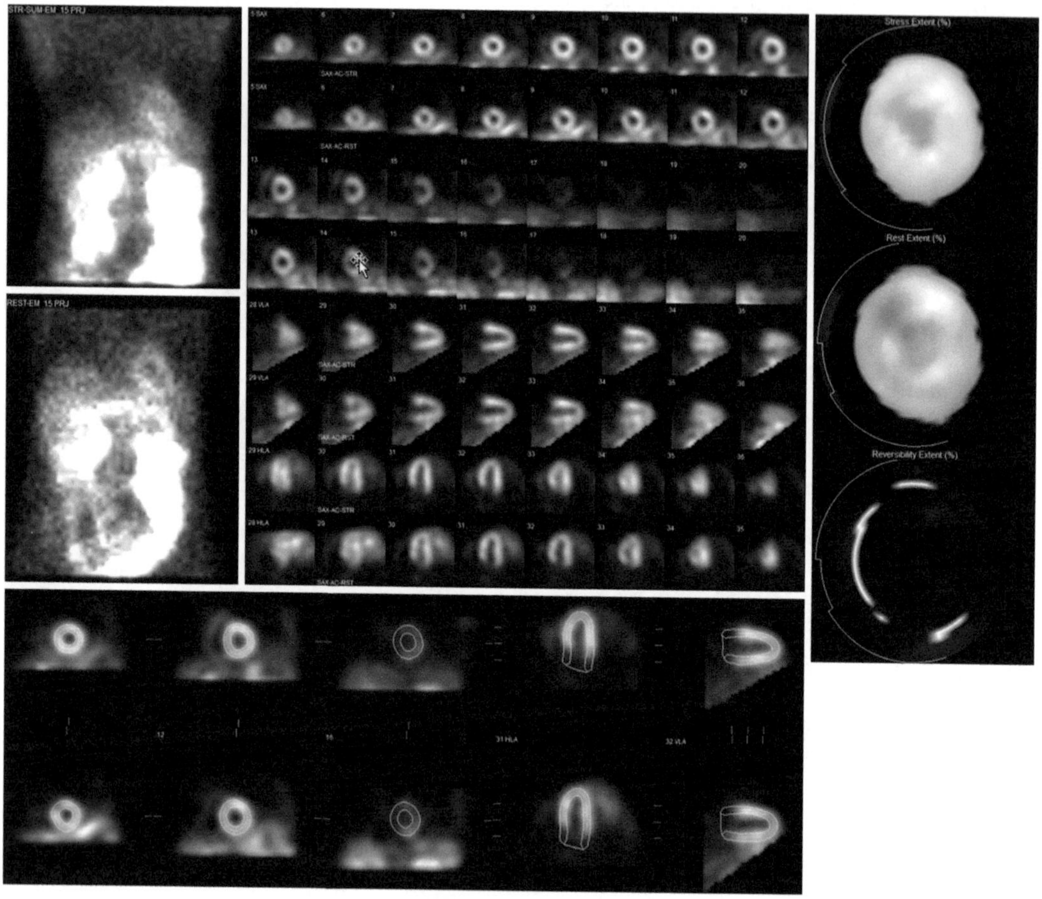

A. Contact the referring physician and recommend performing a viability study
B. Notify the referring physician of a large prior myocardial infarct
C. Request that the patient return to have the study repeated after discontinuation of beta-blockers
D. Adjust the software settings to optimize edge detection

Answer: D

The raw data, short- and long-axis images obtained during stress and at rest are normal. No perfusion abnormalities are identified. The computer analysis, however, finds an annular area of absent perfusion on the bullseye images. There is no known anatomic distribution of vasculature that if occluded can cause this appearance. The finding must be artefactual.

The regions of interest clearly extend beyond the anatomic borders of the left ventricle. When reviewing the computer-generated analysis, this will create an annular artifact at the base of the heart. In this case, that artifact estimates that 34% of the myocardium is abnormal. This is a finding that clearly contradicts the findings of the short- and long-axis images. When such a discrepancy occurs, it is imperative for the interpreter to identify the cause.

While in this case, the appearance is that of a nonreversible perfusion abnormality involving the base of the heart, more correct edge detection on the resting images might give the appearance of reversibility and lead to thoughts of intervention. The distribution, however, does not follow anatomic vascular patterns of the LAD, RCA, or circumflex vessels. Reconciliation of the visual findings and computer findings are an essential component of myocardial perfusion interpretation.

Case 5

41. An 80-year-old male presents for a myocardial perfusion study with a history of newly depressed ejection fraction. He is 5′6″ and weighs 205 lbs. with a BMI of 34. On review of the attenuated-corrected images, the findings are remarkable for:

    A. Chest soft tissue artifact
    B. Right coronary artery ischemia
    C. Large hiatal hernia
    D. RV dilatation

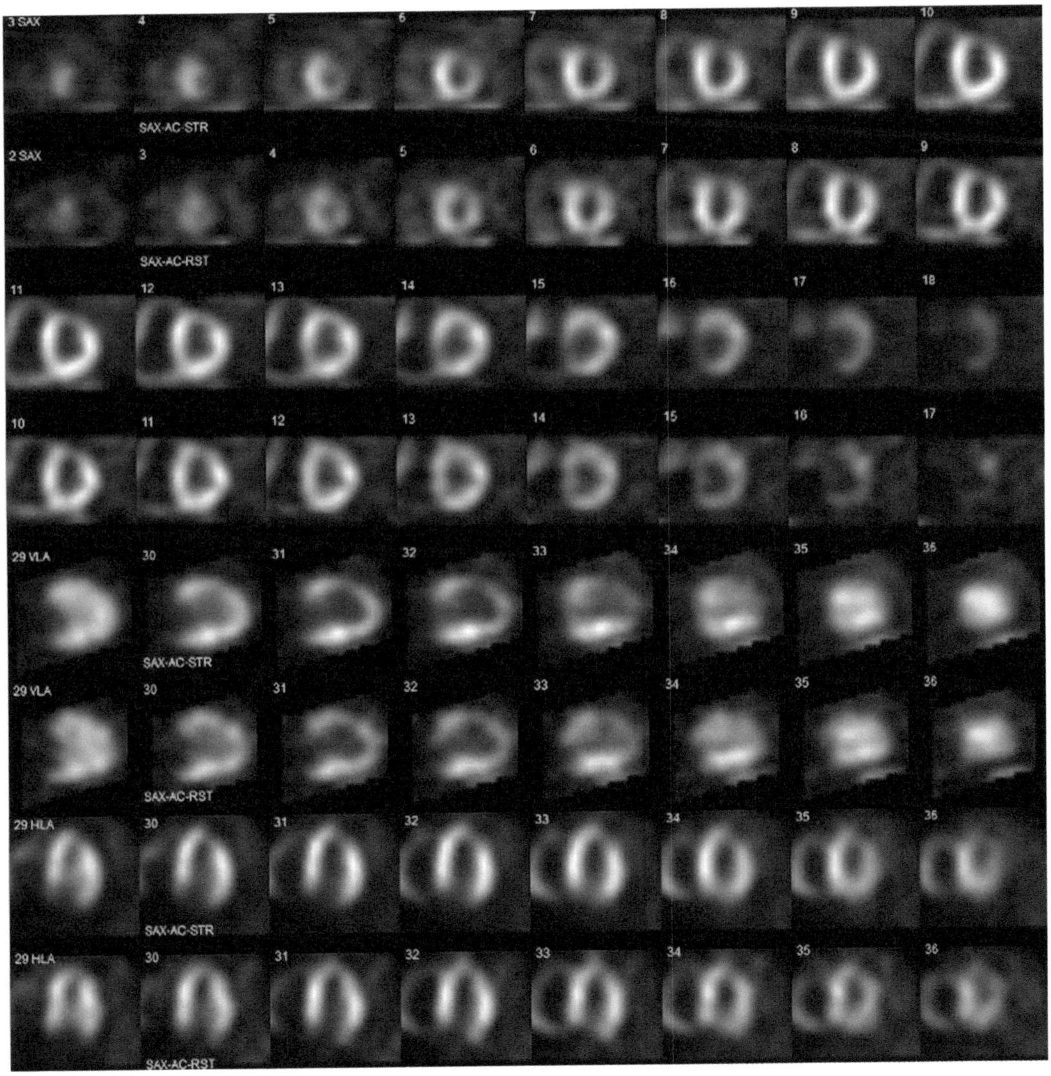

Answer: D: RV dilatation

When reviewing a myocardial perfusion study, it is important to not only evaluate the left ventricle but to review the right ventricle as well since this can indicate underlying pathology. The right ventricular end-diastolic volume is often slightly larger than that of the left ventricle, but in this case can be seen to be significantly larger, raising questions of underlying pathology.

The study is performed with attenuation correction so that a larger BMI should have little effect on soft tissue attenuation. Additionally, males less frequently have soft tissue attenuation over the anterior wall and more commonly have abdominal or diaphragmatic artifacts.

No significant right coronary artery distribution perfusion abnormality is present. Hiatal hernia, when present can be seen at the base of the myocardium.

42. An asymmetrically enlarged right ventricle seen on a myocardial perfusion study can be seen in cases of:

A. Mitral regurgitation
B. Pulmonary embolism
C. Ruptured chordae tendineae
D. Rheumatic heart disease

Answer: B: Pulmonary embolism

Patients may present to the physician's office with a history of shortness of breath. Based on the history provided, the physician may be led towards a cardiac etiology and a myocardial perfusion study ordered. The causes of shortness of breath are many, but a dilated right ventricle may be secondary to tricuspid regurgitation, pulmonary hypertension, or even acute or chronic pulmonary embolism. Unexpected right ventricular dilatation should prompt a review of the patient history for pulmonary or cardiac etiologies. If pulmonary embolism is considered a possibility, a follow-up CT or a V/Q study should be performed.

Mitral regurgitation, rupture of the chordae tendineae, and rheumatic heart disease most frequently involve the mitral valve with resulting left ventricular consequences.

Case 6 A 84-year-old male with a history of coronary artery disease, s/p stent to mid LAD 3 years ago, atrial fibrillation, s/p biventricular ICD, presented with dyspnea on exertion. He underwent myocardial perfusion pharmacological stress testing with Regadenoson. Images are shown below.

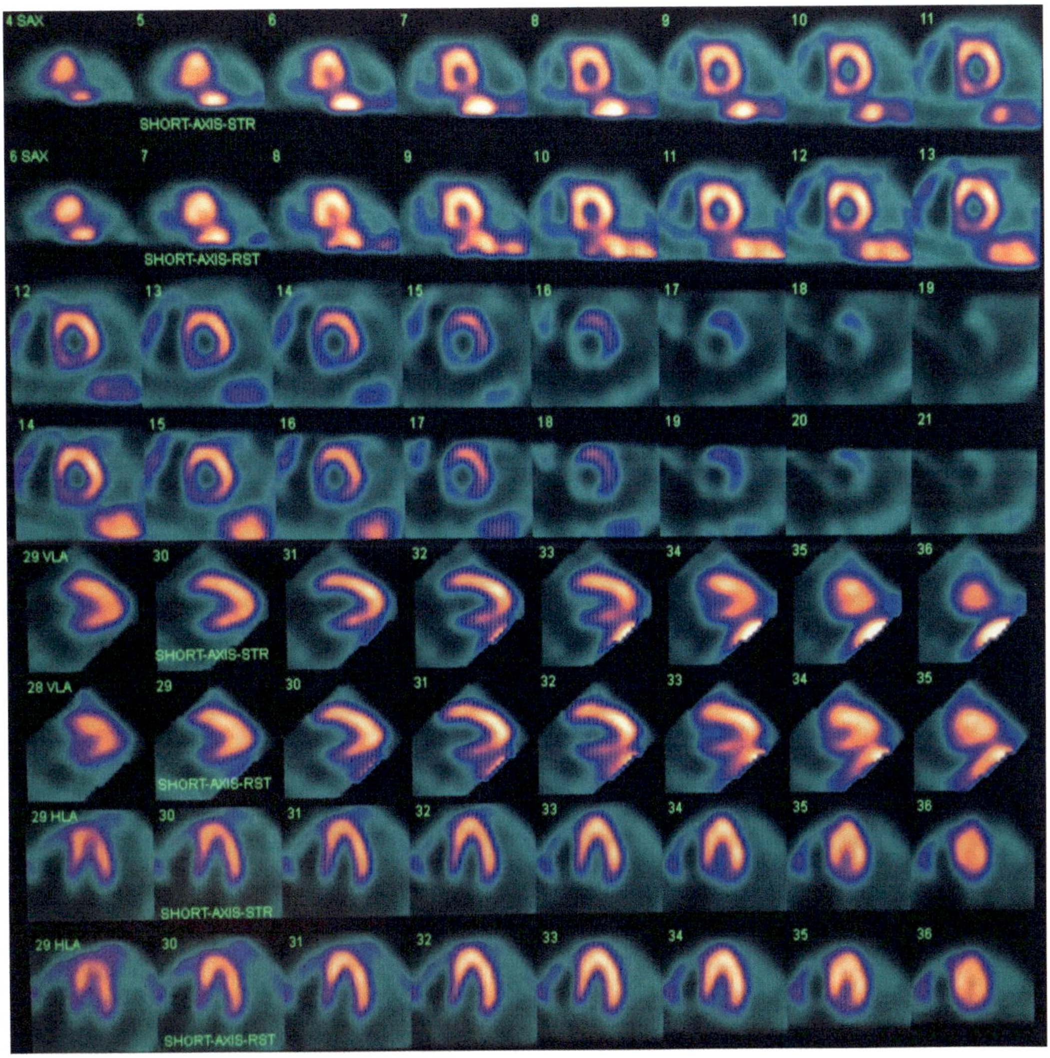

Ejection fraction is calculated at 67%.

43. Upon interpreting of this study you conclude:

   A. Repeat stress images
   B. Repeat rest images
   C. Perform prone images
   D. Urgently refer patient to cardiac catheterization

Answer: C

Stress and rest perfusion images demonstrate decreased counts in the inferior wall. While these findings might represent a nonreversible perfusion defect in the inferior wall, diaphragmatic attenuation artifact should be put into consideration in this male patient. Prone imaging can differentiate one entity from the other, because when prone, the heart will fall towards the anterior thorax, thus reducing the area of the heart obscured by the *diaphragm.* The quality of stress and rest images are diagnostic and there is no need to repeat them. Before sending patient to the cardiac catheterization, differentiation between true perfusion defect and artifact should be made.

Prone myocardial perfusion stress images were performed in this patient (described in prior case scenario). Images are shown below.

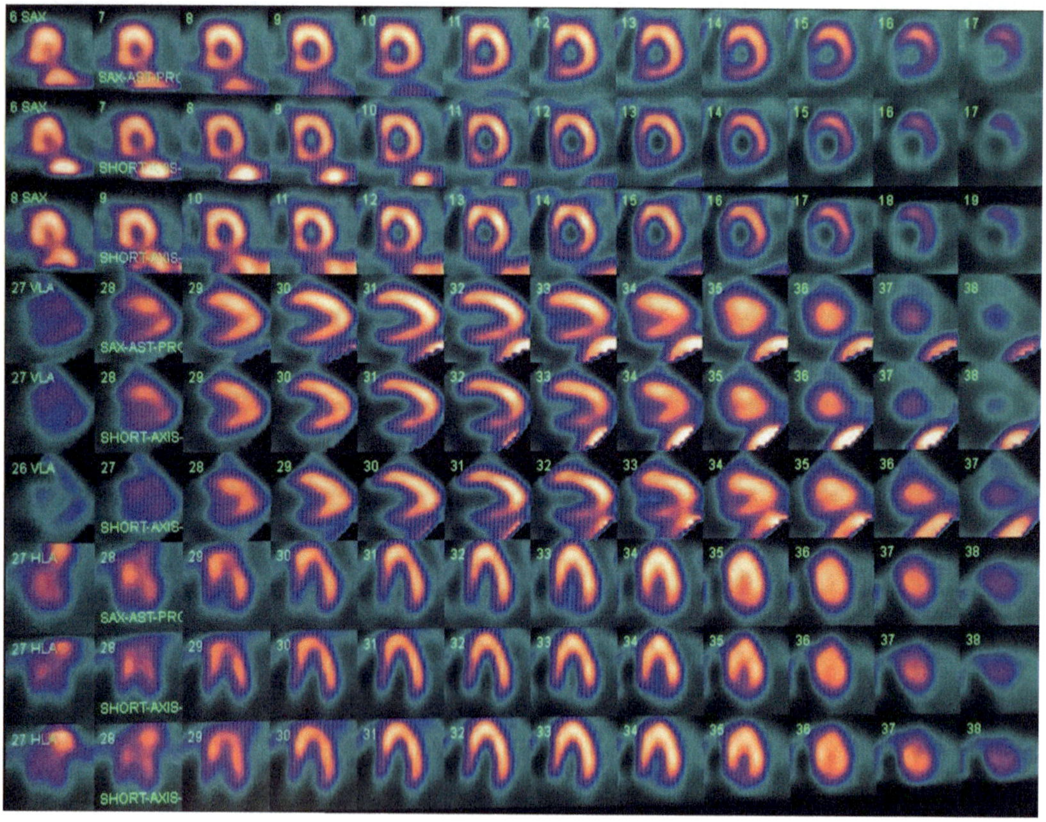

Image 1

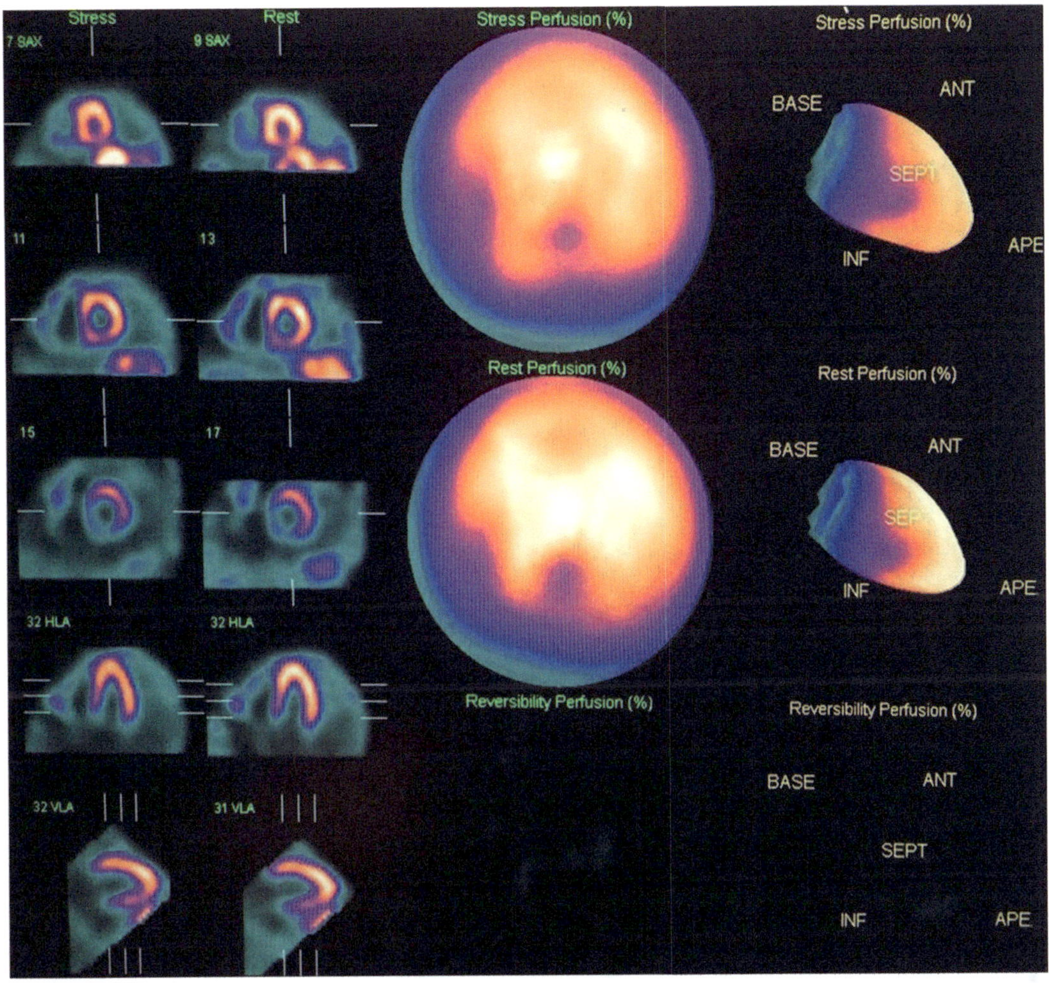
Stress
Rest
7 SAX
9 SAX
11
13
15
17
32 HLA
32 HLA
32 VLA
31 VLA
Stress Perfusion (%)
Rest Perfusion (%)
Reversibility Perfusion (%)
Stress Perfusion (%)
BASE
ANT
SEPT
INF
APE
Rest Perfusion (%)
BASE
ANT
SEPT
INF
APE
Reversibility Perfusion (%)
BASE
ANT
SEPT
INF
APE

Image 2

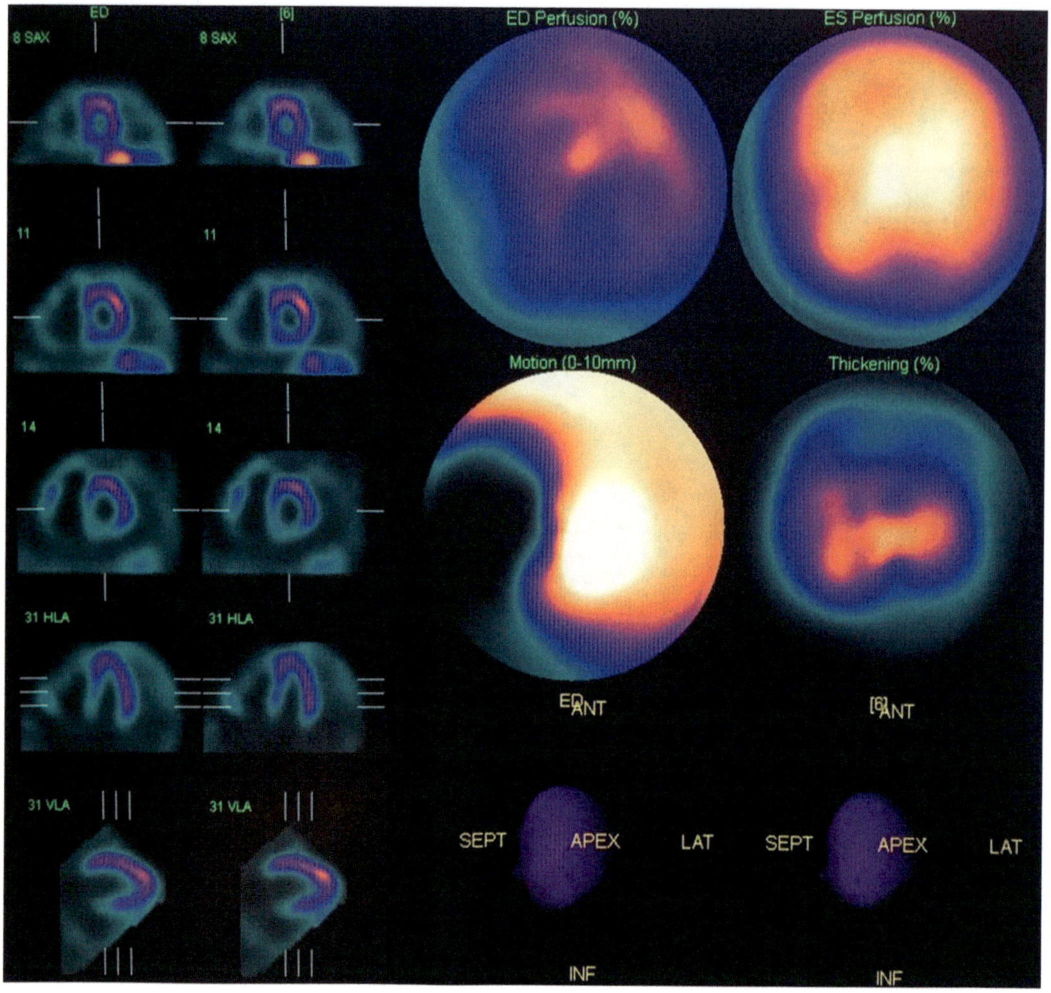

Image 3

44. Upon interpreting the study you conclude:

 A. Septal wall ischemia
 B. Attenuation artifact but no perfusion abnormalities
 C. Significantly impaired perfusion in mid LAD territory, which is suggestive of stent occlusion
 D. Left circumflex disease

Answer: B

Stress and rest perfusion images (image 1) demonstrate reduced counts in the inferior wall, although improved on prone images.

Perfusion stress and rest polar maps (image 2), which compare patient data to gender-matched normal limits accentuate the relative decrease in tracer concentration in the mid-to-basal inferior wall.

Gated tomograms displayed in polar maps (image 3) demonstrate essentially normal inferior wall motion and wall thickening, favoring inferior wall diaphragmatic attenuation artifact rather than scar. There is, however, abnormal septal wall motion. Given normal perfusion to septal wall, it is likely artifactual secondary to ventricular pacing. The lateral wall, which is typically supplied by the left circumflex artery demonstrates normal perfusion.

Case 7

45. What is the imaging paradigm recommended by the ASNC and the SNMMI in the management of cardiac sarcoidosis?

    A. Resting perfusion PET (or SPECT) followed by ^{18}F-FDG PET/CT
    B. Resting perfusion PET (or SPECT) followed by stress PET (or SPECT)
    C. Resting ^{82}Rb PET/CT followed by ^{99}Tc-Tetrofosmin stress SPECT
    D. Resting ^{18}F-FDG PET/CT followed by stress ^{18}F-FDG PET/CT

    Answer: A
    The recommended imaging protocol endorsed by ASNC and SNMMI involves the combination of a resting perfusion study (SPECT or PET) followed by ^{18}F-FDG PET/CT to assess for sites of active inflammation [1].

46. What is the most common pattern of cardiac involvement in symptomatic patients with cardiac sarcoidosis?

    A. Basal ventricular septum
    B. Right Ventricle
    C. Left ventricle free wall
    D. Papillary muscles

    Answer: A
    The formation of noncaseating-granulomas is the main pathophysiologic process in cardiac sarcoidosis. Studies performed postmortem have demonstrated that these granulomas may involve any part of the heart. However, the most commonly involved area is the basal interventricular septum, particularly in symptomatic patients. The natural course of the disease involves three successive stages, including; edema, inflammation, and fibrosis, leading to scarring. Other areas of the heart that may also be involved are the left ventricular free wall, the papillary muscles, and the right ventricle [2].
    It is very important to confirm the suspected areas of active involvement based on the known EKG abnormalities as well as the referring cardiologist's impression. Ideally, this should be performed prior to finalizing the report.

47. In a patient with cardiac sarcoidosis, what pattern of ^{18}F-FDG and ^{99m}Tc-Tetrofosmin distribution is associated with a higher risk for sudden cardiac death and/or VT?

    A. Increased ^{18}F-FDG uptake and increased ^{99m}Tc-Tetrofosmin
    B. Increased ^{18}F-FDG uptake and decreased ^{99m}Tc-Tetrofosmin
    C. Decreased ^{18}F-FDG uptake and increased ^{99m}Tc-Tetrofosmin
    D. Decreased ^{18}F-FDG uptake and decreased ^{99m}Tc-Tetrofosmin

    Answer: B
    Studies have demonstrated that for patients with known cardiac sarcoidosis and preserved left ventricular ejection fraction, a positive PET/CT in combination with perfusion defects was associated with a higher rate of death or sustained ventricular tachycardia [2, 3].

48. What is the most important clinical predictor of mortality for a patient with cardiac sarcoidosis?

    A. The presence of FDG uptake in the right ventricle.
    B. The presence of fixed perfusion defects prior to FDG PET/CT
    C. The extent of left ventricular systolic dysfunction
    D. The extent of left ventricular involvement seen on FDG PET/CT

Answer: C

Recent studies have reported that findings in advanced cardiac imaging, including late gadolinium enhancement by cardiac MR and active myocardial inflammation by ^{18}F-FDG PET/CT are predictors of adverse clinical outcomes [3]. This remains true even among individuals who have preserved left ventricular ejection fraction. Moreover, for a patient with known cardiac sarcoidosis and normal left ventricular ejection fraction, a positive PET/CT in combination with perfusion defects was associated with death or sustained ventricular tachycardia. However, the most important clinical predictor of mortality for this population is the extent of systolic dysfunction [2, 3].

49. What is the characteristic pattern of gadolinium enhancement seen in cardiac MR suggestive of cardiac sarcoidosis?

A. Early gadolinium enhancement, even without late enhancement
B. Late gadolinium enhancement, even without early enhancement
C. Both early and late gadolinium enhancement
D. No gadolinium enhancement

Answer: B

Many patients with suspected cardiac sarcoidosis are initially assessed and even followed up by contrast-enhanced cardiac MR. Because of its higher spatial resolution, it provides valuable information that can potentially determine the stage of the disease. Early gadolinium enhancement can be seen in areas of active inflammation, in addition to wall motion abnormalities. The chronic phase is characterized by delayed gadolinium enhancement secondary to fibrotic changes. Wall thinning is usually seen in the chronic phase. However, the two phases of the disease commonly overlap, and the imaging findings can be difficult to differentiate. In addition, the imaging findings also overlap with those findings seen in myocardial infarction. Despite these difficulties, the strongest imaging hallmark for cardiac sarcoidosis in cardiac MR imaging has been defined as delayed gadolinium enhancement [4].

References

1. Panithaya C et al. Joint SNMMI–ASNC expert consensus document on the role of 18F-FDG PET/CT in cardiac sarcoid detection and therapy monitoring. J Nucl Cardiol. 2017;24:1741–58.
2. Edward H et al. Cardiac sarcoidosis-state of the art review. Cardiovasc Diagn Ther. 2016;6(1):50–63.
3. Chiu C-Z, Nakatani S, Zhang G et al. Prevention of left ventricular remodeling by long-term corticosteroid therapy in patients with cardiac sarcoidosis. Am J Cardiol. 2005;95:143–6.
4. Schatka I, Bengel, Advanced imaging of cardiac sarcoidosis. J Nucl Med 2014; 55:99–106.

Endocrine System

Medhat Sam Gabriel, Bital Savir-Baruch, Robert H. Wagner, Bruch J. Barron, Davide Bova, Aamna Hassan, Stephanie Rodriguez-Santiago, and Yekaterina Kucerova

Case 1

1. A 56-year-old male with long-standing history of GERD, presents with epigastric/RUQ pain, bloating, distention, and occasional diarrhea. Vitals are normal. CT enterography shows no acute inflammatory changes. However, a right adrenal nodule is incidentally noted. Dedicated dynamic contrast-enhanced CT of the adrenal gland shows 1.7 × 2.1 cm right adrenal mass, with marked enhancement and rapid washout, stable in size for 3 years. What is the next step in management?

M. S. Gabriel (✉) · B. Savir-Baruch · R. H. Wagner · D. Bova
Division of Nuclear Medicine, Department of Radiology, Loyola University Medical Center, Maywood, IL, USA
e-mail: sgabriel@lumc.edu; Bital.savir-Baruch@lumc.edu; rwagner@lumc.edu; dbova@lumc.edu

B. J. Barron
Division of Nuclear Medicine and Molecular Imaging, Department of Radiology and Imaging Sciences, Atlanta, GA, USA
e-mail: bjbarro@emory.edu

A. Hassan
Department of Nuclear Medicine, Shaukat Khanum Memorial Cancer Hospital and Research Center, Lahore, Punjab, Pakistan

S. Rodriguez-Santiago
Department of Nuclear Medicine, C.W. Bill Young VA Medical Center, Bay Pines, FL, USA
e-mail: stephanie.rodriguezsantiago@va.gov

Y. Kucerova
Division of Nuclear Medicine, Department of Radiology, Loyola University Medical Center, Maywood, IL, USA

Department of Radiology, Cook County Healthy, Chicago, IL, USA
e-mail: yekaterina.kucerova@cookcountyhealth.org

M. S. Gabriel, B. Savir-Baruch (eds.), *RadTool Nuclear Medicine MCQs*,
https://doi.org/10.1007/978-3-030-69281-0_5

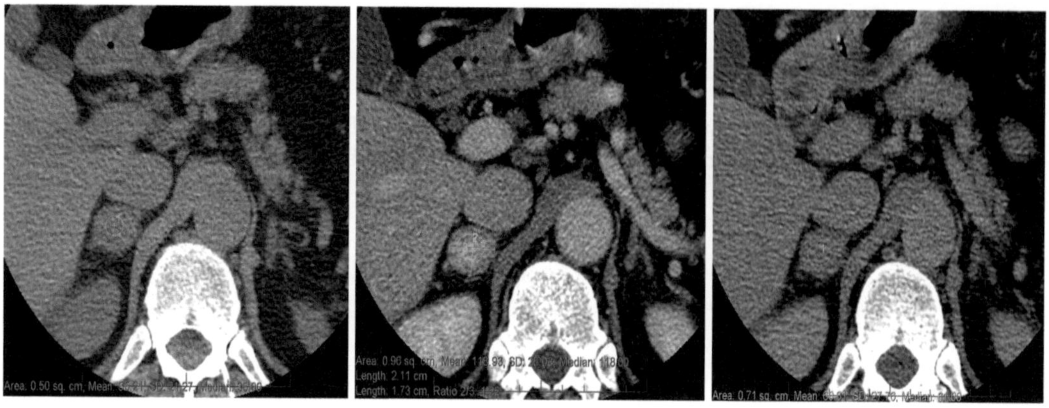

A. Follow up in 12 months for interval stability
B. Biochemical assay
C. MRI abdomen
D. PET/CT

Answer: B

Incidental adrenal lesion (IAL), or incidentaloma, is an adrenal mass found on CT that is usually 1 cm or larger in size. Once an IAL has been detected, it should be further characterized as benign or malignant and functioning or nonfunctioning. Functioning cortical adenomas and pheochromocytomas (PCCs) are best characterized by biochemical assay.

2. Biochemical assay revealed elevated 24 h Norepinephrine and Normetanephrine levels, concerning for pheochromocytoma. Which of the following is the next most appropriate study?

A. Tumor localization OctreoScan
B. F18-FDG PET/CT
C. I-123 MIBG scintigraphy
D. MRI Abdomen

Answer: C

Pheochromocytoma (PCC), a rare catecholamine-secreting tumor, usually manifests clinically, but up to 10% are detected incidentally. On CT, most lesions are smooth and round, with attenuation similar to that of soft tissue. They avidly enhance after intravenous administration of contrast medium and washout characteristics typically similar to those of malignant adrenal lesion. Uncommonly, PCCs show the opposite, with washout values similar to benign disease [1]. Iodine 123–MIBG or, less commonly used nowadays, iodine 131–MIBG is concentrated within PCCs with overall sensitivity of 85–88% and specificity of 70–100% [2].

I-123 MIBG with SPECT/CT demonstrated more uptake in the left adrenal gland when compared to the right. The findings are now suspicious for bilateral pheochromocytoma.

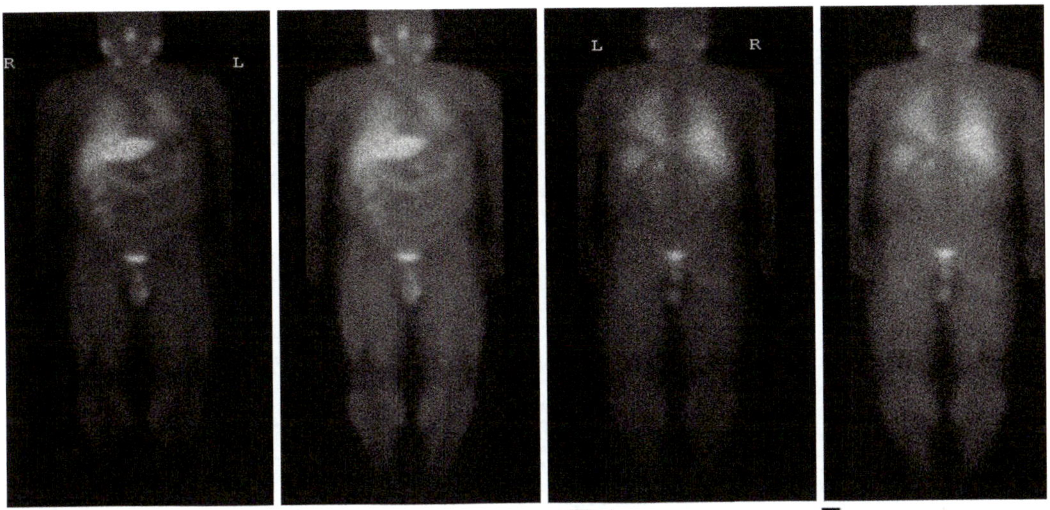

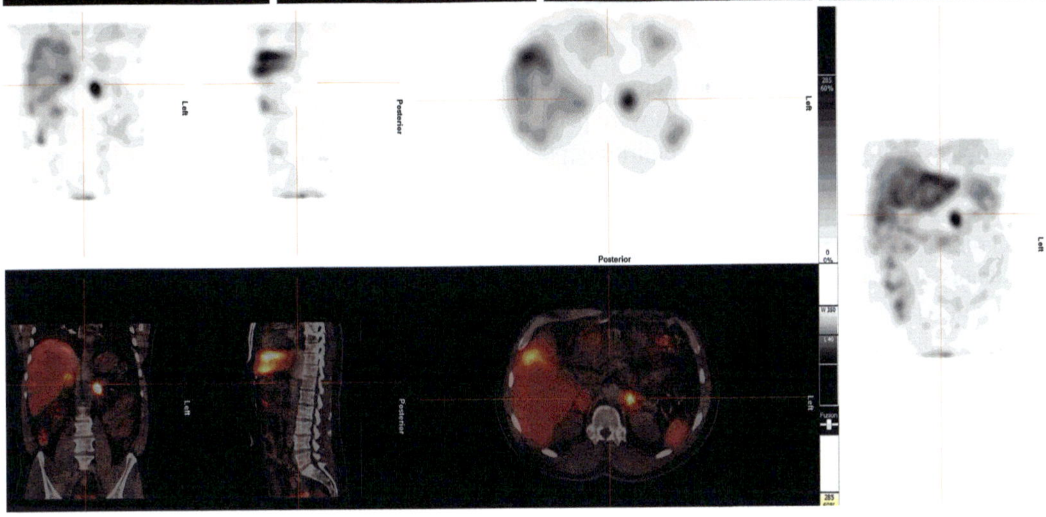

3. Prior to sending this patient to surgery, which of the following imaging studies is reasonable to localize additional catecholamine producing adrenal tissue?

A. MRI adrenal
B. F18-FDG PET/CT
C. OctreoScan scintigraphy with SPECT/CT
D. I- 131 MIBG

Answer: C

PCCs have been shown to express somatostatin receptors and scanning with Indium-111 labeled Pentetreotide (^{111}In-Pentetreotide scan or simply OctreoScan) is useful in the diagnosis of malignant disease, especially in the MIBG negative cases [3]. However, its low sensitivity makes it only supplement to MIBG scintigraphy in suspected metastatic disease. OctreoScan appears more sensitive than ^{18}FDG-PET for well-differentiated NETs (neuroendocrine tumors), while ^{18}FDG-PET demonstrates superior sensitivity in poorly differentiated NETs.

OctreoScan with SPECT/CT was done in this patient, but it was negative. Below are the planner images.

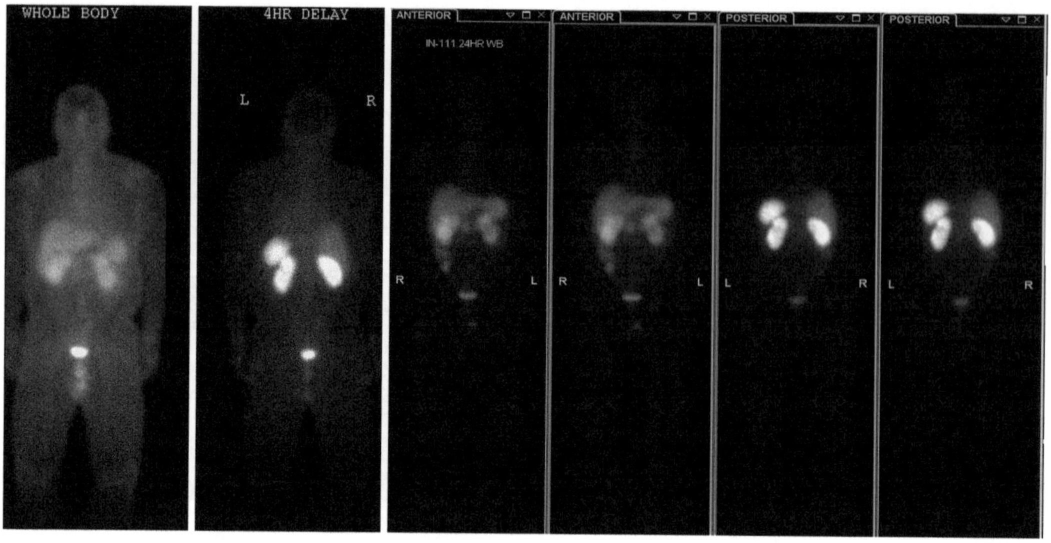

Patient underwent left adrenalectomy (Pathology—Adrenal gland with Medullary Hyperplasia). Postoperative metanephrine levels were persistently elevated and the patient eventually had right adrenalectomy with pathology confirming pheochromocytoma. Additional testing for RET proto-oncogene was negative.

4. Ga-68 DOTATATE has been recently approved by the FDA for the diagnosis of Neuroendocrine tumors and has been shown to have a higher diagnostic value, when compared to conventional scintigraphic techniques. Which of the following is true regarding this novel radiotracer?

   A. Gallium-68 is a generator-eluted, short-lived radionuclide decaying 89% through beta emission.
   B. Ga 68 DOTATATE physiological high uptake is noted in the spleen, adrenals, and liver with variable uptake in brain and prostate.
   C. Short physical half-life of ^{68}Ga ($t_{1/2}$ = 68 min) enables improved dosimetry and repeat imaging, making it ideal for clinical use.
   D. Ga 68 DOTATATE is not useful in diagnosing recurrent NET.

   Answer: C

   Ga-68 is short-lived, a positron-emitting radionuclide with intense physiological uptake in spleen, adrenal glands, liver, pituitary gland and variable uptake in GI, Kidneys, bladder, salivary gland, thyroid. Many recent studies have demonstrated the diagnostic value of this tracer in detecting primary or metastatic and recurrent NET [4, 5].

5. Which of the following statements is true regarding pheochromocytomas?

   A. 10% are solitary
   B. 10% are malignant
   C. 30% are familial
   D. 30% are extra-adrenal (paragangliomas)

   Answer: B

   Approximately 10% of Pheochromocytomas are malignant. Pheochromocytomas are also referred to as the “10% tumors.” 10% are malignant, 10% are familial, 10% are bilateral/multifocal, and 10% are extra-adrenal.

6. Which of the following statements regarding extra-adrenal pheochromocytomas (paragangliomas) is true?

   A. They are more frequent in the elderly
   B. They occur most frequently in the organ of Zuckerkandl
   C. The bladder is affected in approximately 1% of cases
   D. Symptoms are mild

   Answer: C

   Approximately 1% of Pheochromocytomas occurs in the bladder and can cause paroxysmal adrenergic attacks during filling or at micturition. Other symptoms include headaches and intermittent hypertension. Micturition syncope is most frequently "idiopathic." Extra-adrenal pheochromocytomas are more frequent in the pediatric population and occur more frequently along the para-aortic sympathetic chain.

References

1. Szolar DH, Melvyn Korobkin M, Reittner P, et al. Adrenocortical carcinomas and adrenal pheochromocytomas: mass and enhancement loss evaluation at delayed contrast-enhanced CT. Radiology. 2005;234:479–85. Bhatia.
2. Ismail KSMM, Sahdev A, et al., 123I-metaiodobenzylguanidine (MIBG) scintigraphy for the detection of adrenal and extra-adrenal phaeochromocytomas: CT and MRI correlation. Clin Endocrinol. (Oxford), 2008;69:181–8.
3. van der Harst E, de Herder WW, Bruining HA, et al., [(123)I]metaiodobenzylguanidine and [(111) In]octreotide uptake in benign and malignant pheochromocytomas. J Clin Endocrinol Metab. 2001;86:685–93.
4. Mojtahedi A, Thamake S, Tworowska I, Ranganathan D, Delpassand ES. The value of ^{68}Ga-DOTATATE PET/CT in diagnosis and management of neuroendocrine tumors compared to current FDA approved imaging modalities: a review of literature. Am J Nucl Med Molecul Imaging. 2014;4(5):426-34.
5. Bomanji, et al. The impact of 68Ga-DOTATATE PET/CT imaging on management of patients with neuroendocrine tumors: experience from a national referral center in the United Kingdom. J Nucl Med. 2016;57(1):34–40.

Case 2

7. 33 y/o female presents 13 months after total thyroidectomy for papillary thyroid carcinoma with increasing thyroglobulin levels. Shown below are whole-body images obtained 72 h after oral administration of 185 MBq (5 mCi) of I-131.

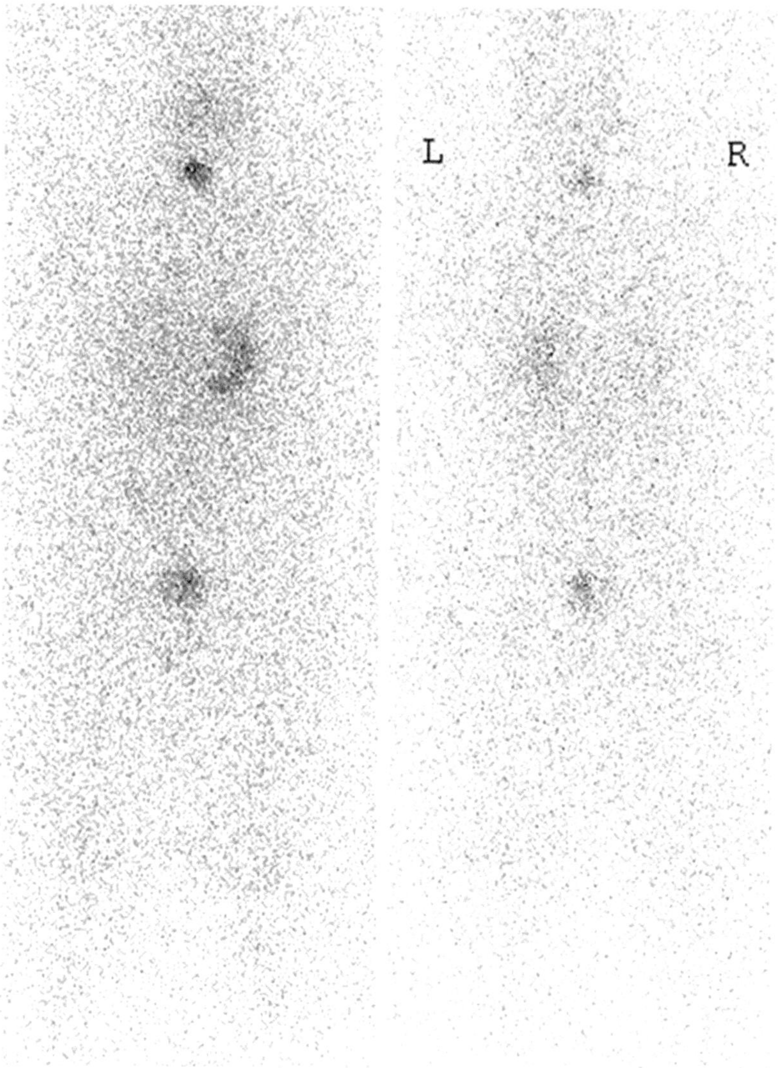

Images show:

A. Negligible residual thyroid parenchyma in the surgical bed
B. Local recurrence of papillary thyroid carcinoma
C. Metastatic linitis plastic from thyroid carcinoma
D. Struma ovarii, previously unsuspected

Answer: A

Uptake in the stomach and bladder is physiologic. Minimal residual activity in the thyroid bed was calculated at 0.02%, within expected limits for a patient following total thyroidectomy.

8. What is the next step in this patient with continuously rising thyroglobulin?

   A. Dynamic perfusion CT of the neck with contrast
   B. I-131 Ablation of residual thyroid tissue
   C. F18-FDG PET-CT
   D. Surgical reexploration

   Answer: B
   The clinical decision was made to administer 150 mCi of I-131 because of the high level of thyroglobulin, positive margins at surgery, lymphovascular invasion, 18/22 positive cervical lymph nodes. Perfusion CECT is a technique recently implemented for localization of parathyroid glands. Surgical reexploration would not likely identify any lesions with such small residual uptake.

9. Patient presents 6 months later with increasing thyroglobulin level now at approximately three times greater than 12 months prior. What is the next step in this patient with continuously rising thyroglobulin?

   A. Surgical reexploration
   B. I-131 Ablation of residual thyroid tissue
   C. F18-FDG PET-CT
   D. Repeat whole-body I-131 scan

   Answer: C
   F18-FDG PET-CT has proven to be a very valuable tool in thyroid carcinomas with poor Iodine uptake. The progressively rising thyroglobulin levels despite successful total thyroidectomy and intervening high dose I-131 ablation highly suggest that the tumor is possibly non-iodine avid.

10. Whole-body images of the PET-CT after F-18 FDG injection are shown below.

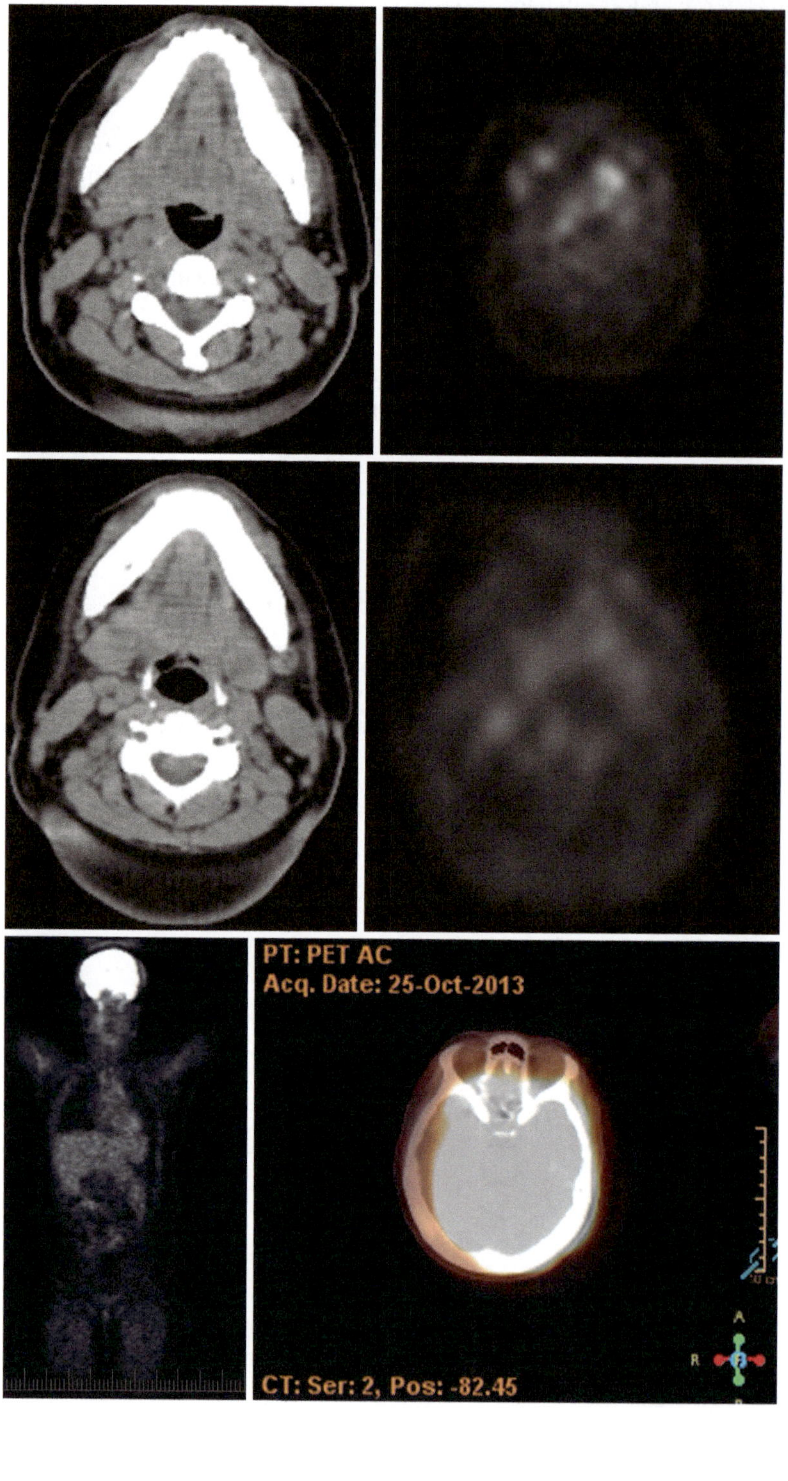

What is your next step?

A. I-131 Ablation of residual thyroid tissue
B. Repeat whole-body I-131 scan to provide anatomic localization of focus of uptake
C. Surgical exploration of the left submandibular region
D. Repeat image acquisition due to motion misregistration

Answer: D

The tumor has already been proven to be unresponsive to I-131 therapy. Blind surgical exploration has been abandoned, especially for solitary lesions of small size. At our institution, we routinely perform dedicated head and neck imaging on all patients with solid tumors of the head and neck in preparation for possible radiotherapy.

11. Separate head and neck acquisition images of the same PET-CT after F-18 FDG injection are shown below. What is your diagnosis?

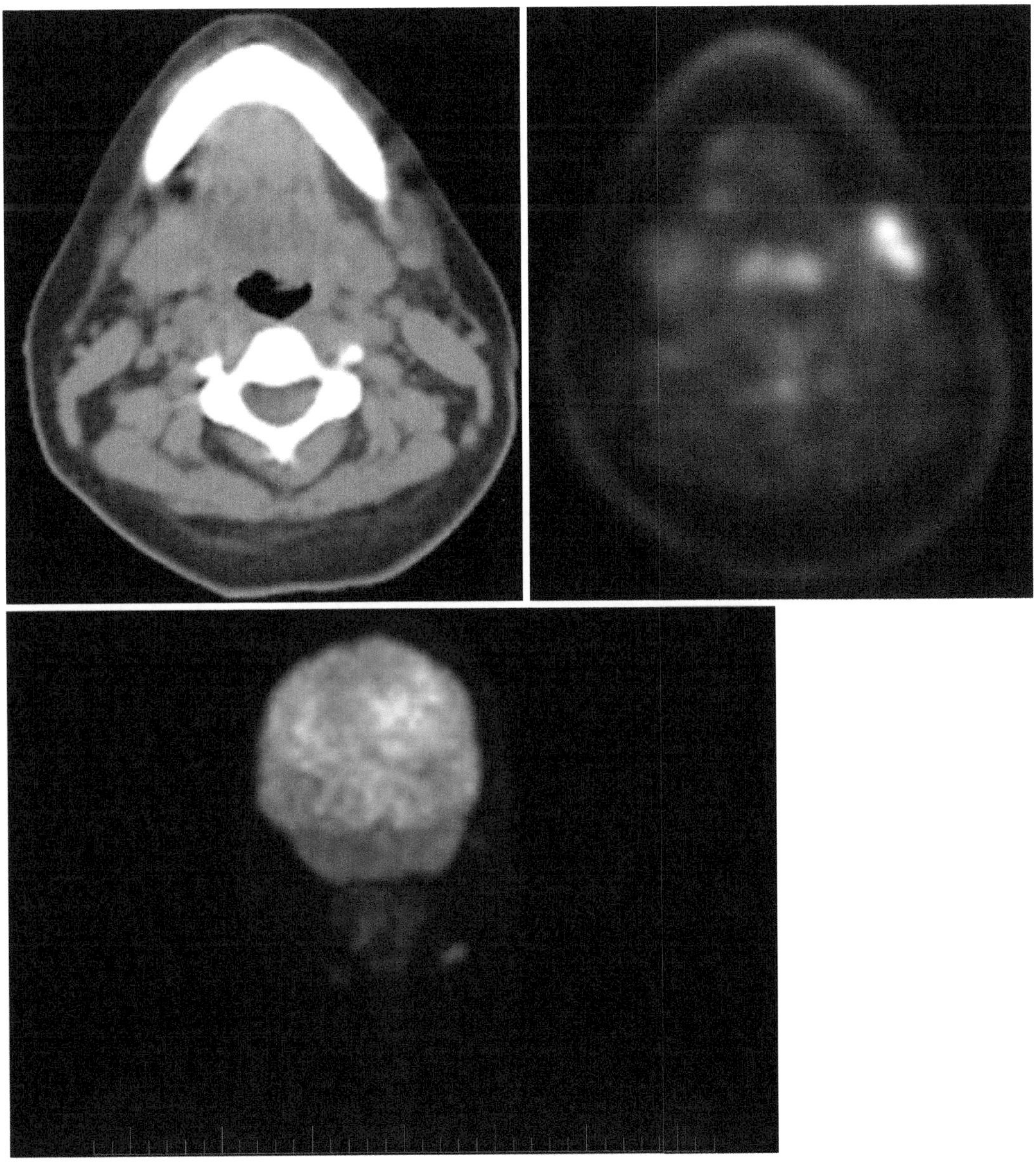

A. Metastatic bilateral cervical lymphadenopathy
B. Reactive non-enlarged left cervical node
C. Normal pharyngeal metabolic activity
D. Recommend MRI of the neck to further characterize

Answer: A

It is always a good rule to remember to look at *all the images on all image sets available, along with reconciling all available imaging and clinical information* with study in object.

*NCCN Clinical Practice Guidelines in Oncology (NCCN Guidelines®)—Thyroid Carcinoma, Version 2.2017—May 17, 2017—Available at NCCN.org.

*Marcus C, Whitworth PW, Surasi DS, Pai SI, Subramanian RM, PET/CT in the management of thyroid cancers. AJR. 2014;202:1316–29.

Case 3

12. A 59-year-old female with history of bilateral breast cancer, hypothyroidism on Levothyroxine, presented for evaluation of primary hyperparathyroidism. Parathyroid scintigraphy was performed. Radiopharmaceutical Tc-99 m Sestamibi was administered IV through the right AC fossa. Thereafter, images were obtained anteriorly over the neck and upper chest immediately and at approximately 2 h later.

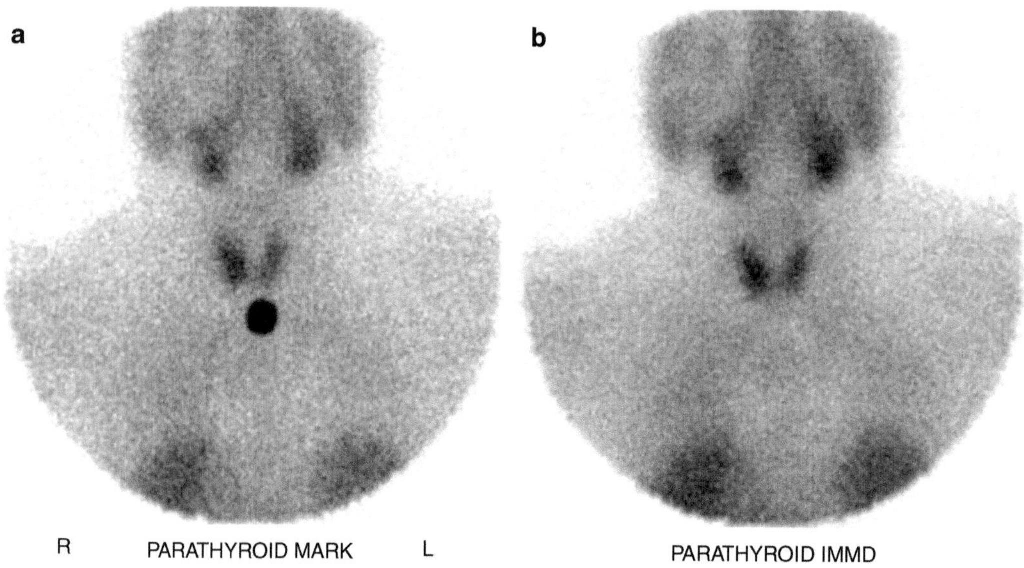

Image 1 (**a** and **b**)

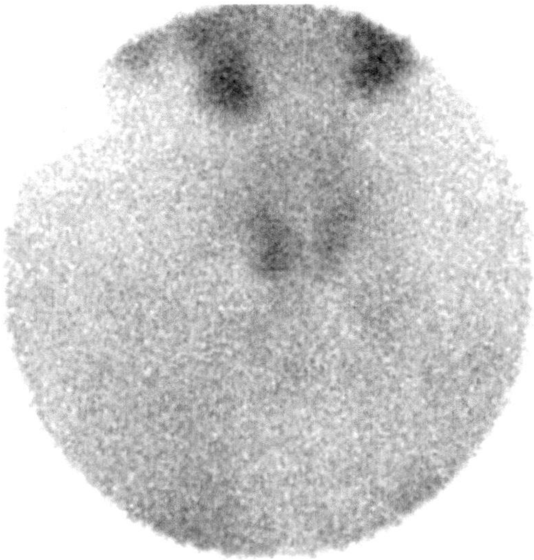

Image 2

Upon evaluation of images, you conclude:

A. The study is consistent with hyperplasia of four parathyroid glands
B. There is a mediastinal uptake, consistent with ectopic parathyroid tissue
C. The scan is normal with no evidence of hyperfunctioning parathyroid tissue
D. Possible parathyroid adenoma, obtain SPECT/CT to aid localization

Answer: D

Immediate pinhole images of the neck (Image 1A with a marker, image 1B without a marker) demonstrate physiological tracer uptake by the thyroid and salivary glands. There is slightly greater tracer accumulation in the region of the right thyroid lobe compared to the left. Delayed image (image 2) demonstrates a subtle area of persistent tracer activity in the inferior pole of the right thyroid lobe. Findings are suspicious for hyperfunctioning parathyroid tissue in the right inferior parathyroid gland. Further evaluation with SPECT/CT may assist in localization.

There is no abnormal mediastinal tracer uptake to suggest the presence of ectopic parathyroid tissue. On immediate images, a marker is placed in the sternal notch for localization.

13. SPECT/CT imaging was performed and shown below:

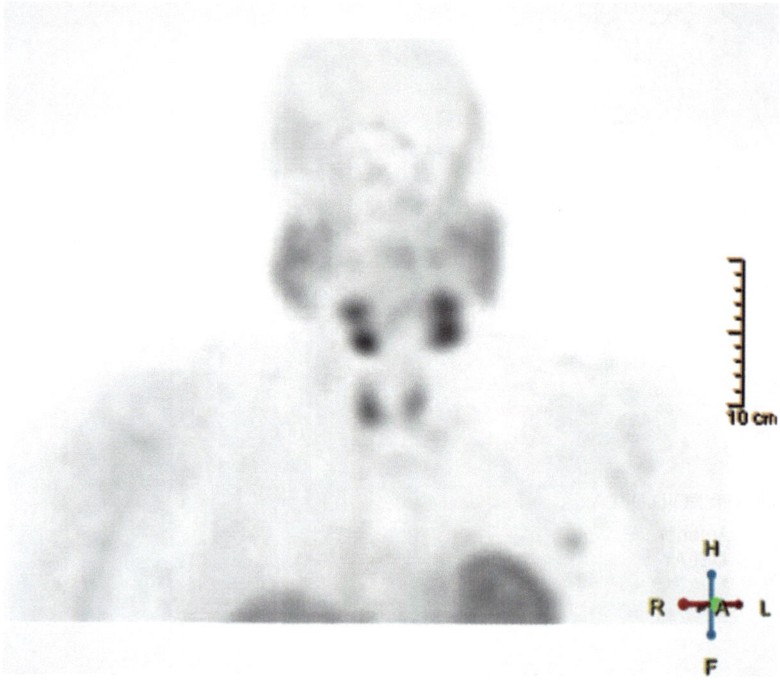

Image 1

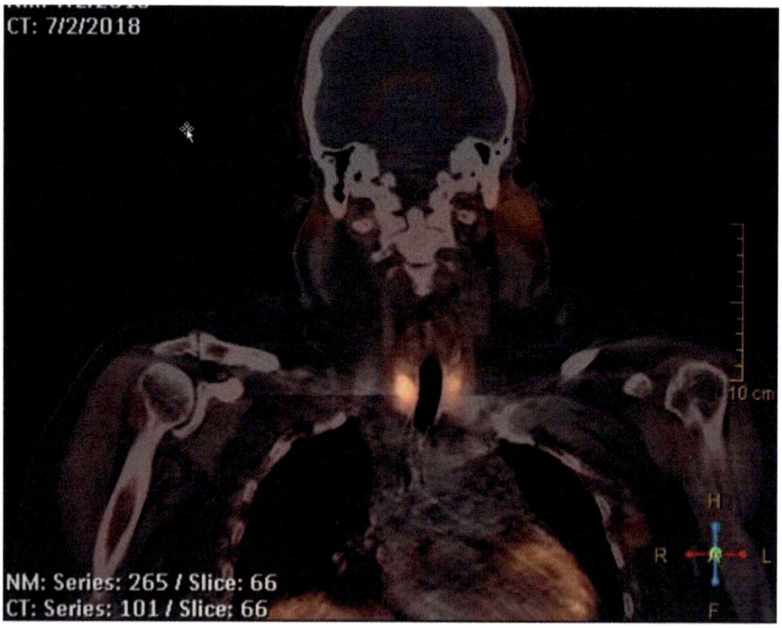

Image 2

Upon evaluation of SPECT/CT images, you conclude:

A. SPECT/CT confirms tracer uptake in four parathyroid glands, consistent with parathyroid glands hyperplasia
B. Mediastinal tracer uptake consistent with ectopic parathyroid tissue
C. Normal parathyroid scan with no evidence of hyperfunctioning parathyroid tissue
D. Focus of tracer uptake in the region of right lower thyroid bed, most suggestive of hyperfunctioning parathyroid tissue

Answer: D

Tc-99 m Sestamibi SPECT-CT images demonstrate asymmetrically increased radiotracer uptake in the region of the right lower thyroid bed. Similar findings were appreciable on planar images and are most suggestive of hyperfunctioning parathyroid tissue at that location.

There is normal physiological tracer uptake within the parotid and submandibular glands. There is no mediastinal tracer uptake to suggest the presence of ectopic parathyroid tissue. Also, there are no multiple foci of abnormal tracer retention to suggest hyperplasia of multiple parathyroid glands.

14. What additional information should be included in the study report?

A. Incidental finding for lung cancer
B. Incidental finding suspicious for breast cancer recurrence
C. Incidental finding for brain metastasis
D. Incidental finding skeletal metastasis

Answer: B

SPECT images demonstrate a focus of mild radiotracer accumulation in the left axilla (image 1), which corresponding to a prominent lymph node on the low dose CT component of the study (image 2, SPECT/CT image), along the lateral margin of the pectoralis minor. Tc-99 m Sestamibi is a nonspecific malignancy seeking agent, i.e., it accumulates in cancer cells more than normal tissue. Given the patient's history of bilateral breast cancer, this axillary uptake is suspicious for breast cancer recurrence. Further evaluation with biopsy should be suggested. No accumulation in the brain, bones, or lungs suggests malignancy.

*Noussios G, Anagnostis P, Natsis K. Ectopic parathyroid glands and their anatomical, clinical and surgical implications. Exp Clin Endocrinol Diab. 2012;120(10):604–10.

*Ayers RR, Tobin K, Sippel RS, Balentine C, Elfenbein D, Chen H, Schneider DF. Does levothyroxine administration impact parathyroid localization?. J Surg Res. 2015;198(2): 360–5.

*Phitayakorn R, McHenry CR. Incidence and location of ectopic abnormal parathyroid glands. Am J Surg. 2006;191(3):418–23.

*Pamarthy S, Wong KK, Gross M, Lorraine Fig. concise review of parathyroid embryology: pictorial guide to ectopic localization. J Nucl Med.2014;55(supplement 1): 1275.

Case 4

15. This 55-year-old male is referred to you from an outside institution and is s/p resection of an invasive papillary thyroid cancer with nodal involvement and metastasis to C7. He is unable to swallow, has a G-tube in place for nutrition, and has had surgical debridement of C7 with cervical fixation. He is experiencing intermittent pain in his right arm. A recent CT study demonstrates nodules in the lungs.
His radioiodine was administered via the g-tube. The 24 h images are provided to you for interpretation. You decide:
    A. Significant activity remains in the g-tube. The study needs to be repeated with greater flushing of the g-tube after administration.
    B. There is no evidence of residual disease. The patient should resume his thyroid medications and follow up with his endocrinologist.
    C. A dose of 100 mCi I-131 should be given empirically as some metastatic disease can only be seen after a therapeutic dose.
    D. A follow-up PET/CT scan using FDG should be performed.

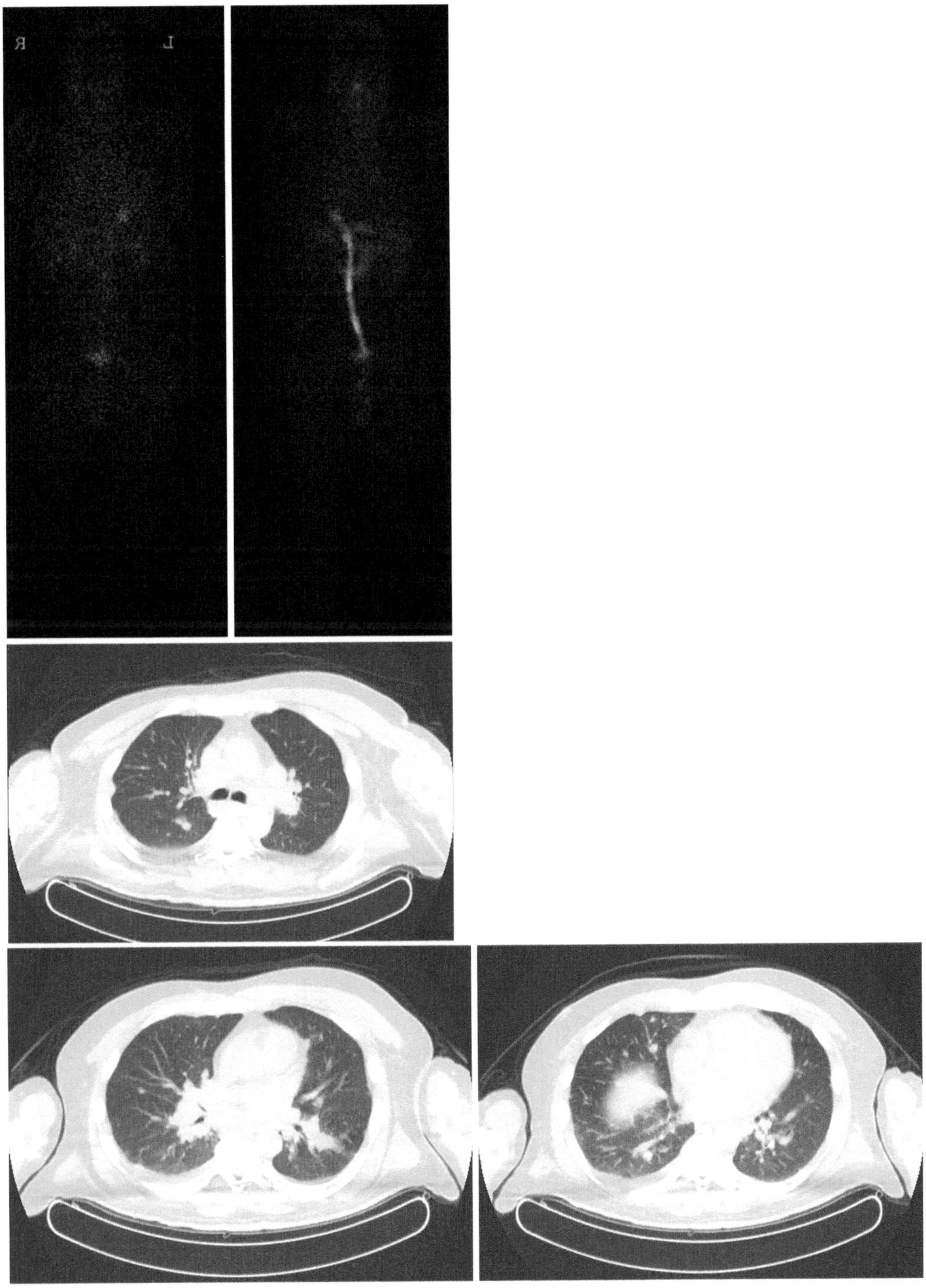

Answer: D

This patient has already had metastatic disease to the bone and has nodules on the CT study that are suspicious for metastatic disease. The scan shows no evidence of iodine avid metastatic disease. Non-iodine avid metastasis is a real concern. While iodine avid thyroid cancer is not frequently seen on FDG PET/CT, the non-iodine avid cancer often demonstrates significant FDG uptake.

Although activity is seen in the g-tube, it was well flushed after administration. Some of the iodine however does remain in the tubing. Sinus, bladder, and bowel activity indicate that a significant amount was absorbed.

16. After performing a FDG PET/CT study on this patient with a history of papillary thyroid cancer, you review the images provided. Your discussion with the referring physician includes a recommendation to:

A. Administer a high dose of I-131 in the hopes to decrease the tumor burden
B. Provide the patient with even higher doses of thyroid hormone replacement in the attempt to slow tumor growth.
C. Consider Ra-223 therapy for bone involvement
D. Suggest consultation with neurosurgery and radiation oncology

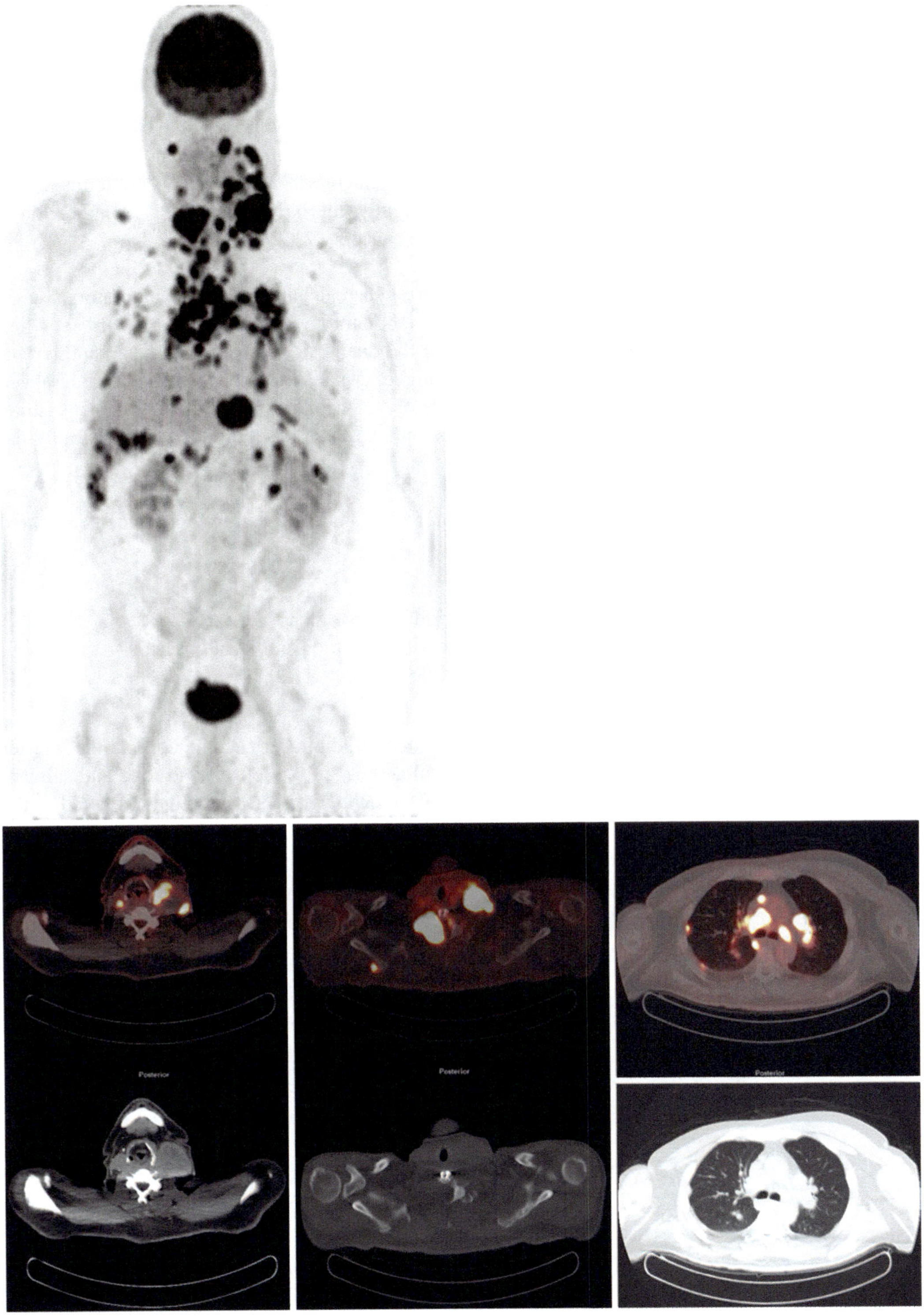
Posterior
Posterior
Posterior

Answer: D

The CT study demonstrates near complete destruction of the right half of the C7 vertebral body. For local control, he would require radiation therapy to this area but may need additional stabilization of his cervical spine.

With non-iodine avid metastatic disease of this extent, radioiodine would have no significant effect. Although thyroid hormone replacement is a part of management, even higher doses would not change the course of the disease or alter the management. Ra-223 therapy is not indicated for bony metastatic disease from thyroid cancer. Currently, the only FDA-approved indication is painful skeletal metastases in patients with prostate cancer.

Case 5 A 72-year-old female patient was found to have hypercalcemia on routine laboratory studies obtained by her primary care physician. Subsequent laboratory workup revealed a corresponding parathyroid hormone level of 227 pg/mL, confirming the diagnosis of a primary hyperparathyroidism. The patient's current symptoms include some fatigue, urinary frequency, and some generalized musculoskeletal aches and pains. She denies any history of nephrolithiasis. A bone density study several years prior showed osteoporosis. No other significant history. She underwent a parathyroid scan (Fig. 1).

17. What should be the most appropriate impression for the findings seen in Fig. 1 when compared to those seen in Fig. 2?

    A. A thyroid nodule
    B. Hyperfunctioning parathyroid tissue of the right inferior gland
    C. Hyperfunctioning right inferior gland and possible ectopic parathyroid tissue
    D. A thyroid nodule and ectopic parathyroid tissue

Answer: C

The early planar images (Fig. 1) demonstrate intense radiotracer accumulation at the inferior aspect of the right thyroid lobe with physiological activity in the left thyroid lobe and salivary glands. An additional unexpected irregular area of moderately increased activity is noted in the

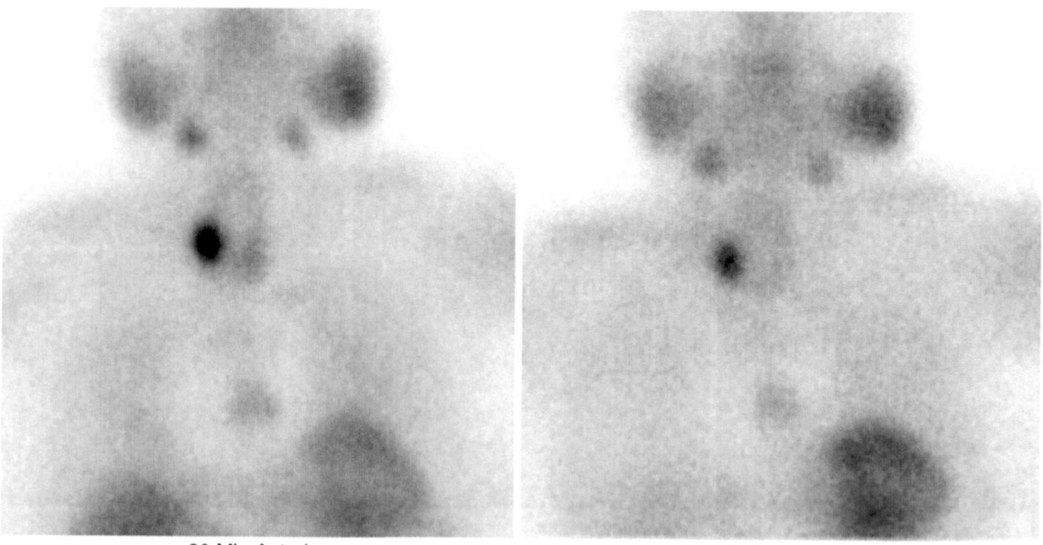

Fig. 1 99mTc Sestamibi parathyroid scan

mid-chest, probably within the mediastinum. Delayed planar images demonstrate washout of activity from the thyroid gland with persistent intense focal activity at the level of the right inferior thyroid pole. Persistent moderate activity is also seen in the mediastinum. The CT images (Fig. 2) demonstrate a large nodule posterior to the right thyroid lobe, which on fused images corresponded to the focal uptake seen in the right neck in the planar images. The findings are consistent with a hyperfunctioning right inferior parathyroid gland.

The differential diagnosis for the anterior mediastinal mass seen on both planar Parathyroid and CT scans includes ectopic parathyroid tissue, thymoma, lymphoma, and others.

18. Considering the findings on Figs. 1 and 2, what is the best next step in the management of this patient?

    A. Surgical resection of the right inferior parathyroid gland.
    B. Perform a contrast-enhanced CT of the neck and thorax.
    C. Prescribe medication including bisphosphonate and furosemide.
    D. Observation, considering the patient's advanced age.

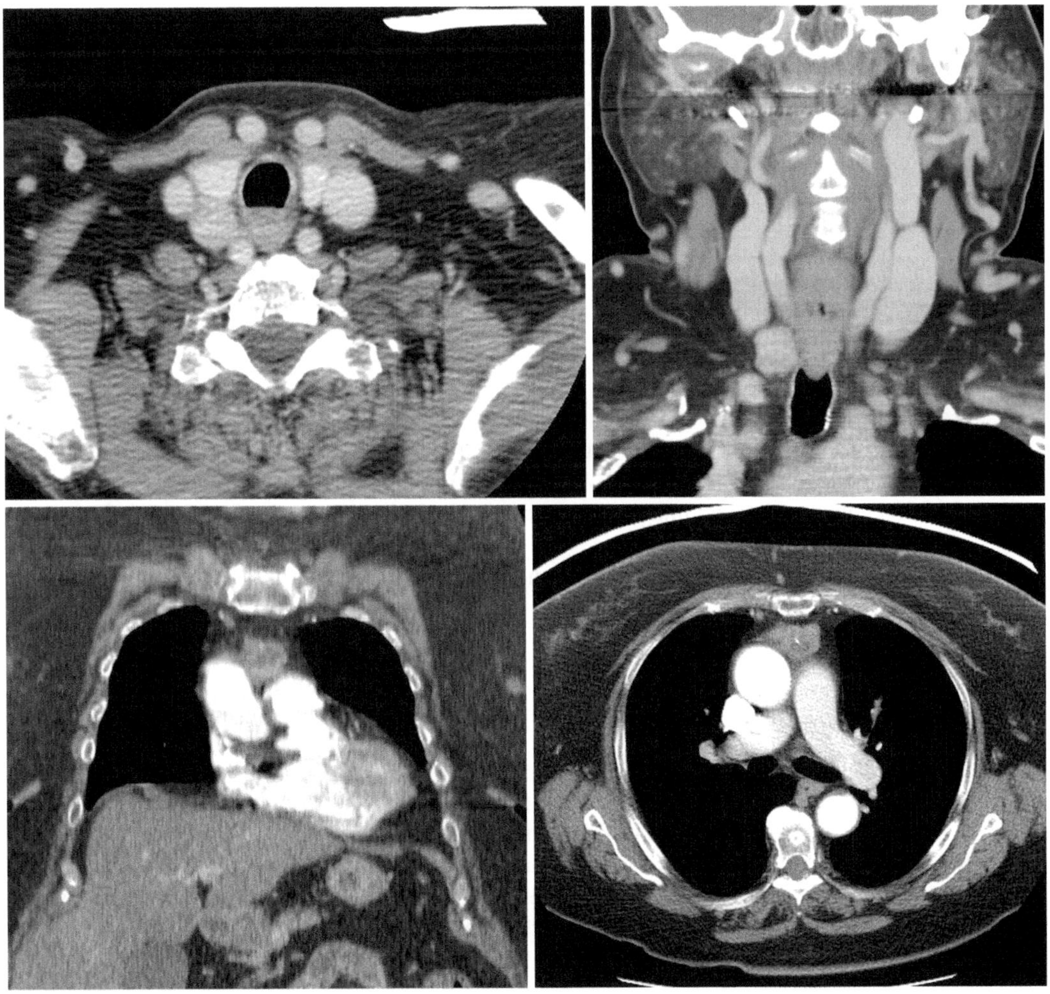

Fig. 2 CT with IV contrast

Answer: A

The curative management of symptomatic patients with primary hyperparathyroidism is surgical resection of the adenoma, provided that they are good candidates for surgery. For asymptomatic patients, currently there is some controversy regarding the most appropriate approach and management of these patients, particularly when hyperparathyroidism is mild. In these cases, the risks versus benefits of parathyroidectomy should always be put into consideration with attention to the patient's age and comorbidities [1].

19. The highest sensitivity for parathyroid scintigraphy is achieved if the adenoma weighs at least?

A. 100 mg
B. 150 mg
C. 200 mg
D. 300 mg

Answer: D

The two superior parathyroid glands are usually located posterior to the midpoles of the thyroid gland (75%) and posterior to the recurrent laryngeal nerve. The two inferior parathyroid glands are commonly located posterior and lateral to the lower poles of the thyroid gland (60%) and anterior to the recurrent laryngeal nerve. Ectopic glands and double adenomas are significantly less common and seen in less than 10% of cases [2]. Normal sized parathyroid glands usually weigh between 35–40 mg each and are typically not detected on scintigraphy. The larger the adenoma the better the diagnostic accuracy on parathyroid scintigraphy, with a detection rate approaching 90% for adenomas larger than 300 mg [2].

20. When using a single isotope, what is the best time to perform the SPECT/CT images?

A. Immediately after injection or following early planar images
B. At 90 min post radiotracer injection
C. At 120 min post radiotracer injection
D. At 180 min post radiotracer injection

Answer: A

Comparisons of diagnostic performance of the traditional planar images with single or dual isotope technique, SPECT, and SPECT/CT demonstrated the superiority of early SPECT/CT in conjunction with any delayed imaging method over dual-phase planar or SPECT [4]

21. The uptake in the mediastinum is likely secondary to:

A. Aortic aneurysm
B. Thymus
C. Physiological uptake
D. Esophageal uptake

Answer: B

Mediastinal mass (thymoma). Thymic neoplasia is the most common cause of anterior mediastinal mass. Parathyroid hormone (PTH)-secreting thymomas are exceedingly rare.

Scanning with technetium Tc 99 m Sestamibi may demonstrate false-positive uptake in a mediastinal thymoma. Thymomas are hypercellular lesions with increased mitochondrial density, which is likely the cause of their Sestamibi uptake. Nonspecific uptake of the Tc 99 mm Sestamibi is also observed in a wide range of benign and malignant tumors.

References

1. Rao DS, et al., Randomized controlled clinical trial of surgery versus no surgery in patients with mild asymptomatic primary hyperparathyroidism. J Clin Endocrinol Metabol. 89(11):5415–22
2. Ziessman HA, et al., Nuclear medicine: the requisites. fourth Edition. Philadelphia, PA. Elsevier Saunders; 2014.
3. Greenspan BS, et al., SNM practice guideline for parathyroid scintigraphy 4.0*. Soc Nucl Med Molecul Imag. 2012. DOI: https://doi.org/10.2967/jnmt.112.105122
4. Lavely WC, et al., Comparison of SPECT/CT, SPECT, and planar imaging with single- and dual-phase 99mTc-sestamibi parathyroid scintigraphy. J Nucl Med. 2007;48:1084–9.

Case 6

22. Expression of the multidrug resistance (MDR)-related P-glycoprotein (P-GP) by parathyroid tumor will likely result in which of the following:

    A. Successful treatment with Chemotherapy.
    B. Successful treatment with monoclonal antibody therapy.
    C. False-negative result of Tc-99 m sestamibi parathyroid scan secondary to Tc-99 m sestamibi efflux from the tumor.
    D. Greater Tc-99 m sestamibi uptake by the parathyroid tumor.

    Answer: C

    Tc-99 m Sestamibi scintigraphy is a useful method in detecting parathyroid adenomas. The size of the tumor is a factor in the detection of the abnormal tissue on Tc-99 m Sestamibi scan. In addition, some drug metabolism transporters, such as p-glycoprotein (P-GP), may decrease 99mTc-sestamibi accumulation in the parathyroid glands and result in a negative study. On the other hand, low or no expression of P-GP will result in positive scans.

    The typical management of hyperfunctioning parathyroid tissue is surgical resection. Chemotherapy and monoclonal antibody therapy have no role in the therapy of parathyroid tumors.

23. A 64-year-old male with elevated calcium and parathyroid hormone levels underwent Tc-99 m sestamibi scintigraphy, demonstrating hyperfunctioning parathyroid tissue. During surgical removal of abnormal parathyroid tissue, what would be the best test to confirm successful resection?

    A. PTH blood serum assay in 3 days to allow serum PTH levels to normalize
    B. Intraoperative parathyroid scintigraphy
    C. Parathyroid scintigraphy in 3 days
    D. PTH intraoperative blood serum assay

    Answer: D

    Surgical resection of abnormal parathyroid gland or glands is the most effective method for cure. The surgical approach is usually through the removal of the adenoma or rarely double adenomas. For hyperplasia, 3½ glands are usually removed. Alternatively, all 4 glands are removed followed by implanting approximately 30 mg of parathyroid tissue in the forearm. Intraoperative PTH levels are measured before and shortly (usually within 15 min) after surgical resection of suspected tissues. This technique is feasible because the half-life of the parathyroid hormone is short, in the range of 3–5 min, allowing for a rapid decrease in the hormone level while the patient is still in surgery. The rule is that 50% decrease in PTH is needed to call the surgery successful. If the PTH level does not decrease by 50% or more, the surgeon needs to search for additional hyperfunctioning area or areas. The procedure is easy where blood samples are collected just before surgery and after resection. If the levels are not decreased at or below 50% of the original value, further measurements are needed at certain time intervals, such as every 10 min or so.

Parathyroid scintigraphy is a useful preoperative localization study. Intraoperative scintigraphy is not feasible. Some surgeons prefer injecting a dose of Tc-99 m Sestamibi approximately 2 h before surgery and use a handheld radioactive probe during surgery to localize the hyperfunctioning parathyroid tissue, a technique similar to lymphoscintigraphy.

PTH blood serum assay and parathyroid scintigraphy in 3 days are not practical or useful since the missed hyperfunctioning parathyroid tissue will require additional surgery.

*Cook GJR, Maisey MN, Britton KE, Chengazi V. Clinical nuclear medicine Fourth Edition.
*UpToDate. https://www.uptodate.com.

Case 7

24. A nuclear medicine study was performed; the image is shown below.

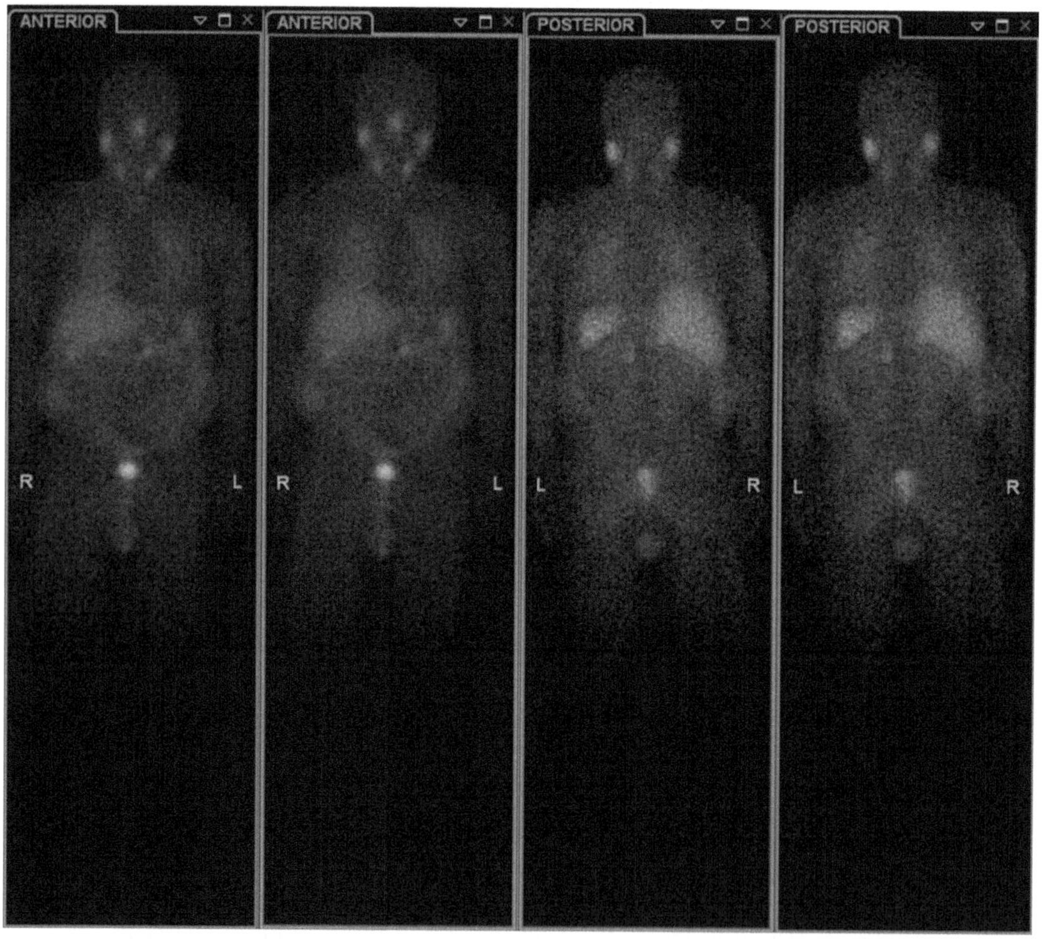

What is the physical half-life of the radiotracer used in the study above?

A. 68 min
B. 6 h
C. 13 h
D. 67 h

Answer: C

I-123 MIBG scan is a scintigraphic study that uses metaiodobenzylguanidine labeled to iodine-123. I-123 has a physical half-life of 13.3 h. It is a norepinephrine analog that localizes in cytoplasmic storage vesicles in presynaptic adrenergic nerves. Therefore, organs with rich adrenergic innervation, such as the heart, salivary glands, spleen, liver, kidneys, and thyroid gland, can be seen on the scan. It is also excreted by the kidneys and the colon.

Indium-111-DTPA-pentetreotide (In-111 OctreoScan) is a somatostatin analog used to evaluate neuroendocrine tumors. In-111 has a half-life of 67 h (2.8 days). In-111 pentetreotide may be used in pheochromocytoma, particularly if extra-adrenal disease is expected but I-123 MIBG is more sensitive.

Ga-68 DOTATATE PET agent with a half-life 68 min. It accumulates in cells with high expression of somatostatin receptors (SSTRs), particularly SSTR2.

Technetium agents have a half-life of 6 h.

25. What is an appropriate patient preparation for the nuclear medicine study shown above?

A. Fasting for at least 6 h
B. Administration of SSKI
C. Checking for WBC level
D. Administration of cimetidine

Answer: B

Thyroid blockade with (nonradioactive) potassium iodide is indicated for nuclear medicine scintigraphy with I-123 MIBG to guard against free I-123 in I-123 MIBG accumulation in the thyroid. This can be done with saturated potassium iodide solutions (SSKI) or potassium iodide-iodine (Lugol solution). In case of allergy to iodine, potassium perchlorate can be given.

F-18 FDG PET/CT is commonly used in the evaluation of several types of cancer which requires fasting for at least 6 h.

Indium-111 Oxine-Labeled WBC scan is used to detect infection and inflammation. White blood cells (mostly neutrophils) are drawn from the patient, tagged with the radioisotope Indium-111 (oxine is a lipid-soluble complex that chelates indium), then radiolabeled leucocytes are reinjected into the patient. Adequate radiolabeling and imaging require more than $5000/mm^3$ peripheral leukocytes, although it can sometimes be performed with counts as low as $3000/mm^3$.

In patients with GI bleeding where Meckel diverticulum is suspected to be the cause, the administration of H2 blockers such as cimetidine increases the accuracy of the study. It minimizes abnormal secretion of tracer by the gastric mucosa, and therefore decreases false-positive studies.

26. Upon evaluation of the nuclear medicine study above, you conclude that:

A. Normal scan
B. Abnormal uptake in the head
C. Abnormal uptake in the mid-abdomen
D. Abnormal uptake in the pelvis

Answer: C

This patient was evaluated with I-123 MIBG scan for additional characterization of a 1.6 cm left adrenal nodule seen on CT imaging. The scan demonstrates focally increased radiotracer uptake in the region of the left adrenal gland, most consistent with neuroendocrine tumor. There are no other areas of abnormal tracer uptake to suggest the presence of metastatic disease. Salivary glands, lungs, heart, liver, spleen, GI tract, and bladder demonstrate physiological tracer uptake.

Pheochromocytoma is a catecholamine-secreting tumor. Most tumors (90%) arise in the adrenal medulla and are benign and unilateral. About 10% ("rule of 10s") are extra-adrenal (paragangliomas; the most common location being the organ of Zuckerkandl near the origin of the inferior mesenteric artery), about 10% are malignant (particularly if metastasis are seen), about 10% are bilateral and 10% are associated with syndromes such as MEN2 (multiple endocrine neoplasia type 2).

*Boland GW, Blake MA, Holalkere NS, Hahn PF. PET/CT for the characterization of adrenal masses in patients with cancer: qualitative versus quantitative accuracy in 150 consecutive patients. AJR Am J Roentgenol. 2009;192(4):956–62. doi: https://doi.org/10.2214/AJR.08.1431.

*Ilias I, Divgi C, Pacak K. Current role of MIBG in the diagnosis of pheochromocytoma and medullary thyroid cancer. Semin Nucl Med. 2011; 41(5): 364–8.

Case 8

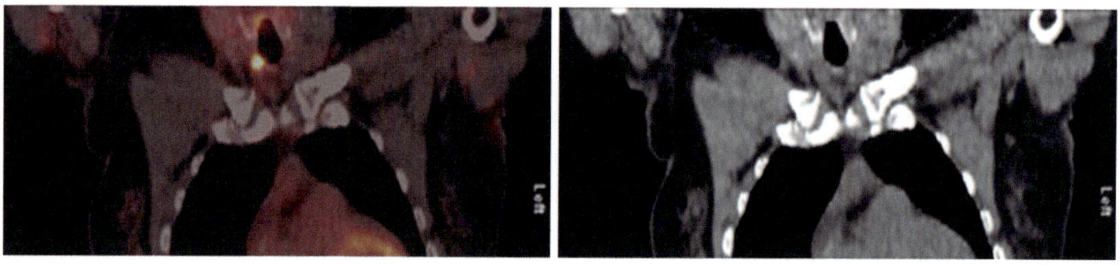

27. ^{18}F-FDG PET/CT performed in an oncology patient. Incidentally, focal ^{18}F-FDG uptake was seen in the inferior pole of the right thyroid lobe. What is the most appropriate next step in management?

    A. Follow up with thyroid ultrasound and FNA when indicated
    B. Follow up with I-123 whole-body scan to evaluate for possible metastasis
    C. Follow up with I-123 thyroid uptake and scan to assess for cold nodules
    D. Total thyroidectomy followed by radioiodine ablation

    Answer: A

    Given the high incidence of thyroid cancer for thyroid nodules demonstrating focal FDG uptake, incidental focal FDG thyroid uptake scan should be followed with a dedicated thyroid ultrasound and fine-needle aspiration if a correlative nodule with suspicious characteristics is identified [1, 2].

28. What is the diagnostic performance of PET/CT for well-differentiated thyroid cancer?

    A. Poor sensitivity
    B. High sensitivity
    C. High positive predictive value
    D. High negative predictive value

Answer: A

Differentiated thyroid cancer has a low glycolytic rate frequently, it is not well seen on FDG PET imaging. High-risk patients diagnosed with differentiated thyroid cancer are better evaluated with I-123 whole-body scan [1, 2]

29. The reported incidence of malignancy within incidentally identified focal ^{18}F-FDG uptake in the thyroid gland is about:

A. 5%
B. 10%
C. 30%
D. 60%

Answer: C

The incidence of focal thyroid uptake in ^{18}F-FDG PET/CT done for other indications has been reported as high as 4%. Given the high incidence of thyroid cancer within these "incidentalomas" (ranging between 15% and 50%), the NCCN guidelines recommend further workup for adequate characterization when focal thyroid uptake is described on ^{18}F-FDG PET/CT done for other oncologic indications [1, 2].

30. What is the most common etiology of diffuse low to moderate level of thyroid uptake seen on Ga-68 DOTATATE PET/CT?

A. Medullary Thyroid Cancer
B. Follicular Variant Thyroid Cancer
C. Physiologic uptake
D. Chronic Thyroiditis

Answer: C

Studies have demonstrated that diffuse uniform ^{18}F-FDG thyroid uptake is not associated with a high incidence of cancer like the more focal uptake. This presentation is most commonly seen in chronic or lymphocytic thyroiditis. In contrast, given the presence of somatostatin receptors in the normal thyroid tissue, diffuse thyroid uptake is commonly seen as part of the physiologic distribution of Ga-68 DOTATATE [5].

References

1. Nishimori H et al. Incidental thyroid "PETomas": clinical significance and novel description of the self-resolving variant of focal FDG-PET thyroid uptake. Can J Surg. 2011;54(2): 83–8.
2. Bae JS et al. Incidental thyroid lesions detected by FDG-PET/CT: prevalence and risk of thyroid cancer. World J Surg Oncol. 2009:63.
3. Chen W et al. Evaluation of thyroid FDG uptake incidentally identified on FDG-PET/CT imaging. Nucl Med Commun. 2009;30(3):240–4.
4. Vaisman F et al. A new appraisal of iodine refractory thyroid cancer. Soc Endocrinol. 2015;22: R301–10
5. Pavel N et al. The rate and clinical significance of incidental thyroid uptake as detected by gallium-68 DOTATATE positron emission tomography/computed tomography. Thyroid. 2016;26(6):831–5.

Skeletal System

Medhat Sam Gabriel, Bital Savir-Baruch, Robert H. Wagner, Bruch J. Barron, Davide Bova, Aamna Hassan, Stephanie Rodriguez-Santiago, and Yekaterina Kucerova

Bone 2020

1. A 38-year-old female with breast cancer had a bone scan. The scan findings are most consistent with:

M. S. Gabriel (✉) · B. Savir-Baruch · R. H. Wagner · D. Bova
Division of Nuclear Medicine, Department of Radiology, Loyola University Medical Center, Maywood, IL, USA
e-mail: sgabriel@lumc.edu; Bital.savir-Baruch@lumc.edu; rwagner@lumc.edu; dbova@lumc.edu

B. J. Barron
Division of Nuclear Medicine and Molecular Imaging, Department of Radiology and Imaging Sciences, Atlanta, GA, USA
e-mail: bjbarro@emory.edu

A. Hassan
Department of Nuclear Medicine, Shaukat Khanum Memorial Cancer Hospital and Research Center, Lahore, Punjab, Pakistan

S. Rodriguez-Santiago
Department of Nuclear Medicine, C.W. Bill Young VA Medical Center, Bay Pines, FL, USA
e-mail: stephanie.rodriguezsantiago@va.gov

Y. Kucerova
Division of Nuclear Medicine, Department of Radiology, Loyola University Medical Center, Maywood, IL, USA

Department of Radiology, Cook County Healthy, Chicago, IL, USA
e-mail: yekaterina.kucerova@cookcountyhealth.org

M. S. Gabriel, B. Savir-Baruch (eds.), *RadTool Nuclear Medicine MCQs*,
https://doi.org/10.1007/978-3-030-69281-0_6

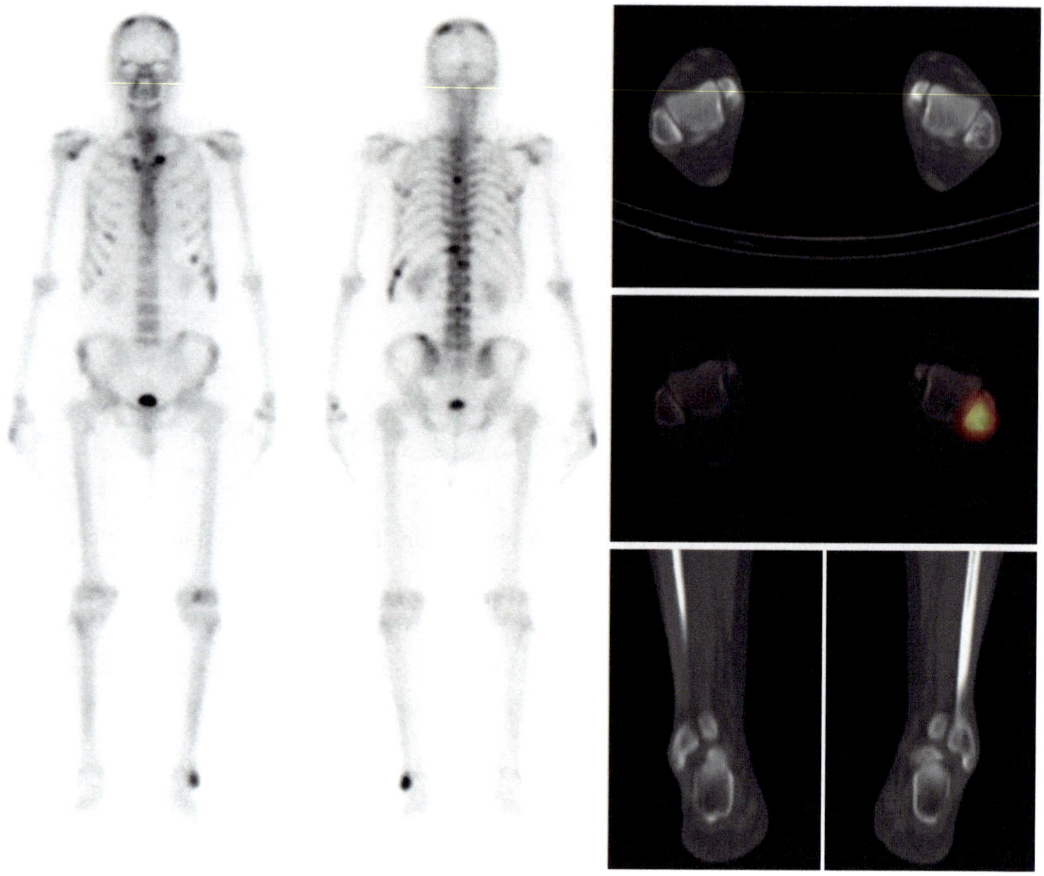

(a) Metastatic lesion in the left fibula
(b) Gout
(c) Melorheostosis of the fibula
(d) Fracture of the distal tibia

Correct answer: A. Metastatic lesion in the left fibula.

The randomly distributed foci of increased radiotracer uptake in the axial skeleton are most consistent with metastases. The distal location of the lesion in the left ankle is unusual and atypical. However, the SPECT/CT images demonstrate a mixed lytic and sclerotic area, which is typical of breast cancer metastases.

Gout does give joint inflammation and remodeling but it typically affects the big toe or the knee joint.

Melorrheostosis usually affects the distal end of long bones with cortical thickening and dense sclerosis on CT images. It is classically described as "dripping candle wax appearance."

Fractures typically show increased uptake on bone scan. However, there is no evidence of fracture in the distal tibia or fibula on SPECT/CT.

2. A 72-year-old gentleman with a recent diagnosis of prostate cancer (PSA 239 ng/dL) underwent a staging bone scan. Based on the scan findings what is the most likely cause of uptake in the left ankle is:

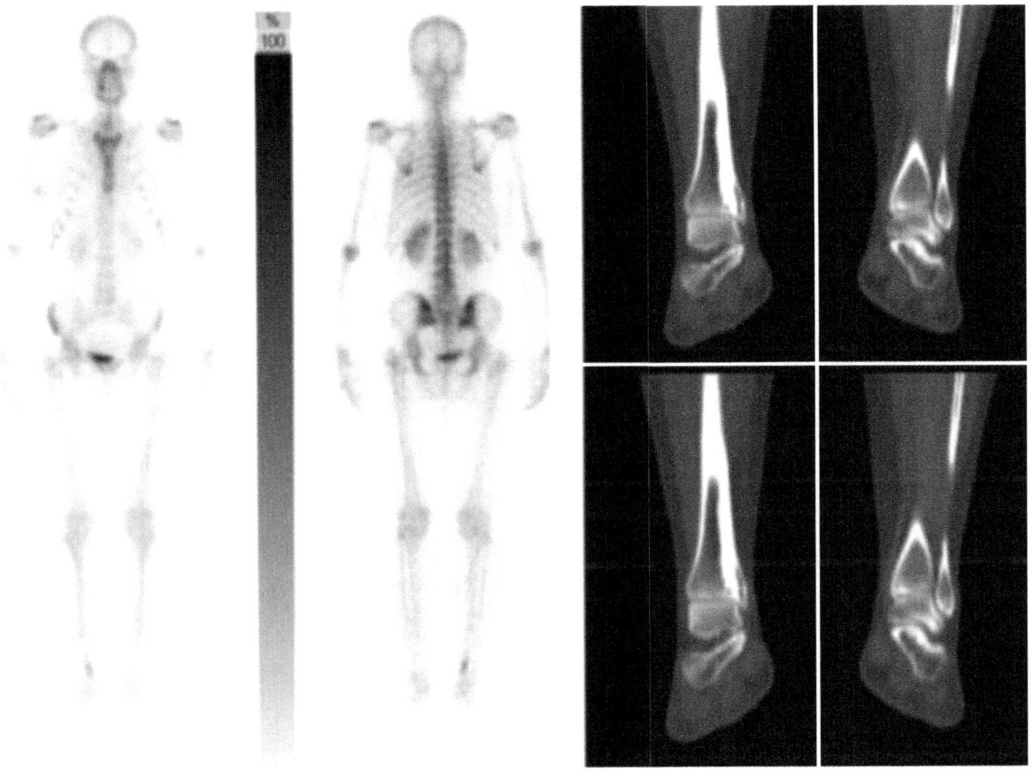

(a) Osseous metastasis
(b) Osteopoikilosis
(c) Fracture of the ankle
(d) Melorheostosis

Correct answer: D. Melorheostosis

Melorheostosis is in the differential for sclerotic MDP avid lesions in the appendicular skeleton. However, classically they are linear lesions giving a candle wax dripping appearance. Melorheostosis is usually seen as asymmetric cortical hyperostosis in contiguous bones. Bone scan appearances of melorheostosis has been described as moderately increased tracer uptake in an asymmetrical linear pattern. It can be polyostotic, involving contiguous bones across the joints along the sclerotomal distribution.

Although the PSA is significantly increased, there is no evidence of osseous metastases on the bone scintigraphy. Prostate cancer metastases are typically osteoblastic, and hence bone scan is the ideal modality to show them.

Osteopoikilosis also gives randomly distributed sclerotic densities, which do not demonstrate significant increase in MDP uptake. These are randomly distributed throughout the skeleton, particularly near the joints and may be innumerable.

Reference: Hassan A, Khalid M, Khawar S. Detection of melorheostosis in a young lady with upper limb pain on three phase bone scintigram and SPECT/CT Clinical Cases in Mineral and Bone Metabolism 2016; 3(1):53–5.

4. This 64-year-old female presents with a history of left-sided upper abdominal pain. A CT study demonstrates findings consistent with a left renal mass. A subsequent bone scan is performed to evaluate for metastatic disease. The findings on bone scan:

   A. Demonstrate the presence of the left renal mass inferiorly but is otherwise normal.
   B. Reliably exclude the presence of osseous metastatic disease.
   C. Suggest that the scan was performed too early and should be repeated in 1–2 h
   D. Suggest the presence of tumor growth into the inferior vena cava.

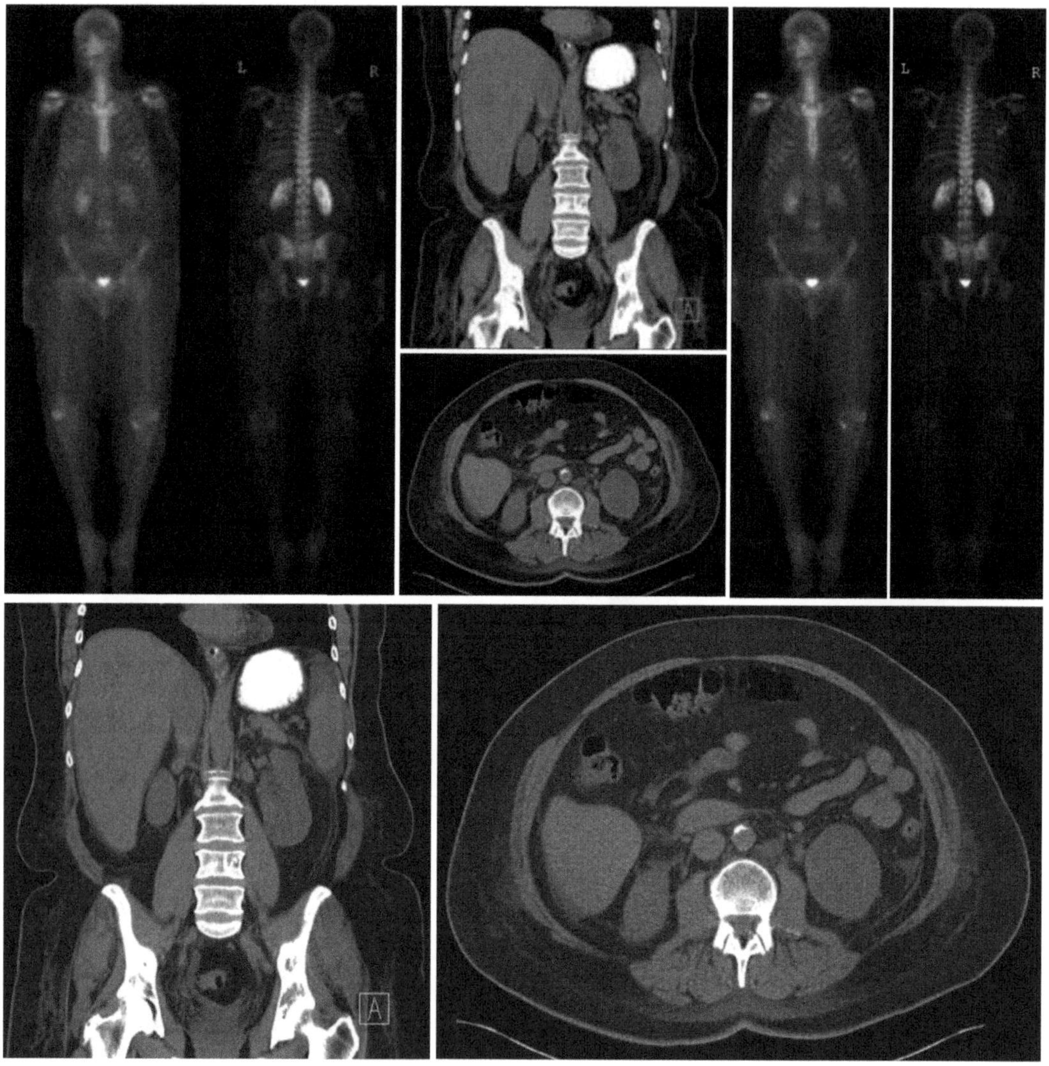

Answer: D

There is significant retention of the radiotracer in the soft tissues of the lower half of the body beginning at the level of the inferior ribs through the lower extremities. The visible upper extremities, base of neck, and chest demonstrate good clearance of tracer. These findings are described as the "Fisherman's Waders" sign and can be seen in cases of inferior vena cava obstruction.

The renal mass can best be seen in the anterior view at the inferior aspect of the kidney.

Metastatic disease from renal cell carcinoma may be subtle or show decreased, rather than increased, uptake. Frequently, these lesions are difficult to identify on bone scan alone.

The scan was performed at the proper time (2–4 h) after injection.

5. In this patient with a suspected renal cell cancer and involvement of the renal vein and the inferior vena cava, further preoperative planning should be considered. When CT findings are equivocal, a good choice for further evaluation would be:

   A. FDG PET/CT imaging
   B. A Ga-67 citrate study
   C. Magnetic resonance angiography
   D. Angiography of the renal mass

   Answer: C

   Planning for surgery includes evaluation of the involvement by tumor of the vascular structures. A CT study with contrast is often sufficient, but in some patients with equivocal findings or in those who cannot tolerate contrast, magnetic resonance angiography is an excellent alternative choice.

   FDG PET/CT imaging is not considered a standard part of the initial evaluation of renal cell carcinoma or to identify local recurrence. However, it has a better sensitivity for identification of metastatic foci.

   Although Ga-67 citrate uptake has been described in renal cell tumors, it is not currently a part of the standard management because of its low sensitivity.

   Angiography is rarely used to evaluate renal cell tumors.

6. This 71-year-old male had a right hip replacement approximately 2 years prior. He recently fell and now complains of right hip pain. No prior bone scan is available for comparison. The findings on the bone scan suggest:

   A. A rib fracture near the tip of the left scapula
   B. Loosening of the right femoral prosthesis
   C. Horizontal uptake in the sternum suggesting fracture
   D. Uptake in the sacrum consistent with a sacral insufficiency fracture

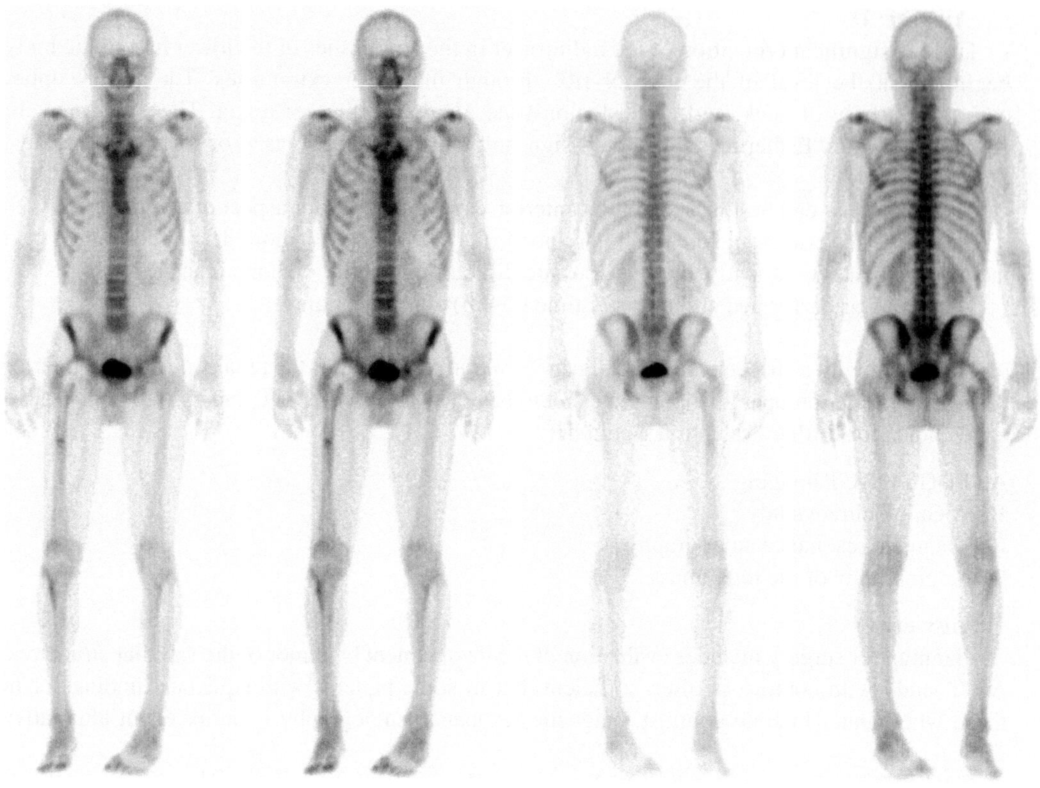

Answer: B

A focal area of increased uptake located at the tip of the prosthesis is suggestive of loosening. In addition, mild uptake is also noted along the length of the prosthesis, further supporting the presence of loosening.

The segmented appearance of the sternum represents ossification sites. This is a normal variant and can persist through adult life.

The focal uptake near the tip of the left scapula is artefactual due to overlapping scapula and rib.

There is no linear uptake across the sacrum extending to the SI joints to suggest a sacral insufficiency fracture, which is typically seen as the letter "H."

7. A 67-year-old male recently fell and is now experiencing persistent burning pain and tenderness, which is worse in his left hand. The bone scan is most consistent with:

   A. Left-hand arthritic changes
   B. Multiple sites of osteomyelitis
   C. Disuse atrophy of the right hand
   D. Abnormal sympathetic nerve activity

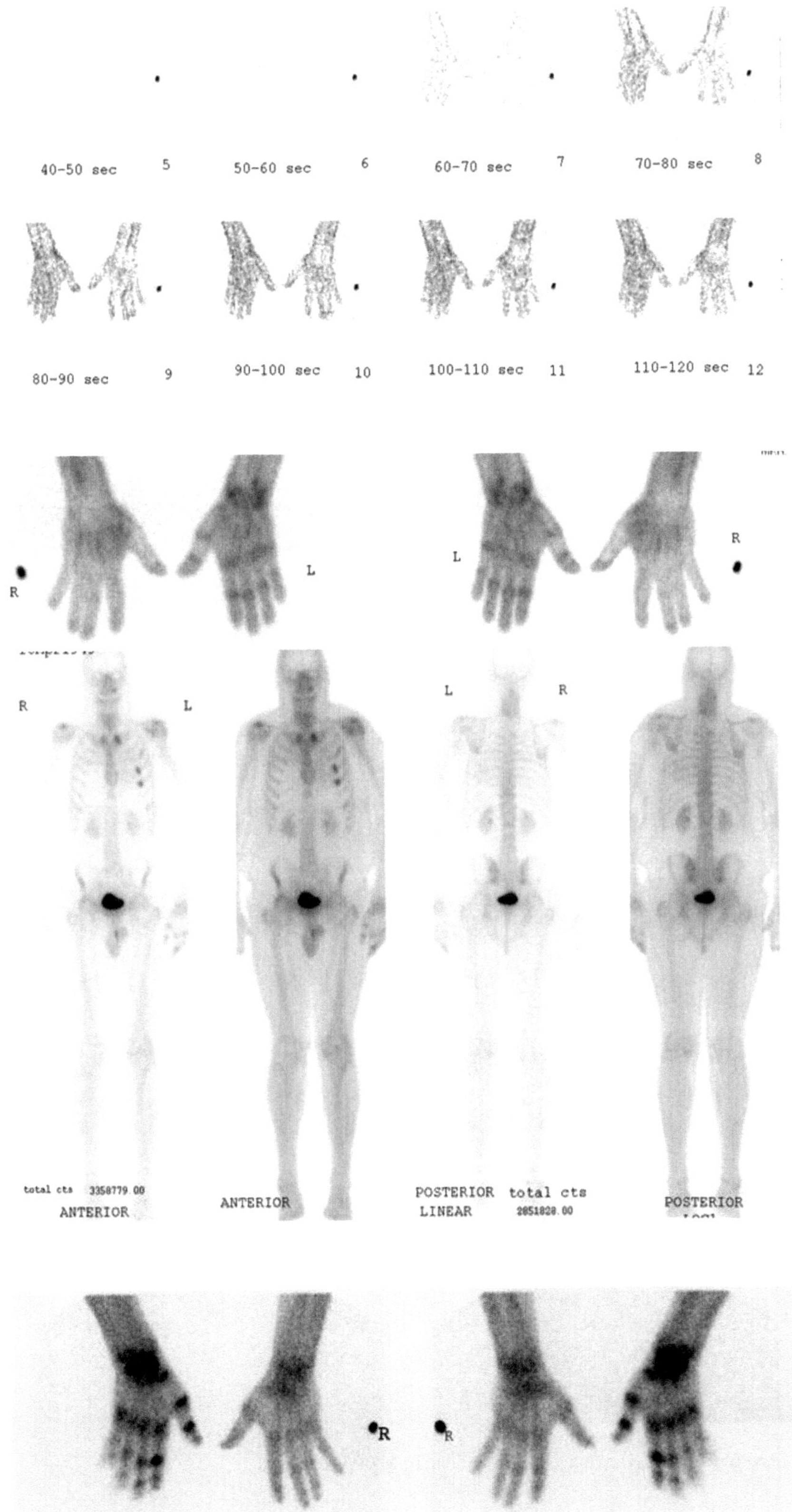
40-50 sec 5
50-60 sec 6
60-70 sec 7
70-80 sec 8
80-90 sec 9
90-100 sec 10
100-110 sec 11
110-120 sec 12
R
L
L
R
R
L
L
R
total cts 3358779.00
ANTERIOR
ANTERIOR
POSTERIOR total cts
LINEAR 2851828.00
POSTERIOR
R
R

Answer: D

Complex Regional Pain Syndrome (CRPS), previously known as reflex sympathetic dystrophy or Sudek's atrophy often occurs after trauma or stroke, although occasionally it is idiopathic. Pain usually lasts for a long time and is believed to result from abnormal sympathetic nerve activity. On bone scan, it is characterized by increased activity on flow, blood pool, and delayed images. Increased uptake is most noted in the periarticular areas.

Multiple areas of osteomyelitis involving all the joints of the left hand would be highly unusual. The right hand appears normal.

Case 1 A 29-year-old female patient is referred for a sacrum MRI (Fig. 1) to further evaluate reports of chronic low back pain, suspicious for sacroiliitis. A 6 × 12 mm oval-shaped lesion within the right ischium which contains a small area of suspected fat signal centrally is described. Subsequently, a CT of the pelvis is performed and demonstrates the lesion to be stable in size and showing mixed density with central lucency probably representing a nidus as could be seen with osteoid osteoma. A bone scan with a SPECT/CT of the pelvis is performed (Figs. 2 and 3).

Fig. 1 AX T1 images

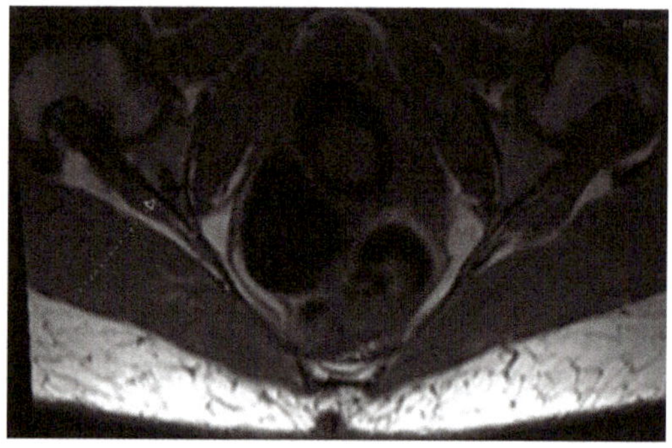

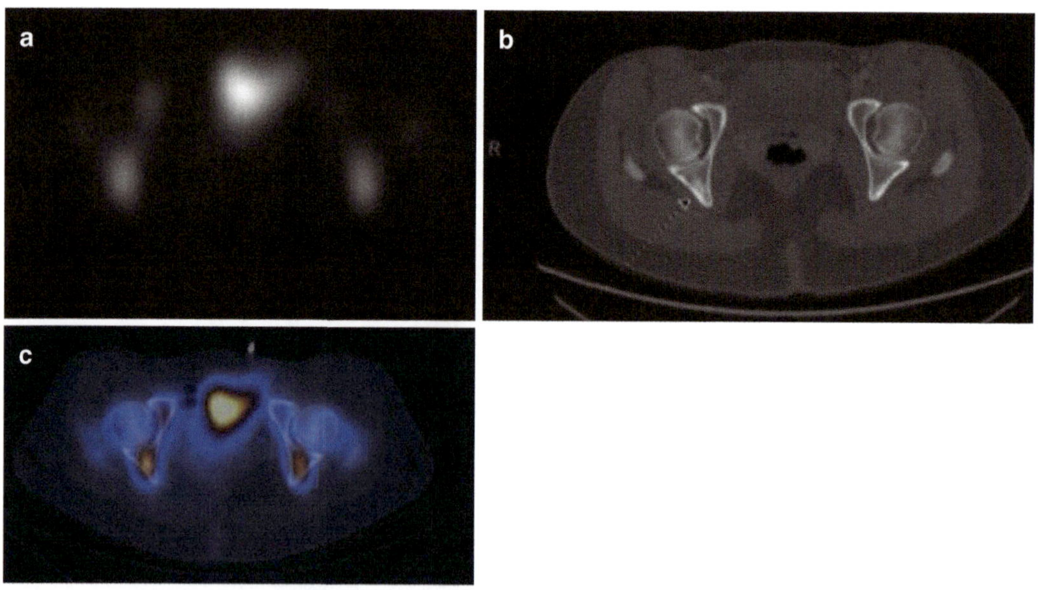

Fig. 2 **a** 99mTc MDP SPECT; **b** CT axial images; **c** Fused 99mTc MDP/CT images

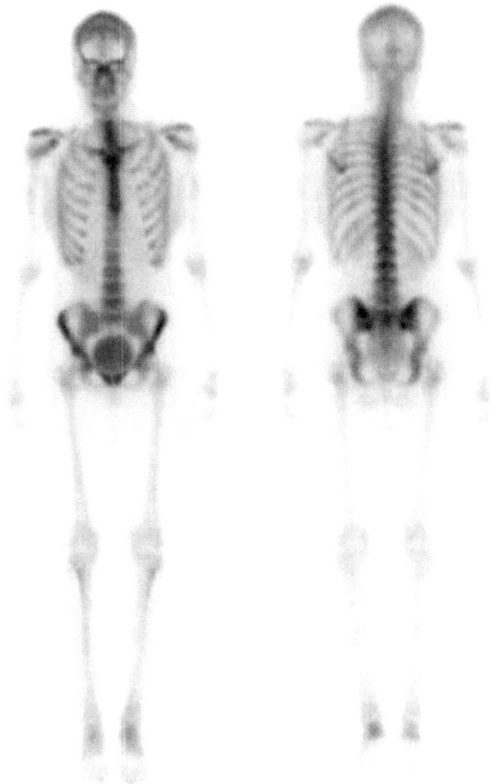

Fig. 3 Planar 99mTc MDP Bone Scan

8. Based on the radiographic and scintigraphic findings shown above,

   (a) The lesion is most consistent with osteoid osteoma
   (b) Metastatic disease is highly suspected
   (c) The findings are typical for enchondroma
   (d) The lesions are likely of a benign etiology

   Answer: d
   The combination of magnetic resonance imaging findings and the lack of increased uptake suggest a benign etiology. Commonly osteoid osteomas demonstrate increased MDP uptake. The whole-body scan demonstrates no evidence of a blastic metastatic process. Enchondromas demonstrate variable levels of uptake but usually present as MDP avid cystic lesions in the bones of the hands.

9. Metastatic lesions in the distal extremities are most commonly seen with what type of cancer?

   (a) Lung cancer
   (b) Prostate Cancer
   (c) Hepatocellular carcinoma
   (d) Breast cancer

Answer: a

Metastatic lesions in the distal long bones of the appendicular skeleton are less common than in the axial skeleton, particularly for prostate cancer. Bone scans are significantly less commonly performed for the evaluation of metastasis in patients with lung cancer. This is in part due to the advent of ^{18}F-FDG PET/CT and its utility in the evaluation of this population. Lung tumors have a significantly higher rate of vasculature invasion and arterial metastasis. As a result, metastatic lesions in the distal extremities are more commonly seen in this group of patients.

10. How often bone scans in the initial evaluation of patients with stage 1 prostate adenocarcinoma with PSA levels below 8 ng/mL show positive findings of bone remodeling as the result of metastatic lesions?

    (a) 1–5%
    (b) 10–20%
    (c) 30–40%
    (d) 50–70%

    Answer: a

    It is important to remember that not all patients with a diagnosis of prostate adenocarcinoma benefit from bone scintigraphic studies. The routine use of bone scans for newly diagnosed patients with low-risk prostate adenocarcinoma as a part of their baseline workup is not recommended. The likelihood of a true positive bone scan for osseous metastatic disease correlates with the Gleason score and PSA levels. The incidence of bone metastases appreciable in bone scans for patients with low-risk prostate adenocarcinoma has been extensively documented. Osseous metastatic disease in patients with PSA level below 10 ng/mL and a Gleason score of less than 7 is uncommon, with positive scans in this population seen in less than 5%.

11. Tc-99 m MDP bone scan has the lowest sensitivity for which of the following diseases?

    (a) Prostate cancer
    (b) Breast Cancer
    (c) Multiple Myeloma
    (d) Lung Cancer

    Answer: c

    Tc-99 m MDP bone scans are highly sensitive for blastic osseous lesions and less sensitive for lytic or lucent lesions. The permeative osseous pattern that can be seen with multiple myeloma in skeletal surveys may be seen as photon deficient areas in the bone scan. However, the sensitivity of the bone scan for the detection of these lesions in this group of patients is limited. In contrast, PET imaging has a high sensitivity for lytic lesions.

12. In a patient with prostate adenocarcinoma, a single focus of uptake is seen in the right anterior fifth rib. What is the probability of malignancy for the single focus of uptake in a rib?

    (a) About 30%
    (b) About 15%
    (c) About 5%
    (d) About 1%

    Answer: b
    A solitary focus of rib uptake is most commonly of benign and likely posttraumatic etiology, even for a patient with known malignancy. The likelihood of a solitary metastatic focus of uptake in the ribs in this population is between 10–20%. The probability of metastasis as the cause for solitary osseous findings varies per location. Foci of uptake in the sternum, spine, and pelvis have the highest probability in the range of 60–75%. Skull uptake can be variable and many times challenging to interpret. However, a focus of increased uptake in the skull in a patient with known malignancy has approximately 40–50% change of representing osseous remodeling secondary to changes induced by a metastatic lesion.

References

1. Donohoe KJ et al. Society of nuclear medicine procedure guideline for bone scintigraphy. Soc Nucl Med Molecul Imag. 2003.
2. Donohoe KJ et al. Appropriate use criteria for bone scintigraphy in prostate and breast cancer. Soc Nucl Med Molecul Imag. 2017.
3. Mohler JL et al. Prostate adenocarcinoma. National comprehensive cancer center clinical practice guidelines in oncology. Version 3.2018–June 21, 2018

13. What is the chance of developing sarcoma in this patient?

    (a) 25%
    (b) 10%
    (c) 5%
    (d) 1%

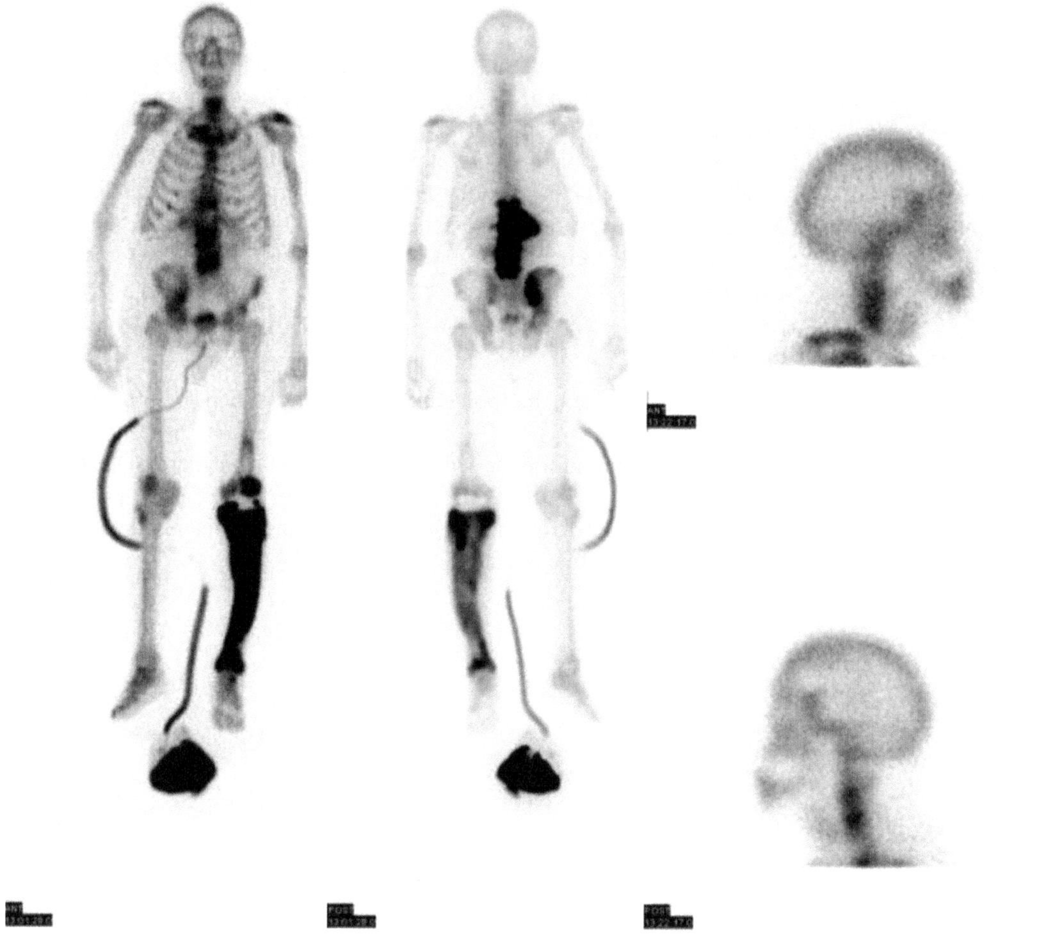

Correct answer: D

The patient has Paget disease in the left tibia, right iliac bone, and spine. The disease affects elderly people and is usually benign. However, in about 1% of patients, aggressive sarcomatous transformation occurs.

Case 2 49-year-old female on maintenance therapy for breast carcinoma metastatic to the spine presents for surveillance PET-CT. Mammogram showed a 1.9 cm mass with satellite nodules. US-guided core biopsy demonstrated Infiltrating ductal carcinoma and the patient was scheduled for mastectomy with chemoradiation. A whole-body bone scan for initial staging was obtained.

14. Based on the images below, what is the most likely diagnosis?

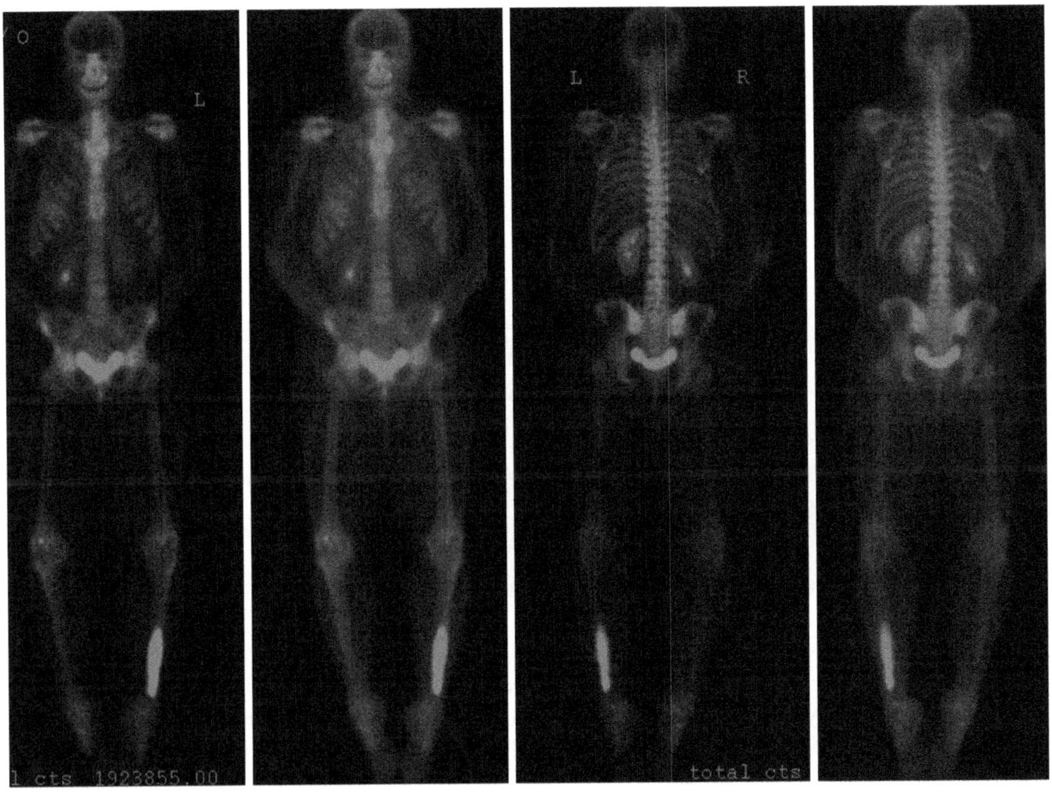

A. Metastatic osseous disease
B. Paget's Disease
C. Fibrous Dysplasia
D. Osteoid Osteoma

Answer: C

Fibrous dysplasia is a benign intramedullary fibro-osseous lesion and may be monostotic (more common) or polyostotic. The degree of increased tracer uptake is relatively high, rivaling that seen in Paget's disease. Suggestive features of Fibrous Dysplasia are younger age, ground-glass radiographic appearance, and pattern of distribution. When Paget's involves a long bone it invariably extends to at least one end of the bone, whereas fibrous dysplasia only rarely involves the epiphysis. The pattern of uptake in metastatic disease is more focal than linear and for the most it affects the axial and proximal appendicular skeleton.

15. What additional imaging can establish the diagnosis to aid the scintigraphy findings in this patient?

    A. MRI
    B. X-Ray
    C. Ga 67 scintigraphy
    D. Tc-99 m SPECT/CT

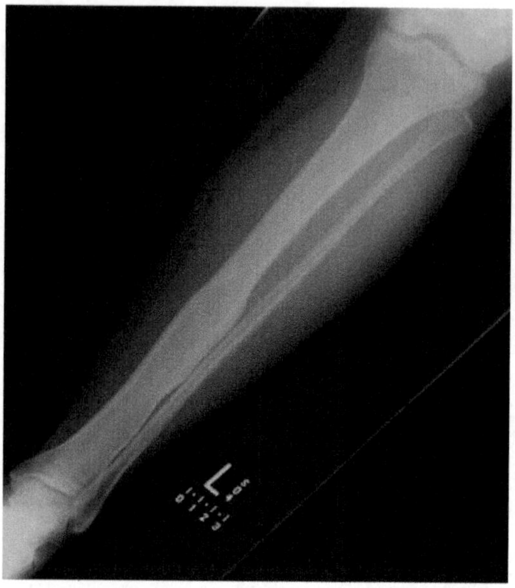

Answer: B

Radiographs are the initial and most cost-effective confirmatory test in these cases. The normal bone is replaced by tissue that is more radiolucent, with a grayish "ground-glass" pattern that is similar to the density of cancellous bone but is homogeneous, with no visible trabecular pattern, generally sparing the epiphysis. The lesion characteristically is surrounded by a distinct rim or shell of reactive bone that is defined more sharply on its inner border than on its outer border, where it may fade gradually into normal cancellous bone. GA 67 scintigraphy is used in diagnosing osteomyelitis of the spine, as well as a problem-solving tool in other more indolent infectious processes of the skeleton, when other faster imaging modalities (MRI, CT, Tagged Leukocyte Imaging) are nonconclusive or degraded by artifact, such as that caused by metal hardware. MRI is a second line morphological imaging tool, but is best performed after radiographs, as it tends to be less specific in appearance: it provides exquisite anatomic detail and spatial distribution, as well as the ability to exclude other diagnoses (lack of chondroid matrix, lack of aggressive periosteal reactions, etc.). Tc-99 m SPECT/CT would not provide additional information in this patient as the lesion is brightly hypercaptating on the planar images.

16. Regarding fibrous dysplasia, which of the following statements is true?

    A. It is caused by a defective gene
    B. It is more frequently diagnosed in elderly patients
    C. It is caused by repeated microtrauma
    D. It is more frequent in the Caucasian population

    Answer: A

    It is believed to be caused by a defective gene in the bone and connective tissue cell lineage, more frequently diagnosed in children and young adults and does not appear to demonstrate gender or race preference.

Reference

1. Fitzpatrick KA, Taljanovic MS, Speer DP et al. Imaging findings of fibrous dysplasia with histopathologic and intraoperative correlation. AJR Am J Roentgenol. 2004;182(6):1389–98.

Case 3

17. A 55-year-old female treated approximately 3 years before for infiltrating ductal carcinoma of the left breast, with a negative sentinel node, presents for surveillance bone scan. The patient also has recent episodes of bloody nipple discharge.

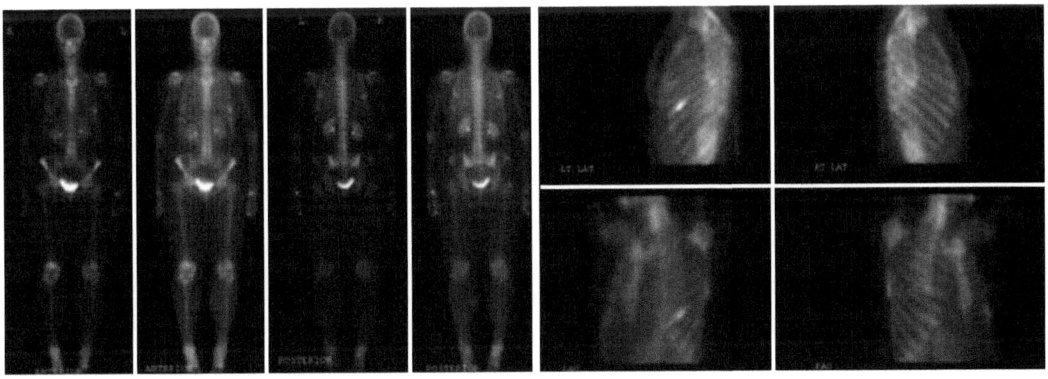

What is your recommendation?

A. Treatment for metastatic disease
B. Correlate with CT obtained on the same day
C. Recommend MRI to further characterize
D. Recommend radiographic rib series to look for fractures

Answer: B

While in general radiographic examinations are the first diagnostic test in skeletal disease, the availability of a recent CT and its greater contrast resolution obviate the need for additional radiographs. In addition, radiographs provide greater diagnostic accuracy in the appendicular skeleton than in the axial skeleton and the ribs; the lesser sensitivity is due to superimposition of numerous structures over the spine, pelvis, and ribs.

18. CT with bone windows at the level of the left rib uptake is shown below. What is your diagnosis?

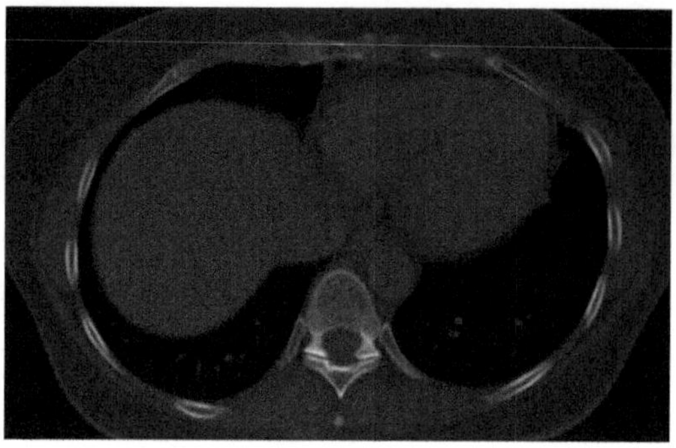

A. No evidence of fracture, consistent with metastatic disease.
B. CT-guided percutaneous core needle biopsy
C. Further evaluation with MRI
D. Follow-up in 6 months to allow posttraumatic changes to resolve

Answer: C

While the elongated morphology on bone scan is highly suspicious for metastasis, the history of limited stage disease 3 years prior and its solitary nature are confusing. Patient provides no history of trauma upon questioning but does admit to left chest pain upon deep inspiration. CT-guided biopsy cannot be performed without a visible target. Follow-up of potential metastatic disease is not advisable.

19. The medical oncologist ordered a PET to restage the disease and further characterize the rib lesion. Select images below show:

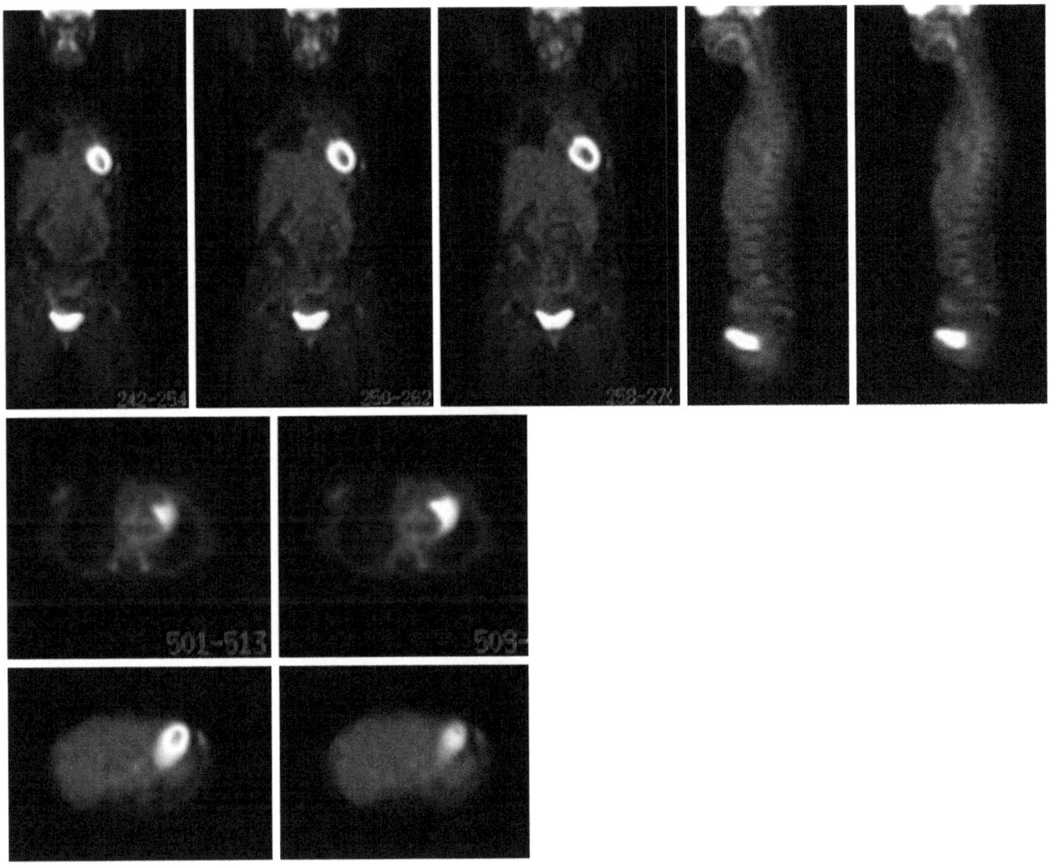

A. Infiltrating lobular carcinoma of the right breast
B. Diffuse metastatic disease to the liver
C. Hypermetabolic lesion in the left sixth rib suspicious for metastasis
D. Cardiac sarcoidosis

Answer: C

Uptake in the right breast is physiologic and lack of uptake in the left breast is secondary to prior radiation therapy. Uptake in the liver and myocardium is physiologic.

20. Below are select images from MRI of the ribs (axial T1 and FST2, oblique sagittal FST2).

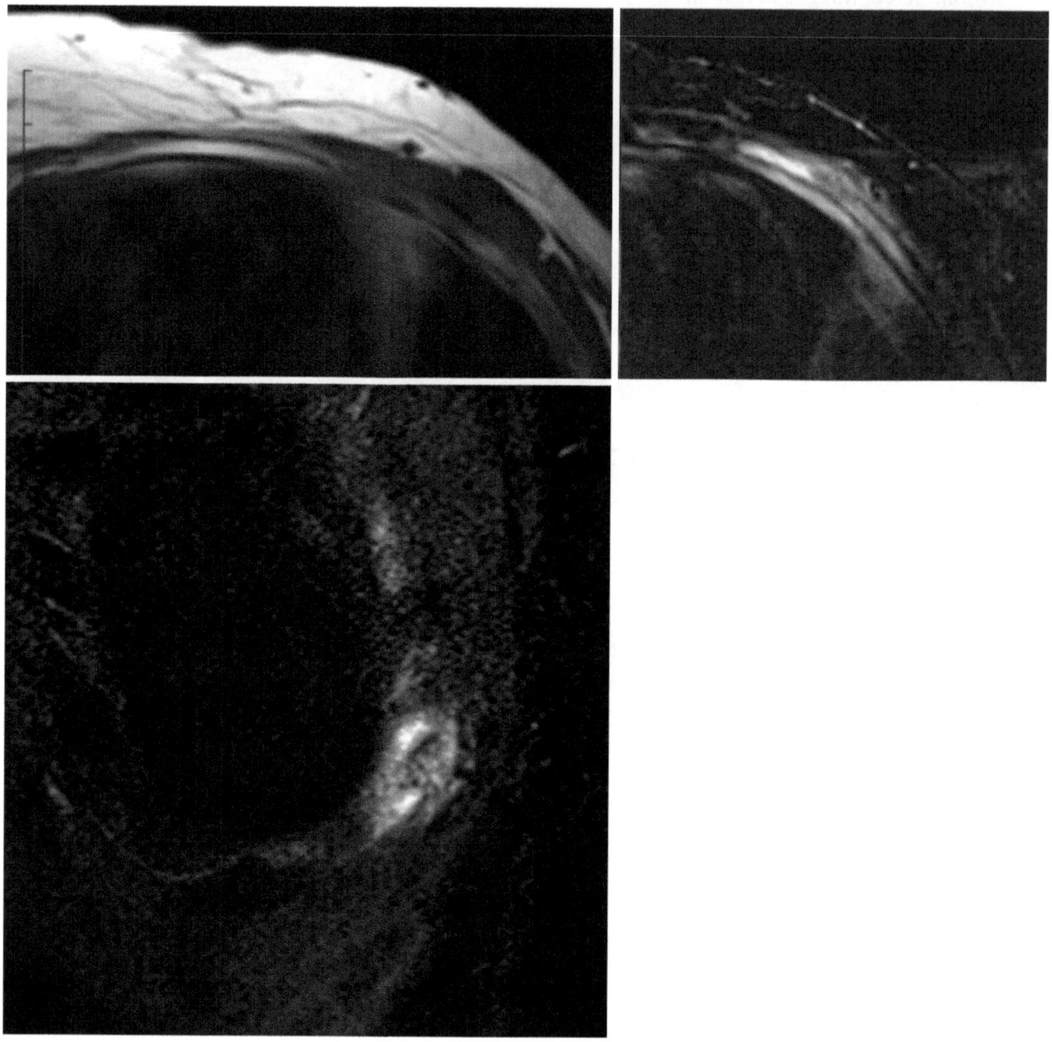

What is your diagnosis?

A. Trauma
B. Metastatic disease
C. Fibrous dysplasia
D. Indeterminate

Answer: D

The lesion demonstrates markedly increased T2 signal, marked marrow edema and adjacent edema, greater than expected for neoplasm. However, there is neither fracture or traumatic deformity, cortical interruption, nor history of trauma. The patient underwent surgical resection of the anterior left sixth rib immediately after bone scan and intraoperative localization with a Gamma Probe by thoracic surgeon and nuclear medicine physician team. Pathology demonstrated "BONE WITH DIFFUSE MARROW REPLACEMENT BY FIBROSIS WITH BONE AND STROMAL NECROSIS CONSISTENT WITH FRACTURE CALLUS."

Case 4

21. A 10-year-old female presents for follow-up Whole Body Bone Scan (WBBS) after chemotherapy of known metastatic telangiectatic osteosarcoma of the left humerus. Below are the current and the two prior studies obtained 4 and 6 months prior, as well as select images from the time of diagnosis.

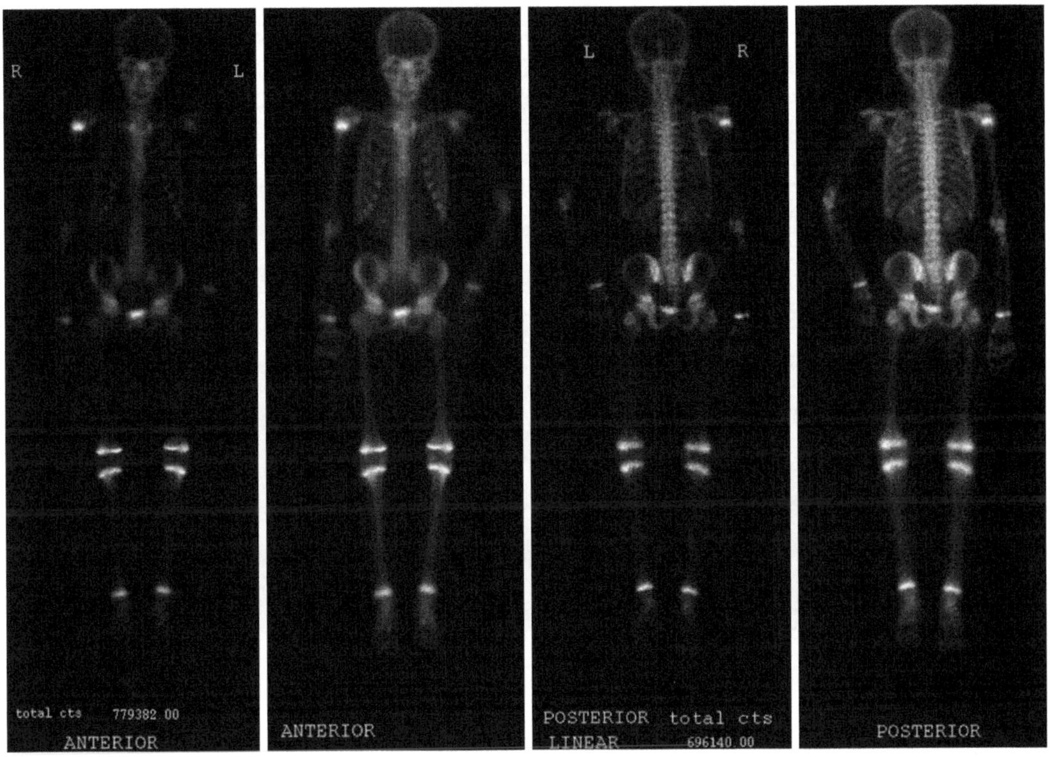

Current Study

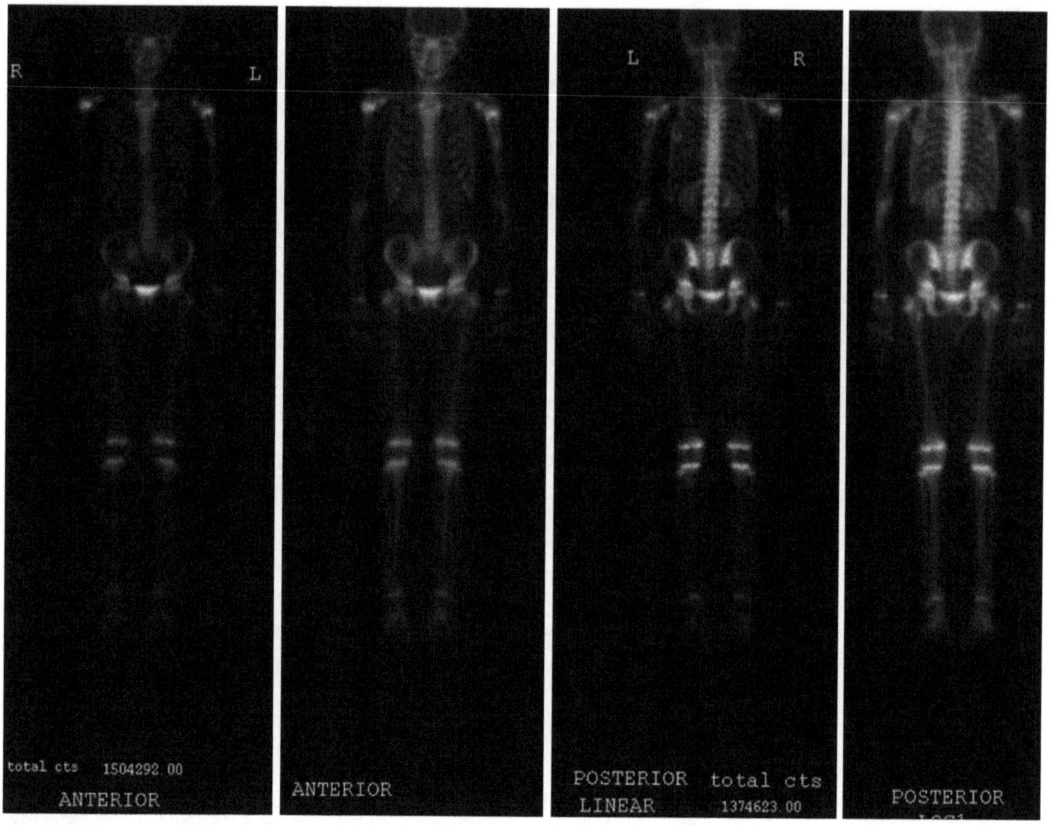

4 months prior

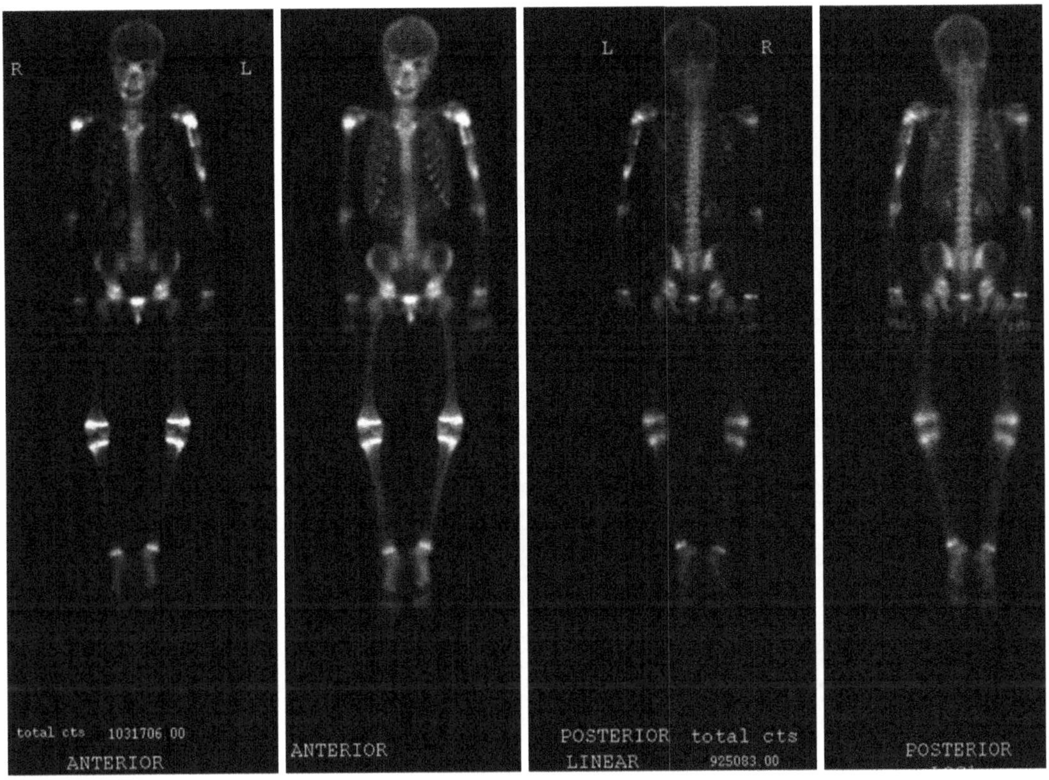

6 months prior, at presentation

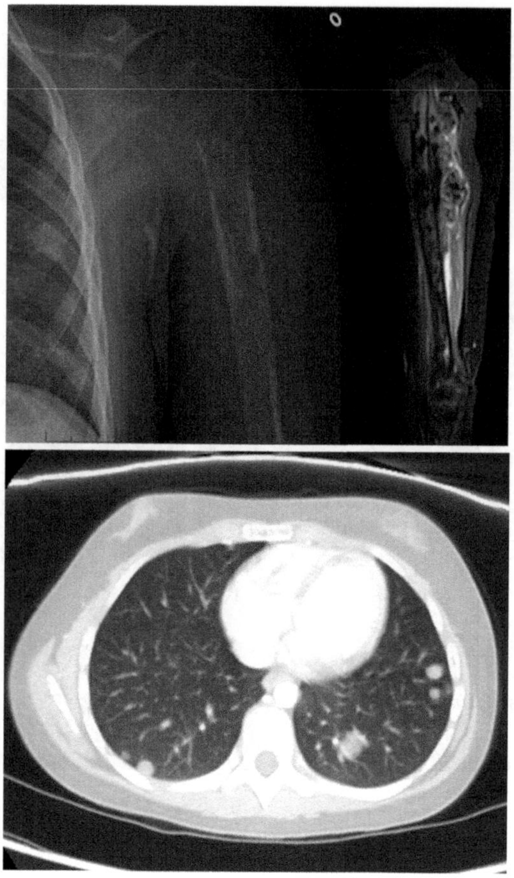

Imaging at time of presentation: WBBS, Humeral radiograph, Fat-Suppressed T2 sagittal images of the left humerus, along with pulmonary contrast-enhanced CT.

What is your interpretation?

A. S/p surgical excision. Follow-up in 12 months for interval stability
B. Favorable response to chemotherapy. No suggestion of new neoplastic foci.
C. Progression of disease
D. Findings compatible with AVN of the humerus

Answer: B

There has been progressive resolution of heterogeneous uptake in the proximal 2/3 of the left humerus.

22. Subsequent imaging obtained later that day (not available at time of interpretation) is provided below.

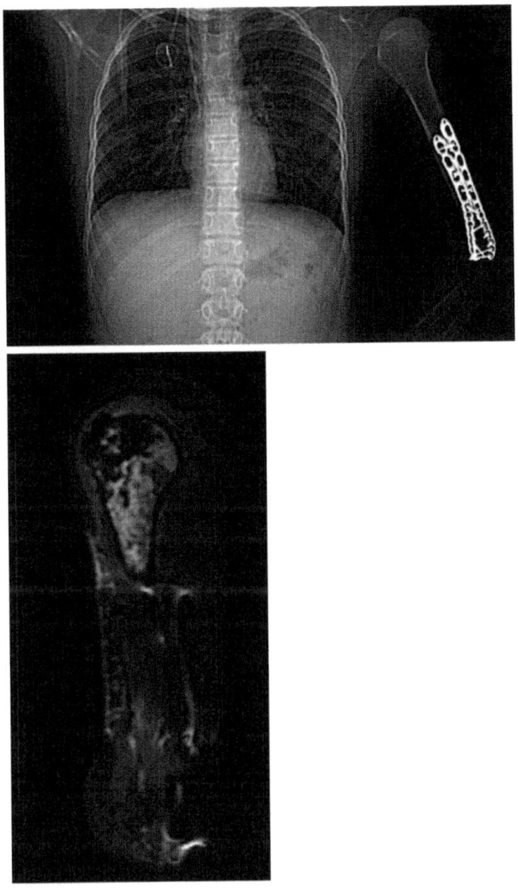

Scout of CT of Thorax and Inversion Recovery MR sagittal image of the left humerus.

What is your interpretation?

A. S/p surgical excision. Follow-up in 12 months for interval stability
B. Favorable response to chemotherapy. No suggestion of new neoplastic foci.
C. Progression of disease
D. Findings compatible with AVN of the humerus

Answer: D

Upon further investigation, we came to realize that the patient had undergone surgical resection of the distal 3/4 of the left humerus with allograft interposition at an outside institution. There is hardware in the humerus on scout and MRI, implying a major surgery.

Bone scanning has been shown to have a high prognostic value for bone graft viability, with a high incidence of failure in those grafts demonstrating decreased or absent radiotracer localization. Additionally, correlating bone scan findings with other available imaging (CT, MR, X-Rays, etc.) is of paramount importance for a correct interpretation of findings. Also, accurate interpretation of any imaging study cannot be performed without thorough knowledge of the patient's medical and surgical history, a task today made easier by the advent of electronic medical records (EMR), but not always possible.

*Leslie A, Jones AJ, Goddard PR, The influence of clinical information on the reporting of CT by radiologists, Br J Radiol. 2000;73(874):1052–5.

**Cohen MD, Accuracy of information on imaging requisitions: does it matter?, J Am Coll Radiol. 2007;4(9):617–21.

***Droll KP, Prasad V, Ciorau A, Gray BG, McKee MD, The use of postoperative bone scintigraphy to predict graft retention.

Case 5

23. 76-year-old gentleman presents for restaging a prostate adenocarcinoma treated with brachytherapy at an outside institution, now with rising PSA. Whole-body bone scan images are shown below.

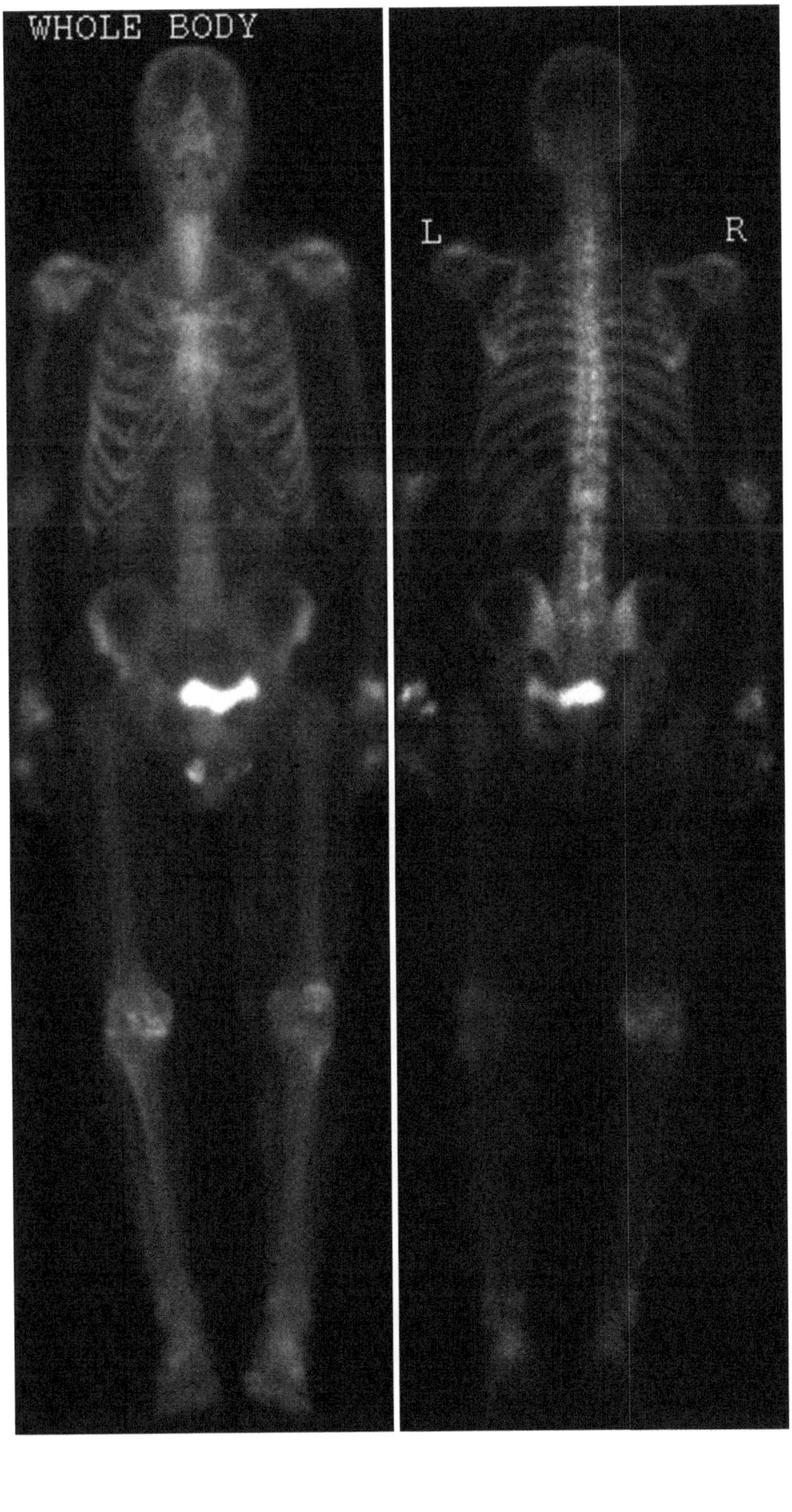

What is your diagnosis and/or recommendation?

A. Urinary contamination: repeat pelvic imaging after wiping.
B. Probable bladder diverticulum: acquire supplemental T.O.D. (Tail on Detector) view after voiding
C. No evidence of metastatic disease
D. Metastatic disease to the pelvic skeleton

Answer: B

There is an asymmetric appearance of the bladder, with faint uptake noted left of the bladder on posterior views, projecting over the expected location of the left superior pubic ramus. The Tail on Detector view is a quick and cheap additional view to discriminate bony uptake from overlapping pelvic structures. Urinary contamination is present on the genitalia, which is too low to really provide concern for overlapping artifact. The presence or absence of metastatic disease question cannot be answered at this time.

24. The patient left the imaging department by the time you interpret this study. What would be your next step?

A. Review the CT of the abdomen and pelvis the patient had after the bone scan.
B. Recommend Pelvic MRI to exclude the possibility of bony metastases.
C. Recommend FDG PET-CT because it is more sensitive for metastatic disease
D. Recommend repeat bone scan with pelvic SPECT

Answer: A

"Correlation with existing imaging studies for ALL patients undergoing bone scintigraphy" has been a quality metric for NM physicians and radiologists in the US for several years, impacting reimbursement cuts ("Merit-Based Incentive Payment System" or MIPS, formerly "Physician Quality Reporting System" or PQRS of Centers for Medicare and Medicaid Services). It is also the correct thing to do for our patients, before embarking on any additional imaging. The CT may provide an answer to our dilemmas or help guide the most appropriate next step. FDG is notoriously insensitive to prostate carcinoma and sclerotic skeletal lesions.

Shown below are select images from the same-day CT with bone windows.

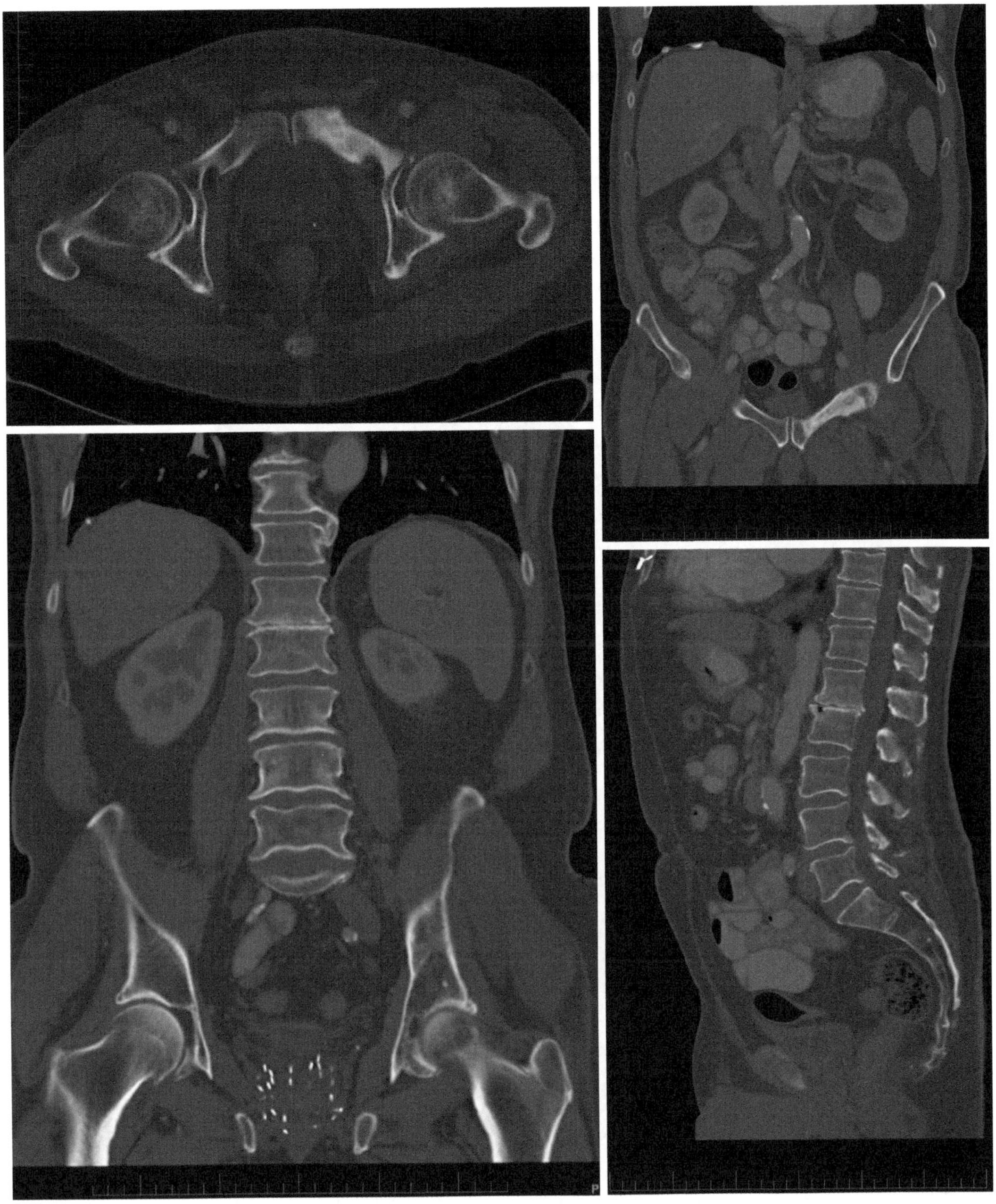

25. What is your diagnosis now?

A. Metastatic disease to the pelvis and lumbar spine.
B. Degenerative changes of the lumbar spine and bladder diverticulum
C. Degenerative changes of the lumbar spine and healed stress fracture of the left superior pubic ramus
D. Solitary metastasis to the left superior pubic ramus and degenerative changes of L1–L2.

Answer: D

Irregularly sclerotic lesion in the left superior pubic ramus accounts for faint linear uptake noted on posterior WBBS view is consistent with metastatic disease. Linear uptake in the superior lumbar spine noted on both anterior and posterior bone scan views corresponds to endplate sclerosis of inferior L1 and superior L2, with associated marked narrowing of the interposed intervertebral disc space, classic features of degenerative changes.

26. Your technologist brings you this bone scan and is unsure why it has this appearance. He insists that the detectors were close to the patient. Quality control for uniformity and resolution was just performed prior to the patient and was unremarkable. You should ask the technologist:

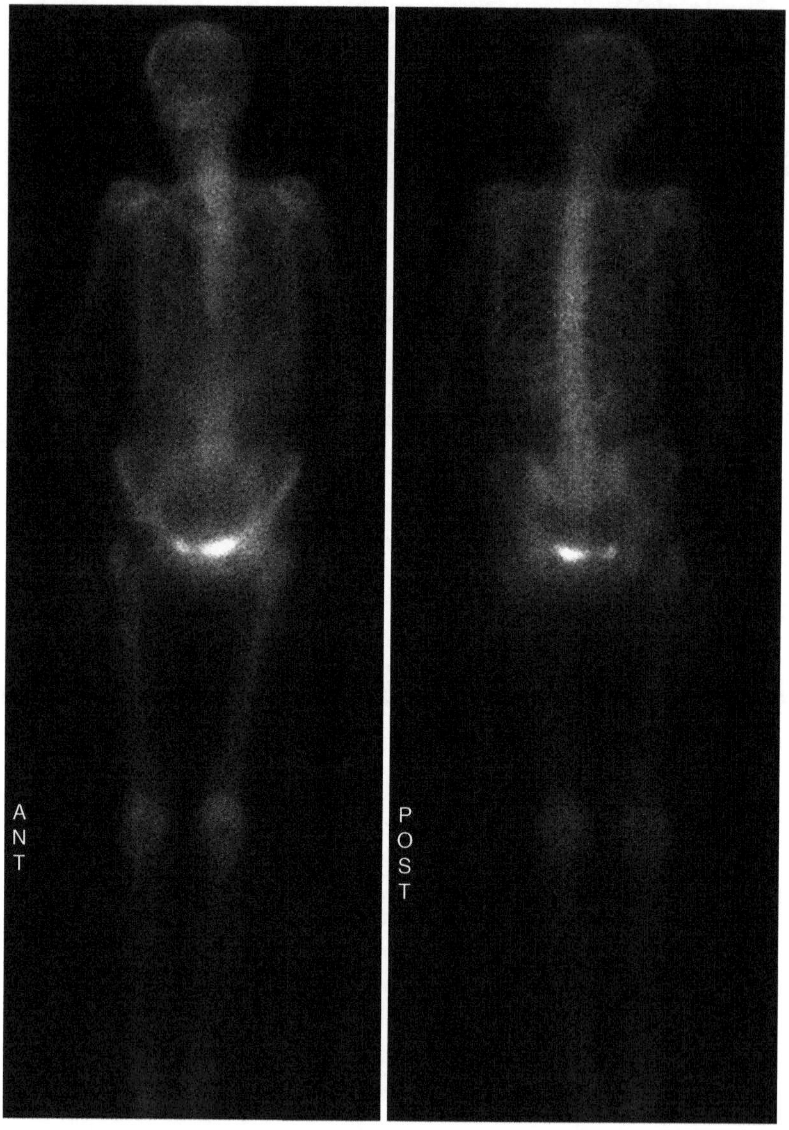

A. To check if the scan speed of the camera was too fast
B. To use the high-resolution collimator instead of the low energy all-purpose collimator
C. To check the photopeak of the camera
D. To contact the radiopharmacy since the wrong radiopharmaceutical may have been injected.

Answer: C

The images represent a bone scan that was acquired off-peak. When performing quality control, a Co57 sheet flood source, is often used. Failure to change the photopeak back to Tc-99 m will result in a scan that is suboptimal in appearance.

Too fast of a scan speed would result in fewer image counts. Either high resolution or low energy all-purpose collimators are acceptable for bone imaging. While an incorrect radiopharmaceutical administration should always be considered as a possibility, in this case, the biodistribution is typical for bone scan agent.

27. A 68-year-old female with a history of lower back pain, bilateral knee pain, and left ankle pain presents from her podiatrist's office who suggested that she has left knee osteomyelitis. She has had a 1 year history of intermittent fevers of unknown origin. Workup has been negative so far. Pain has been progressing in her left knee for 2 weeks. A MRI 1 year ago demonstrated a previously fractured patella and meniscal injury. Based on the bone scan findings:

A. Begin broad-spectrum intravenous antibiotics
B. There is asymmetric patellar uptake, a normal variant.
C. Initiate a metastatic workup to identify the primary tumor
D. Perform a biopsy to evaluate for chronic osteomyelitis

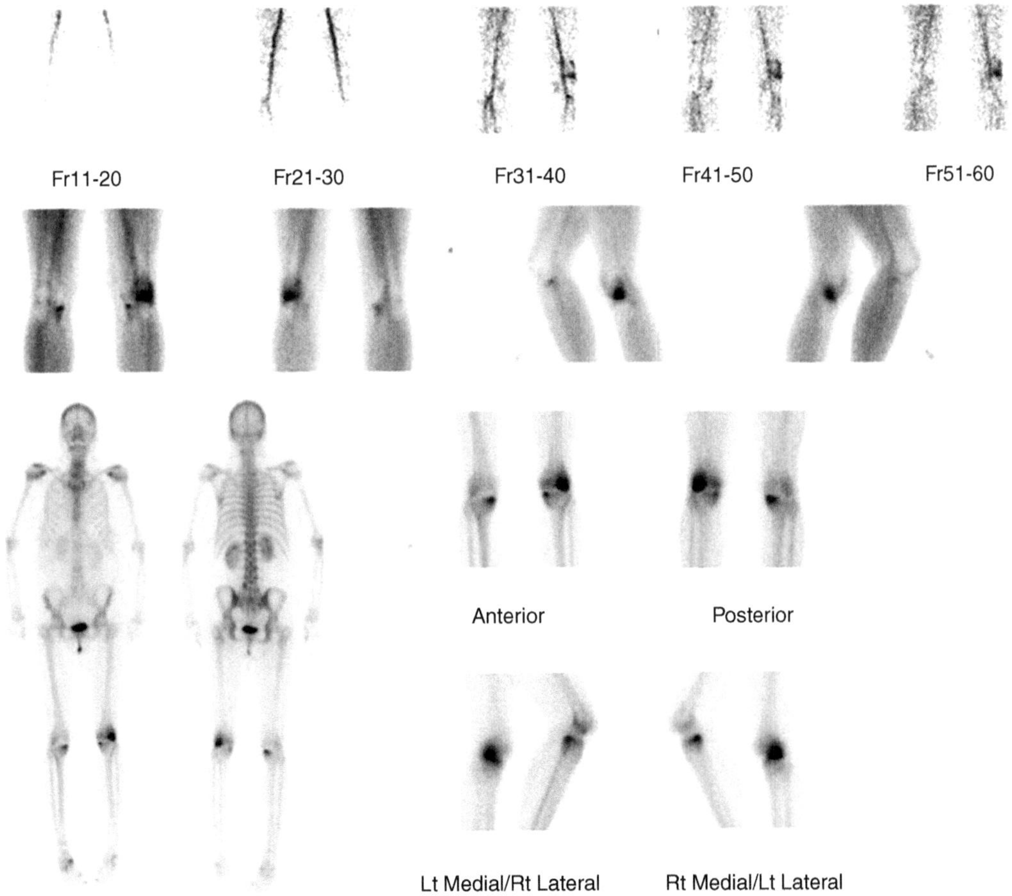

Answer: D

There is uptake seen in all three phases. The scan findings are most suggestive of osteomyelitis. Prior to beginning antibiotics, a biopsy should be performed to identify the pathology and infectious organism.

While asymmetric patellar uptake can be seen as a normal variant, the uptake is clearly in the lateral condyle of the femur. Metastatic disease when it occurs, is more common in the axial skeleton rather than to the extremities.

In this case, the biopsy revealed osteonecrosis of the femoral condyle. No infectious agent was identified. Osteonecrosis can also demonstrate increased uptake on all three phases of a bone scan.

28. This patient presents with a history of elevated alkaline phosphatase on recent lab work. The bone scan demonstrates findings most suggestive of:

A. Paget disease
B. Metastatic disease
C. Renal osteodystrophy
D. Hypertrophic pulmonary osteoarthropathy

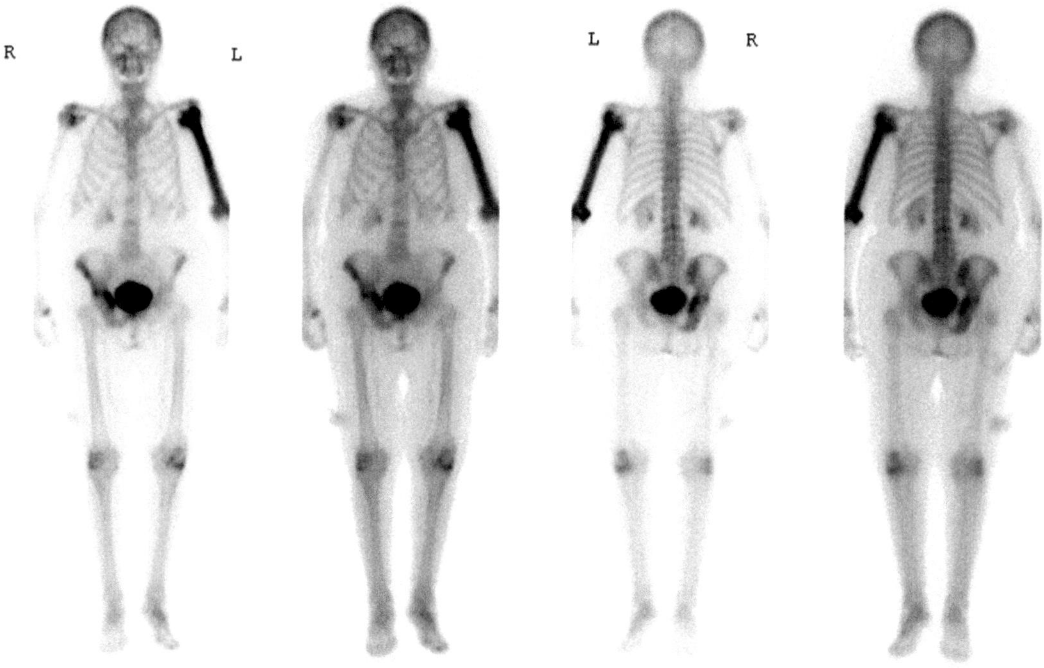

Answer: A

Paget disease when active will produce some of the most intense uptake seen on bone scan. Commonly involving the pelvis, femur, skull, or vertebral bodies, it can affect any bone in the body.

Renal osteodystrophy results in what is often termed the "beautiful bone scan." Characteristically, there is little soft tissue uptake and beading of the costochondral junctions can be seen.

Hypertrophic pulmonary osteoarthropathy is seen in the lower extremities and has a "tram track" appearance.

29. A 52-year-old female with a history of end-stage renal disease, bilateral nephrectomy, and failed renal transplant presents with an elevated alkaline phosphatase level. The referring physician asks you to perform a bone scan to help evaluate for other causes. Your interpretation is:

A. Recent trauma to the chest causing uptake at the costochondral junctions
B. The paucity of soft tissue uptake indicates dialysis between injection and imaging
C. A suspicious focus of uptake is seen in the base of the skull posteriorly
D. The patient likely has renal osteodystrophy

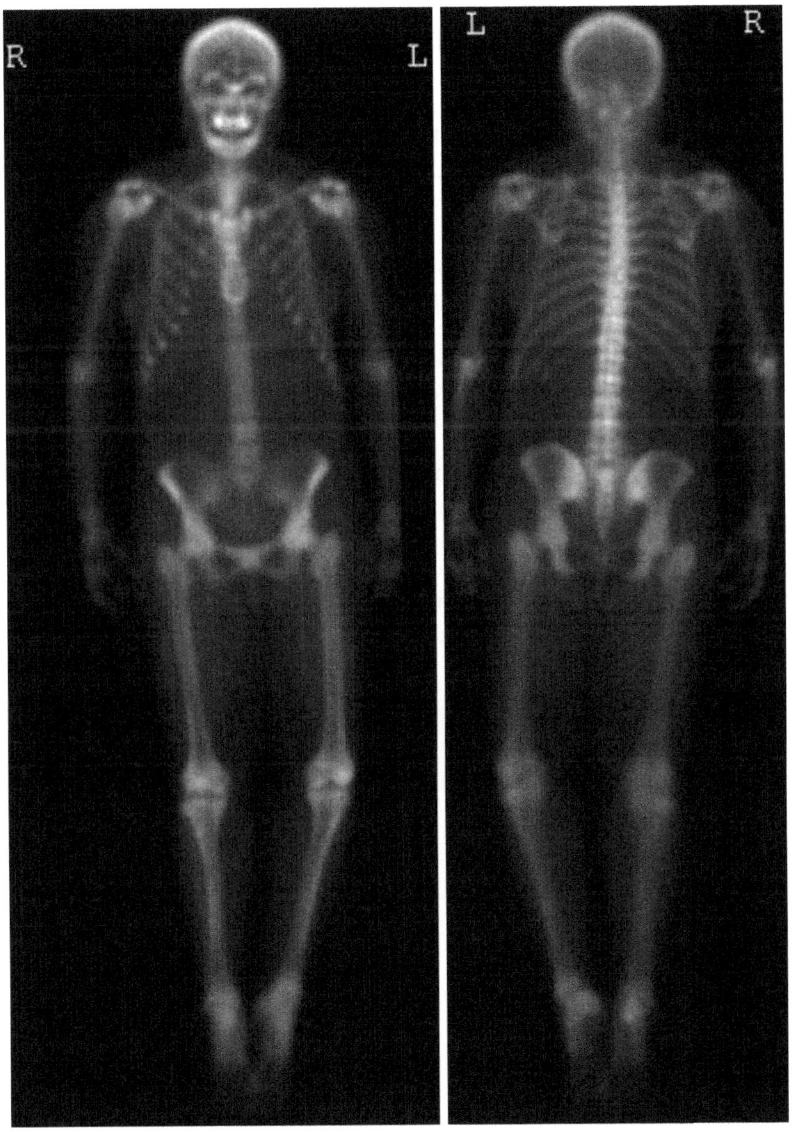

Answer: D

Often associated with chronic kidney disease and dialysis, the changes in the bone are related to altered blood levels of calcium and phosphorus. Absorption of calcium from food is impaired and the body responds by removing it from the bones. Monitoring bone density is important in these patients as osteoporotic fractures are more common.

The bone scan in renal osteodystrophy is sometimes described as the "superscan" of metabolic disease or a "beautiful bone scan." A similar appearance can be seen in hyperparathyroidism. It differs from the "superscan" of metastatic disease in that the distal extremities are less commonly involved with tumor spread.

Common findings on the "superscan" of metabolic disease include "beading" of the costochondral junctions, a high bone to soft tissue ratio of uptake, and occasionally greater uptake in the sternum referred to as a "necktie" sternum.

This patient had no prior history of trauma to the chest and did not undergo dialysis between injection and imaging. The focal uptake in the midline of the posterior skull is the torcular herophili which is the location of the confluence of cranial venous sinuses.

30. This 46-year-old female was diagnosed with breast cancer approximately 1 year ago with a large left breast infiltrating ductal carcinoma with suspicious lymph nodes on MRI. She underwent neoadjuvant chemotherapy and was scheduled for definitive surgery. Prior to her surgical procedure, she began to complain about pain in her chest under her right breast. This bone scan was performed to exclude the presence of metastatic disease. Selected images from the CT study performed on the same day are also provided. You review the images and mention in your report:

    A. The bone scan is normal with the exception of the uptake in the left ankle, consistent with known prior trauma.
    B. An MRI of the ribs should be performed
    C. Correlation with a MRI study of the head is required for a complete evaluation
    D. The changes seen on imaging are due to myositis ossificans

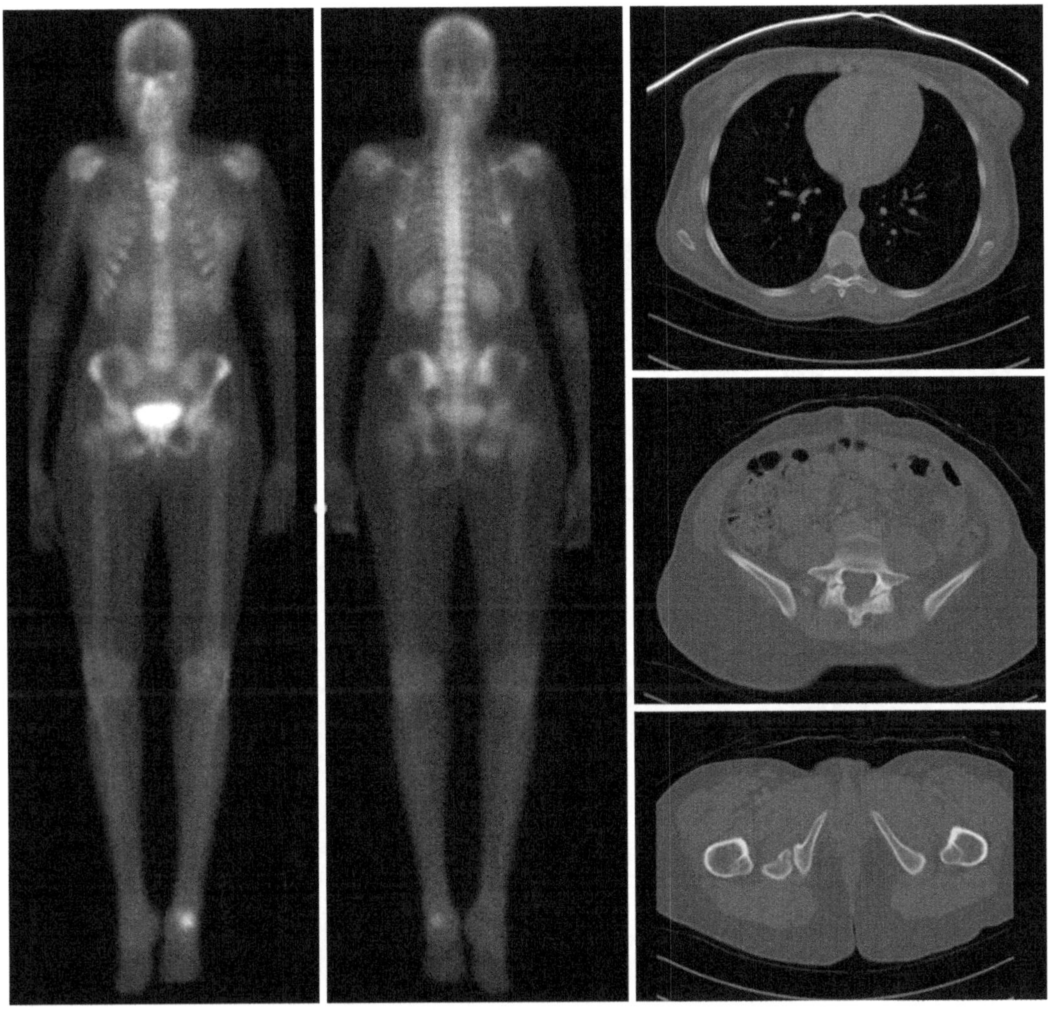

Answer: D

The bone scan demonstrates asymmetric uptake in the right ischium. On the CT study, this can be seen as calcification within the adjacent muscle. The patient had prior trauma to this area which is a common cause of myositis ossificans. Early in the development, increased uptake can be seen in all three phases of a bone scan. Later in the process, it is seen mainly on the delayed images. In this case, confirmation that this is not a metastatic focus is important in management.

The left ankle was injured at the same time but is not the only finding on the bone scan. No abnormality can be seen in the ribs. Bilateral spondylolysis is present on the CT study at L5.

31. You are presented with this bone scan on an elderly male patient who is complaining of right hip pain. The history mentions bilateral hip replacements. After reviewing the study you:
    A. Recommend obtaining a PSA level
    B. Recommend a CT study of the abdomen
    C. Request additional information on kidneys and renal function
    D. Explain that findings are most consistent with loosening of the prosthesis

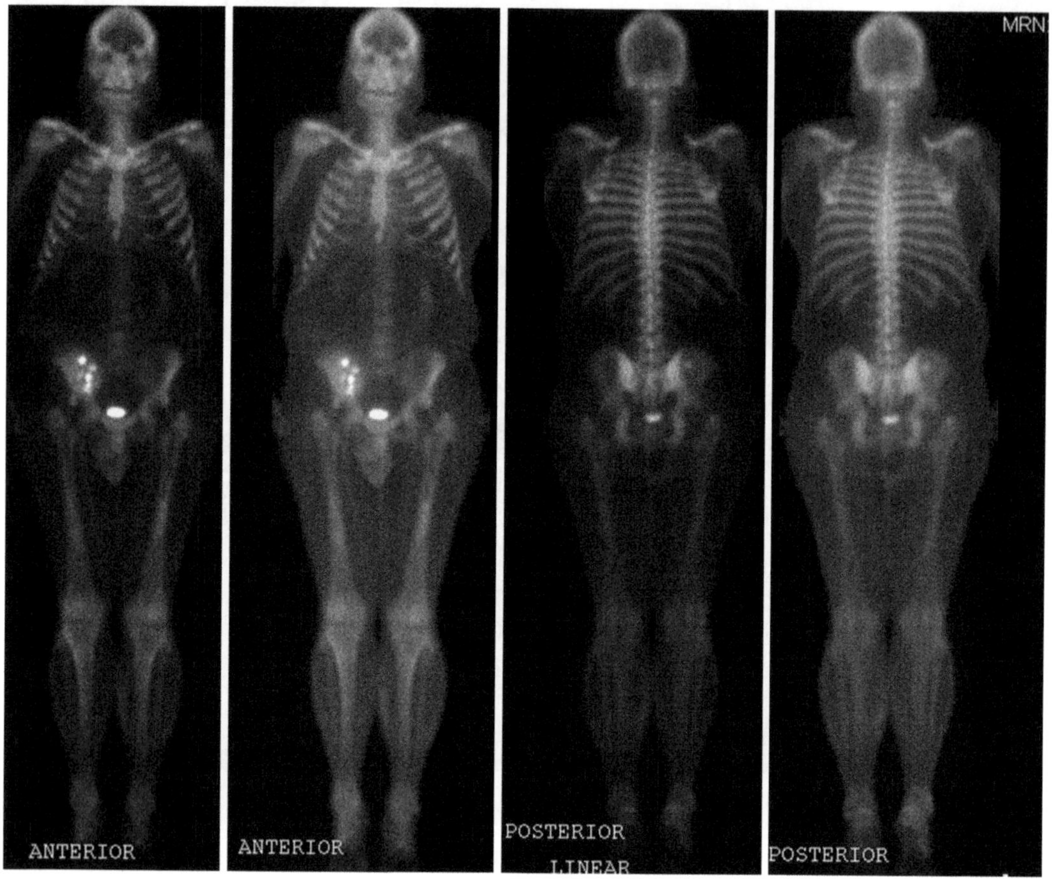

Answer: C

The multiple areas of uptake seen in the right iliac region are seen best on the anterior view and barely seen on the posterior view. This makes it unlikely to represent a bony abnormality. There is no uptake noted in the native kidneys. The uptake in the right iliac fossa is from a renal transplant that was conveniently omitted from the history. Additional information on renal function and review of the history would confirm this finding.

The study illustrates the importance of good history in clarifying the findings on the scan.

Although there are bilateral femoral prostheses present, the findings do not suggest loosening. The patient has no history of cancer and it would be highly unusual for the multiple sites in the right iliac wing to represent metastatic disease without other sites of involvement.

Case 6

32. A diabetic 60-year-old male smoker with a 1 year prior history of distal tibial fracture that was openly reduced and internally fixated. He now presents with swelling and erythema in his leg. There is no drainage, but the swelling improves with elevation. There is no history of fevers or chills, but there is mild tenderness. A bone scan is performed for evaluation.
    The bone scan:

    A. Demonstrates that uptake is only due to postsurgical changes
    B. Demonstrates loosening of the plate and screws
    C. Requires additional imaging to confirm infection
    D. Shows findings consistent with a new fracture

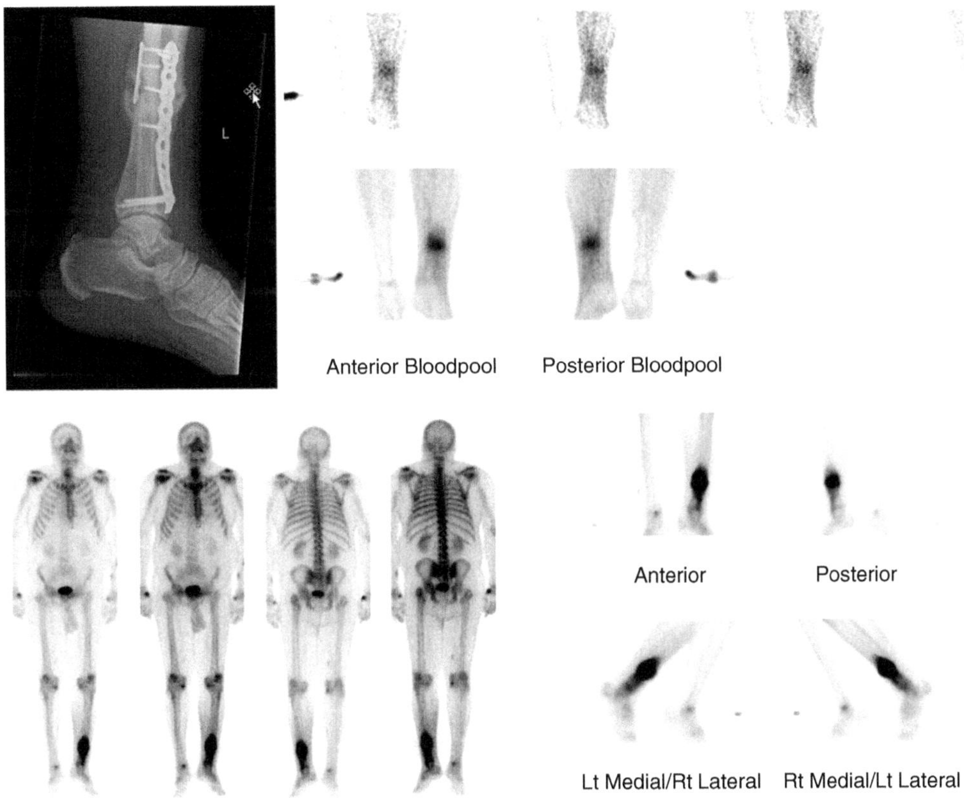

Answer: C

With a prior history of fracture in this weight bearing area, increased uptake would be expected to be present for years. The uptake seen on all three phases however is most suspicious for infection rather than just postsurgical changes. Despite the expected uptake, the bone scan is useful when paired with a Ga67 study which can aid in the determination of infection.

The intense uptake seen in the area would not allow for the determination of new fracture or screw loosening.

33. Subsequent to the bone scan, a Ga67 scan was performed. This demonstrates:

A. Uptake at the fracture site consistent only with prior surgical intervention
B. Shine through activity from the prior technetium study
C. An infectious process is ongoing at the site of the prior fracture
D. There is no osteomyelitis present, only cellulitis

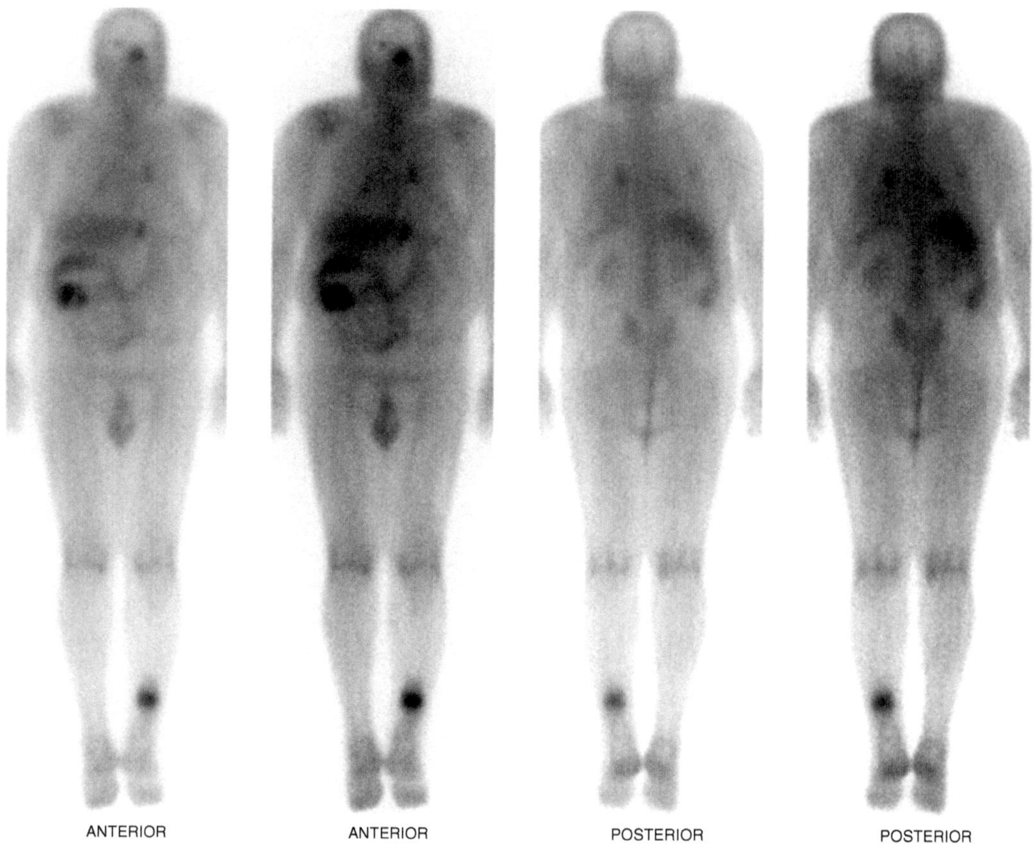

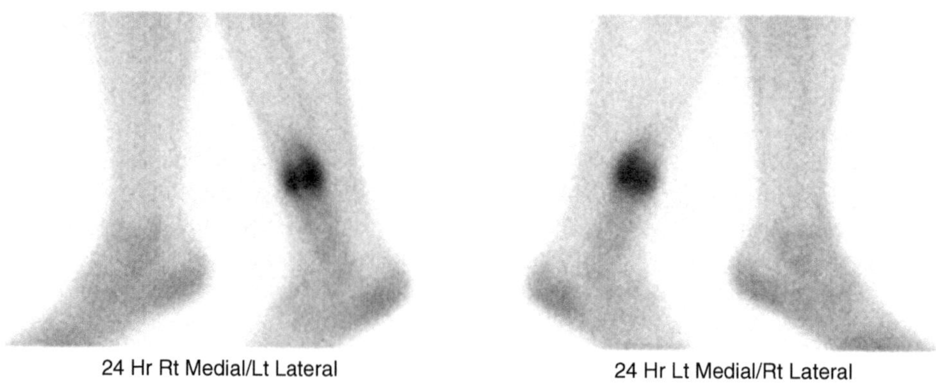

Answer: C

On a gallium scan, a general rule is that uptake at any location that is as great as that of the liver is usually pathologic. The uptake at the fracture is more focal than that of the bone scan and indicates the location of the infection. Note that there is little uptake on the gallium scan from the screws or plate.

Since a gallium scan is performed 24 h (and occasionally at 48 and 72 h) postinjection, there is little if any technetium remaining to cause shine through. Cellulitis would demonstrate a greater degree of soft tissue involvement.

Case 7

34. A 60-year-old female with a history of breast cancer (invasive lobular carcinoma with metastatic axillary lymph nodes), remote chronic compression fracture of T11 presented with acute onset of mid back pain. Radiographs of the lumbar and thoracic spine showed stable T11 compression fracture. DEXA demonstrated the lowest T-score of −1.5. Pain was somewhat improved with pain medications and muscle relaxant, but however never resolved completely. Due to persistent unexplained symptomatology, a whole-body bone scan was performed.

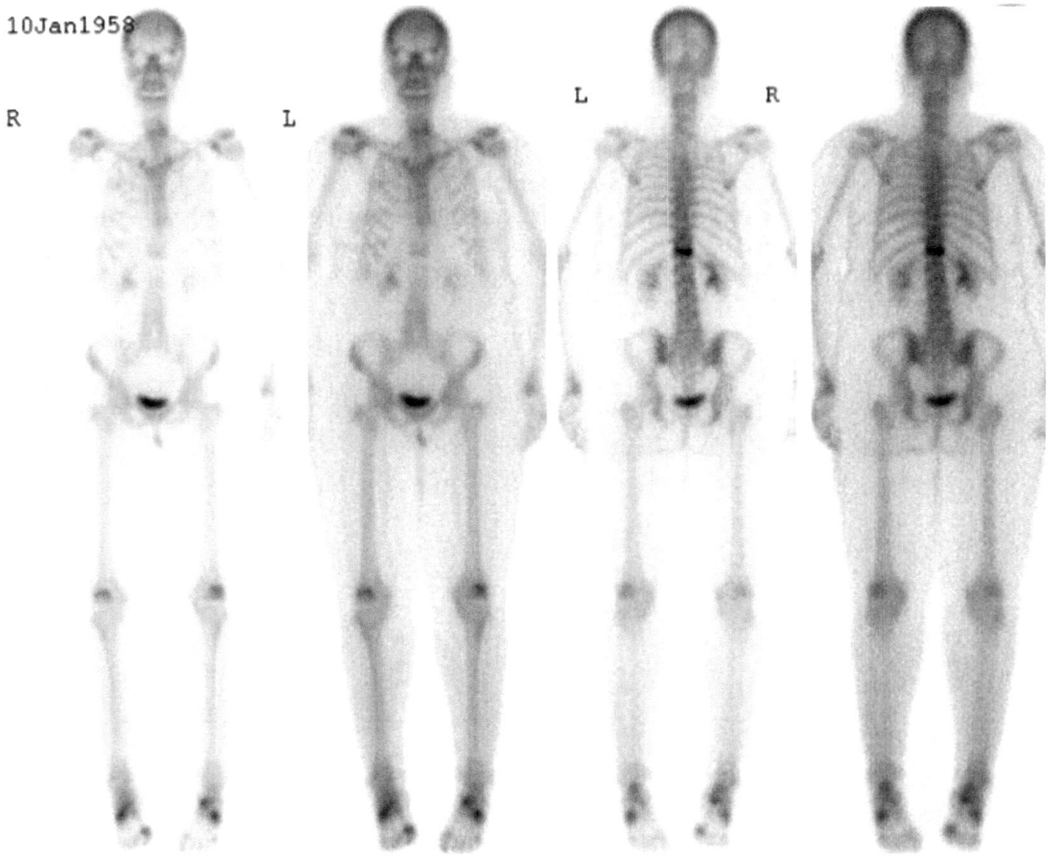

Image 1

Upon evaluation of the images, you conclude that:

A. There is a compression fracture in the lower thoracic spine likely secondary to patient's osteoporosis.
B. Linear uptake in the mid-body is likely secondary to artifact, and you would ask a technologist to repeat the images after removing the overlying object.
C. The uptake in the lower thoracic spine is suspicious and you would recommend further evaluation with additional diagnostic imaging.
D. Should recommend CT of the pelvis to evaluate incidental finding of tracer uptake in pubic bones.

Answer: C

Whole-body bone scan demonstrates a linear band of increased uptake in the lower thoracic spine vertebral body, consistent with a compression fracture. Bone scan is a very useful modality for the evaluation of osseous pathology (with high sensitivity of 95–100%). In the setting of acute fractures the bone scan is usually abnormal in 80% of patients at 24 h, 95% at 72 h, and 98% at 1 week (longer in elderly). Bone scan also has some limitations: false-negative scans can occur with osteoporosis, as high bone turnover in the osteoporotic site might increase background noise and "obscure" the lesion.

The patient has history of chronic T11 compression fracture. Chronic compression fractures usually demonstrate a gradual decrease in activity when healing occurs, and would not usually present as "hot" uptake on bone scan in a patient with acute symptomatology. Given the acuity of the pain and marked uptake on bone scan in the lower thoracic spine, with a known history of breast cancer a pathological fracture should be suspected. Further evaluation of this area with SPECT/CT would be beneficial for anatomic localization of the lesion and detection of bone destruction. MRI would add valuable information in characterizing a lesion.

Decreased bone mineral density predisposes to compression fractures. Osteoporosis is defined as a BMD 2.5 standard deviations or more below average value for young healthy women (T-score > -2.5 SD). Osteopenia is defined as a BMD 1.0–2.5 standard deviations below average value for young healthy women ($-1.0 <$ T-score < -2.5 SD). This patient has osteopenia (T-score of -1.5) rather than osteoporosis.

Tracer uptake in the pelvis is within the bladder (not within pubic bones), which is commonly seen secondary to normal tracer excretion through the urinary system.

35. In this same patient, MRI of the thoracic spine was performed to further evaluate the bone scan findings, as well as F-18 FDG PET/CT study.

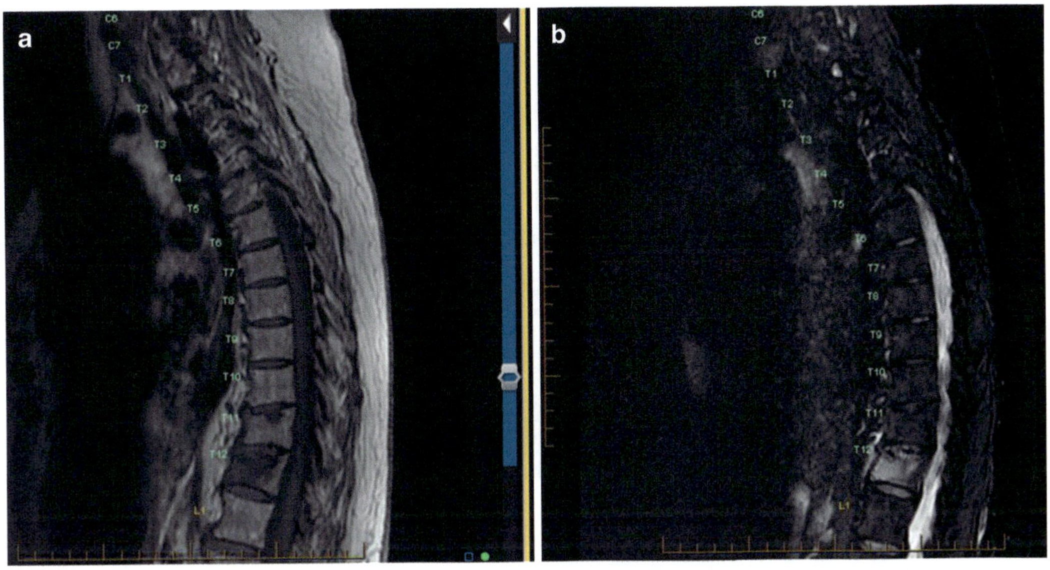

Image 1 **a** and **b**

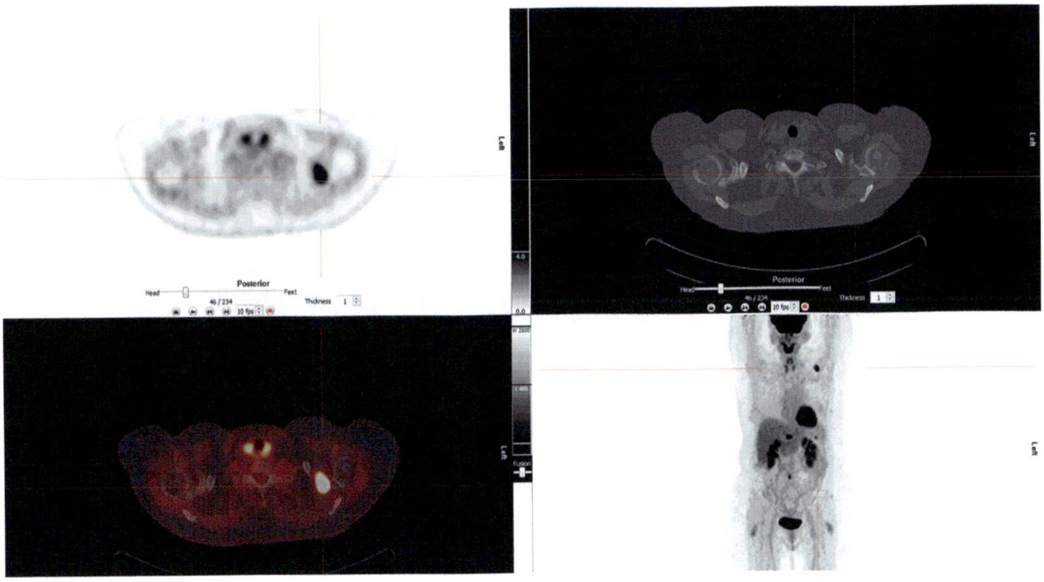

Image 2

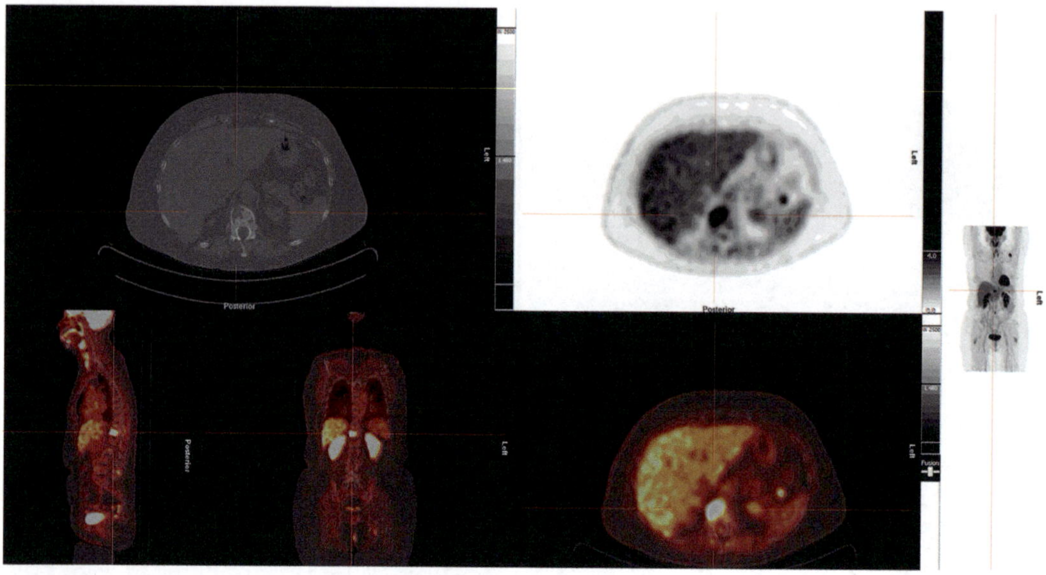

Image 3

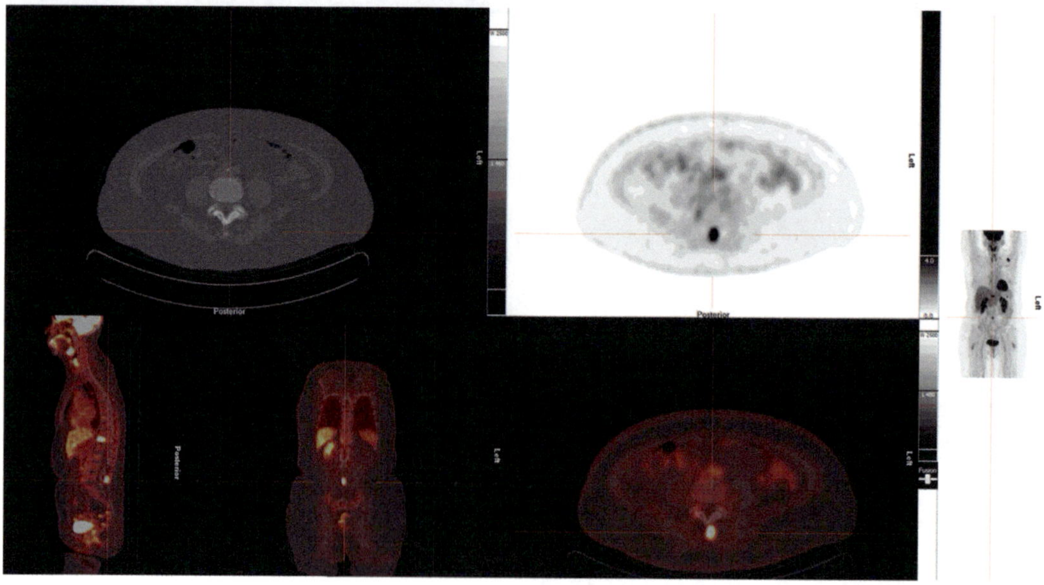

Image 4

Upon evaluation of these additional studies, you conclude that:

A. Compression fracture of T12 is likely secondary to osteoporosis.
B. Uptake in scapula and spinous process of L4 seen on PET/CT images are likely nonspecific because they were not seen on whole-body bone scan.
C. The patient might benefit from palliative radiation therapy.
D. This patient has a thyroid cancer

Answer: C

MRI frequently adds valuable information when characterizing skeletal lesions and their surrounding soft tissues in musculoskeletal disease. In this patient, MRI image 1A (*T1* weighted *image*) demonstrates focal T1 hypointense signal within the T12 vertebral body, compatible with a metastasis. On *T2* weighted *image (image* 1B*), there is a s*uperimposed fracture with associated compression deformity of the T12 vertebral body with T2/STIR hyperintense signal as well and less than 25% loss of anterior height loss and overall appearance of a pathologic compression fracture.

F-18 FDG PET/CT images (Images 2–4) demonstrating hypermetabolic lytic lesions involving the left scapula, T12 vertebral body, and spinous process of L4, with the appearance of metastatic disease. This was found to be secondary to metastatic breast carcinoma. She might benefit from palliative radiation therapy for symptomatic relief of skeletal pain.

Bone scan is sensitive in detecting predominantly osteoblastic sites that are easily seen as areas of increased activity. Lesions that are mostly osteolytic are more difficult to detect because they will appear cold or isointense (in this patient bone scan failed to detect lytic lesions in the left scapula and spinous process of L4). Increased uptake in poorly lytic lesion can be seen if it is associated with aggressive periosteal change (as in this patient with pathological fracture of T12 vertebral body). F-18 FDG PET often can detect involvement from cancers in the marrow or in lesions that are more aggressive thus complementing bone scan in the detection of lytic lesions.

Decreased bone mineral density does predispose to compression fractures. Osteoporosis is defined as BMD 2.5 standard deviations or more below average value for young healthy women (T-score < −2.5 SD). Osteopenia is defined as BMD 1.0–2.5 standard deviations below average value for young healthy women (−1.0 < −T-score < 2.5 SD). As mentioned in the previous question, this patient has osteopenia (T-score of −1.5), rather than osteoporosis.

36. You noticed increased metabolic activity in the thyroid bed on this F-18 FDG PET/CT scan. What is the most likely diagnosis for incidental findings?

A. Subclinical thyroiditis
B. High-grade follicular thyroid carcinoma crossing the midline
C. Metastatic breast carcinoma to the thyroid
D. Metastatic breast carcinoma to bilateral station VI nodes

Answer: A

There is a diffuse bilateral tracer uptake in the thyroid gland. This happens in about 3% of patients imaged with FDG PET and then in about 45% there is a known diagnosis of autoimmune thyroiditis or hypothyroidism. Thyroid carcinoma or metastatic disease usually will demonstrate focal FDG accumulation, rather than diffuse and symmetric.

*Ziessman HA, O'Malley JP, Thrall JH, Nuclear medicine: the requisites.
*Karantanis D, Bogsrud TV, Wiseman GA, Mullan BP, Subramaniam RM, Nathan MA, Peller PJ, Bahn RS, Lowe VJ. Clinical significance of diffusely increased 18F-FDG uptake in the thyroid gland. J Nucl Med. 2007;48(6):896–901.
*StatDx. https://my.statdx.com/.

Case 8

37. Ewing sarcoma most commonly affects which type of bone?

   A. Tubular bones
   B. Flat bones
   C. Irregular bones
   D. Short bones

   Answer: A
   The rare Ewing sarcoma tumors usually present as neoplastic skeletal tumors and less commonly as soft tissue neoplasms. These tumors belong to the now better-defined spectrum of bone tumors known as Ewing sarcoma family of tumors (EFT). Its most common radiographic presentation is of a large destructive lesion in the large tubular bones of the lower extremities [1].

38. When evaluating the imaging workup of a patient with Ewing sarcoma, which of the following findings is considered a poor prognostic factor?

   A. Whole-body ^{99m}Tc-MDP with diffusely increased uptake in the epiphyses.
   B. Whole-body ^{99m}Tc-MDP with two foci of uptake in the left proximal femur.
   C. ^{18}F-FDG PET/CT with two hypermetabolic lesions involving the left proximal humerus and the left lung apex.
   D. ^{18}F-FDG PET/CT with two foci of increased metabolic activity in the left distal femur and mid tibia.

   Answer: C
   The presence or absence of metastatic disease at the time of presentation is the most important indicator of prognosis. Approximately 25% of patients demonstrate metastasis at the time of diagnosis. Another important predictor of outcome is the location of the tumor. Patients with tumors involving the distal extremities have a better prognosis than those with tumors involving the upper extremities or central or pelvic tumors. Patients may present with osseous metastasis at the time of diagnosis. Soft tissue metastasis is most commonly reported in the lungs [1, 2].

39. Which imaging modality better delineates intramedullary and soft tissue invasion in cases of Ewing sarcoma?

   A. PET/CT
   B. Bone scan
   C. Contrast-enhanced computed tomography (CECT)
   D. MRI

   Answer: D
   ^{18}F-FDG PET/CT has demonstrated to be superior to traditional bone scintigraphy for the initial staging and restaging of patients diagnosed with Ewing sarcoma. MRI is significantly superior for the evaluation of intramedullary extension and soft tissue invasion [1].

40. The population most commonly affected by Ewing sarcoma include:

A. Females between10 and 15 years of age
B. Males between 10 and 15 years of age
C. Females between 20 and 25 years of age
D. Males between 20 and 25 years of age

Answer: B
Ewing sarcoma most commonly affects patients in the first three decades of life with a peak prevalence between the ages of 10 and 15 years of age and a modest predilection for males [1].

41. The clinical presentation of Ewing sarcoma best resembles which of the following diseases:

A. Legg-Calve-Parthe disease
B. Sacral insufficiency fracture
C. Slipped capital femoral epiphysis
D. Osteomyelitis

Answer: D
Patients with Ewing sarcoma usually present with complaints of pain and swelling. Leukocytosis can be present. These combinations of nonspecific findings are commonly present with osteomyelitis [1, 2].

42. Ewing's sarcoma metastasis most commonly occurs to which soft tissue organ?

A. Brain
B. Liver
C. Spleen
D. Lungs

Answer: D
Approximately 25% of patients with a new diagnosis of Ewing sarcoma present with metastatic disease. When the disease spreads to the soft tissues, it demonstrates a predilection for the lungs [1].

43. Which of the following patient groups with Ewing sarcoma has a favorable prognosis?

A. Older age and male gender
B. Older age and female gender
C. Younger age and male gender
D. Younger age and female gender

Answer: C
Ewing sarcoma demonstrates a peak prevalence for patients younger than 20 years of age with a sling prevalence for males. Studies of incidence, prevalence, and outcome have demonstrated a better prognosis for male patients diagnosed at an earlier age [2].

44. Which imaging modality can be used to better determine the extent of metastatic disease?

    A. ^{18}F-FDG PET/CT
    B. Whole-body MDP
    C. CECT
    D. Skeletal Survey

    Answer: A

    ^{18}F-FDG PET/CT has demonstrated to be superior to traditional bone scintigraphy for the initial staging and restaging of patients with Ewing Sarcoma. MRI is significantly superior for the evaluation of intramedullary extension and soft tissue invasion [1, 2].

45. Which imaging modality may be superior to assess response to treatment?

    A. ^{18}F-FDG PET/CT
    B. Whole-body MDP
    C. CECT
    D. Skeletal Survey

    Answer: A

    ^{18}F-FDG PET/CT has demonstrated to be superior to traditional bone scintigraphy for the initial staging and restaging of patients diagnosed with Ewing's Sarcoma. MRI is significantly superior for the evaluation of intramedullary extension and soft tissue invasion [1, 2].

References

1. Mark DM et al. From the radiologic pathology archives: ewing sarcoma family of tumors: radiologic-pathologic correlation. Radiographics. 2013;33.803–31.
2. Jimmi H, et al., Imaging features of primary tumors and metastatic patterns of the extraskeletal ewing sarcoma family of tumors in adults: a 17-year experience at a single institution. Korean J Radiol. 2015;16(4):783–90.
3. Mark DM, et al. From the radiologic pathology archives: ewing sarcoma family of tumors: radiologic-pathologic correlation. Radiographics. 2013;33:803–31.
4. Jimmi H, et al., Imaging features of primary tumors and metastatic patterns of the extraskeletal ewing sarcoma family of tumors in adults: a 17-year experience at a single institution. Korean J Radiol. 2015;16(4):783–90.

46. This is a 56-year-old male with diffuse joint pain referred by rheumatology. What diagnosis is represented by this scan?

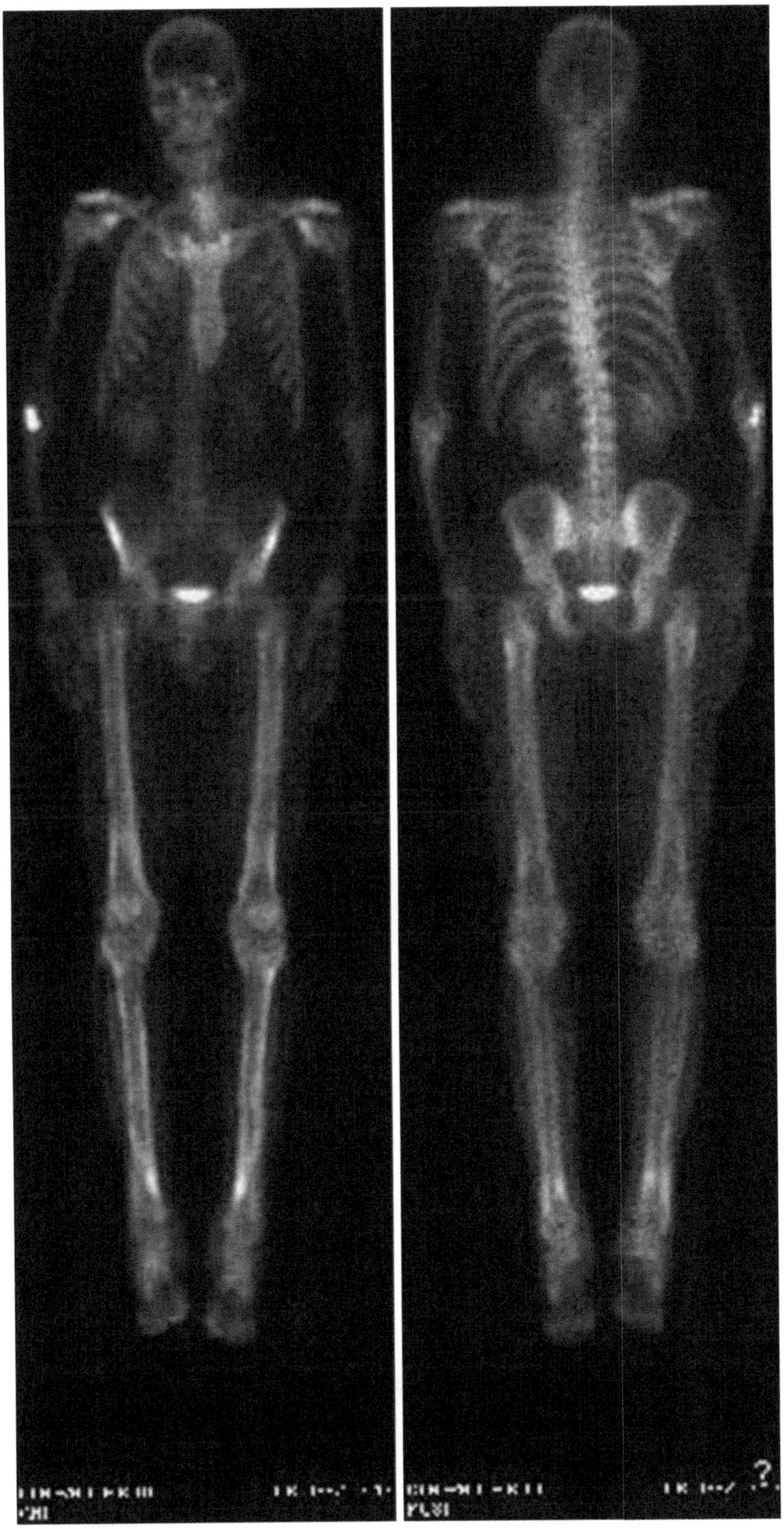

A. Metastatic disease to the bones
B. Hypertrophic osteoarthropathy
C. Rheumatoid arthritis
D. Osteopetrosis

Answer: B

HPO—hypertrophic pulmonary osteoarthropathy.

Osteoarthropathy is often associated with lung cancer but can be seen in other pulmonary disorders. Key features include painful joint swelling, clubbing of the fingers and toes, and periostitis. It is often misdiagnosed as a rheumatologic disorder. One can see a mild increase in activity in the right hemithorax, consistent with hyperemia in the region of the tumor as identified on subsequent chest radiograph.

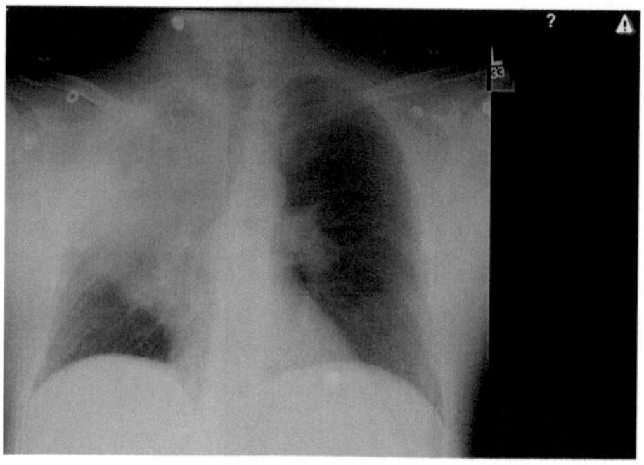

47. Which of the following was the bone-seeking radiopharmaceutical used for the scan?

A. Tc-99 m-Sestamibi
B. Tc-99 m-MDP
C. Tc-99 m-MAA
D. F-18 NaF

Answer: B

There are several bone-seeking radiopharmaceuticals, the most common being Tc-99 m MDP or HEDP. There are others which are useful for palliation therapy of bone metastases. These include Sr-89 chloride, Radium-223 (Xofigo), Samarium 153-EDTMP. In addition, F-18 NaF is a PET tracer used for bone imaging.

48. What is the mechanism of action of the tracer used?

1. Replacement of hydroxyl group in hydroxyapatite
2. Replacement of calcium moiety in hydroxyapatite
3. Replacement of phosphate moiety in hydroxyapatite
4. Accretion onto hydroxyapatite

Answer: D

Hydroxyapatite is the matrix of bone formation. Radiopharmaceuticals used for bone pain palliation, generally replaces the calcium moiety in the crystal and when incorporated into bone, release radiation that affects adjacent nerve endings, NaF replaces the OH moiety of the crystal. The Tc-99 m compounds do not replace any element in the crystal. They are absorbed onto the crystal by a process known as accretion.

Reference

- Cengiz A et al. Hypertrophic pulmonary osteoarthropathy on bone scintigraphy and 18F-fluorodeoxy glucose positron emission tomography/computed tomography in a patient with lung adenocarcinoma. Indian J Nucl Med. 2015;30(3):251–3. https://www.ncbi.nlm.nih.gov/pmc/articles/PMC4479915.

Case 9

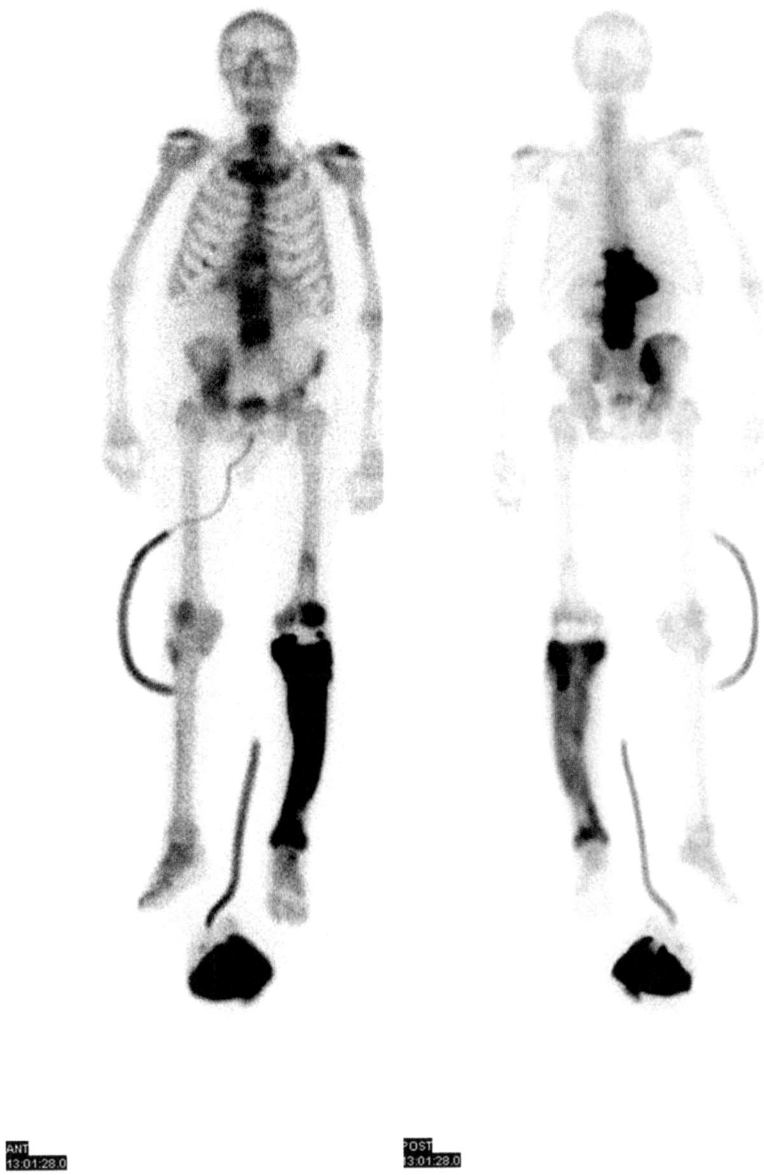

49. What is the most likely diagnosis of this patient?

    A. Multiple fractures
    B. Paget disease
    C. Osteomyelitis
    D. Widespread skeletal metastatic disease

    Answer: B

    The scan demonstrates intense uptake at least in the lower thoracic and upper lumbar spine, right SI joint, and the left tibia. The markedly increased uptake and the regions affected are typical for Paget disease of bone. The disease usually affects the pelvic bone, the skull, the spine, and the long bone. When a long bone is affected, it usually involves one end of the bone only.

50. What is the chance of developing osteogenic sarcoma, given the findings on this scan?

    A. 1%
    B. 5%
    C. 10%
    D. 25%

    Answer: A

    There is less than a 1% chance of developing osteogenic sarcoma in a patient with Paget disease of the bone. They occur mostly in patients with long-standing polyostotic Paget disease and often occur in the seventh decade of life. Paget disease reflects changes in the matrix of bone. Often, it is discovered as part of a workup for elevated alkaline phosphatase.

Reference

- Hanson ME, Seton M, Merchant A. Osteosarcoma in Paget's disease of bone. J Bone Miner Res. 2006;Suppl2:58–63.

Case 10

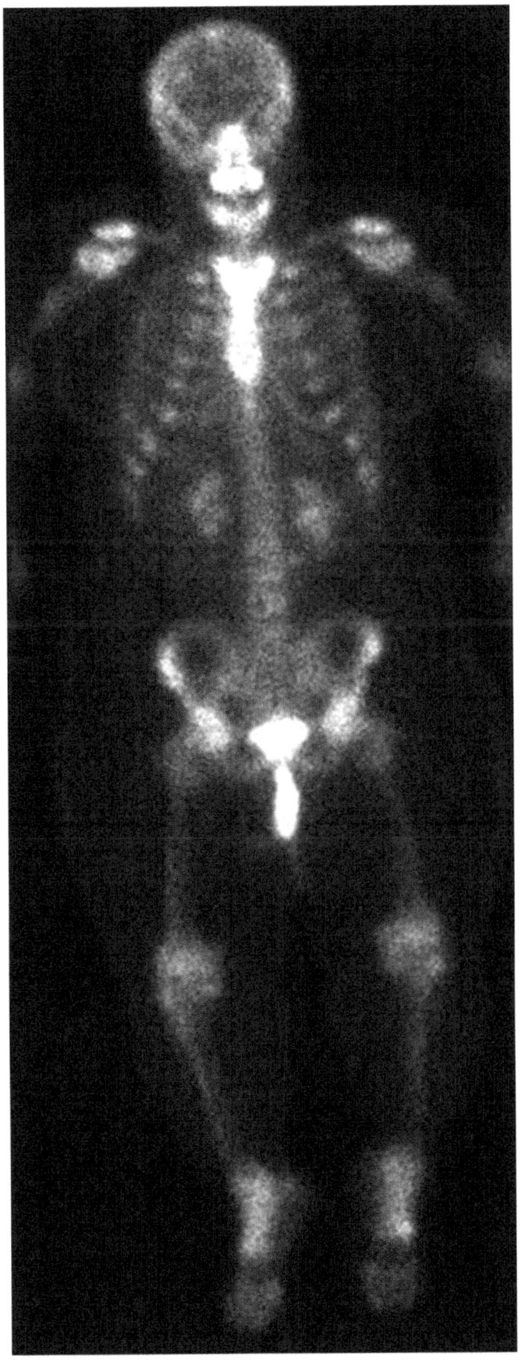

51. What is the key to the diagnosis for this patient?

    A. Costochondral uptake
    B. Upper to lower ratios for each limb
    C. Uptake in the arms
    D. Uptake in the hips

Answer: B

This is the bone scan of a dwarf. The ratios for the tibia to femur and radius to humerus are not normal.

This is a case of an achondroplastic dwarf in whom there is a reduced growth rate of the limbs. A short trunk due to reduced growth rate is seen in spondyloepiphyseal dysplasia. In generalized short stature conditions, both limbs and trunk are shortened.

This can be seen in osteogenesis imperfecta.

Reference

• Chaudhary V, Bano S, Imaging in short stature. Indian J Endocrin Metab. 2012;16(5):692–7.

52. What is the cause of the uptake in the upper arms?

    A. Fractures
    B. Neurofibromas
    C. Brown Tumors
    D. Increased stresses on deltoid insertions.

Answer: D

Increased stress on deltoid insertions can be seen in those doing manual labor for a living or those with abnormal stresses on the upper arms. It can be unilateral or bilateral. Increased stress on muscle insertions can also be seen in long-distance runners and is manifest by shin or thigh splinting on bone scan.

53. What is the most likely diagnosis for the bone scan on an elderly woman?

A. Osteoporosis Circumscripta
B. Hyperparathyroidism
C. Super scan
D. Hyperostosis totalis

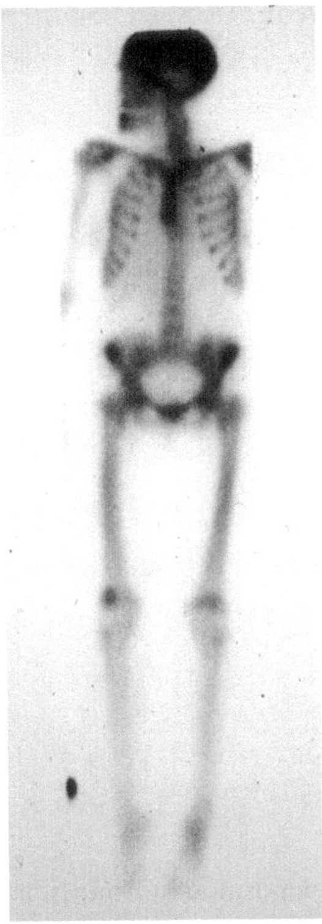

Answer: B

There are several clues that help with the diagnosis. Two subtle clues are the costochondral uptake and patellar uptake, often seen in cases of metabolic bone disease. Osteitis pubis (increased activity in the pubic symphysis) is not uncommon in elderly people with hyperparathyroidism. The intense uptake in the skull and lack of renal activity are key findings. X-ray findings include cortical resorption on the clavicle and phalanges.

54. What is the cause of the abnormality on this bone scan?

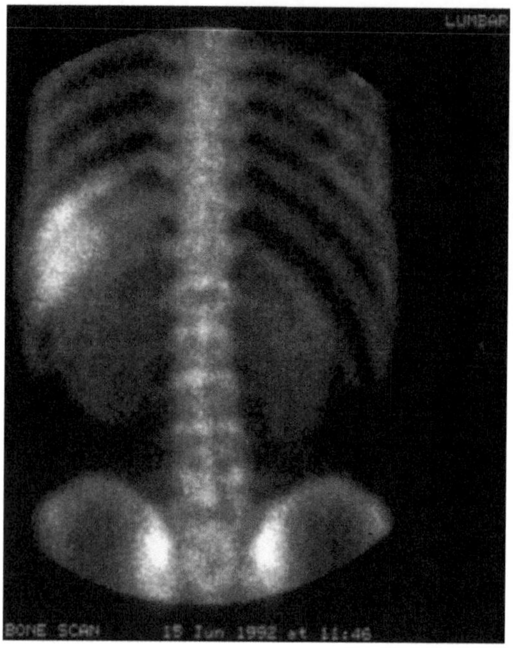

A. Free pertechnetate
B. Rib fracture
C. Increased alumina level
D. Splenic uptake caused by underlying sickle cell disease

Answer: D

In sickle cell disease, there is red blood cell destruction leaving deposition of iron within the spleen. Tc-99 m MDP binds to areas of iron deposition. In this disease, one would not expect uptake on a liver spleen scan as there is functional asplenia. Other disorders such as multiple myeloma may also result in splenic uptake on a bone scan.

DEXA 2020

Case 11

55. An 85-year-old female with no significant past medical history underwent a DXA scan for evaluation of bone mineral density. Based on WHO criteria her bone mineral density was classified as osteopenia. View acquired for femoral BMD show:

DualFemur Bone Density

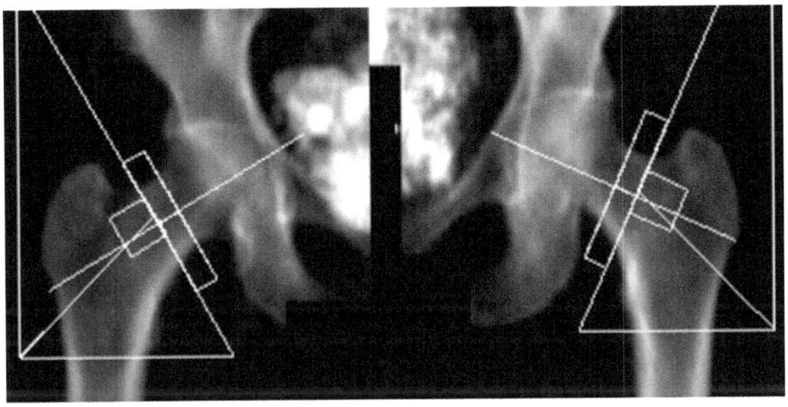

A. Calcified uterine fibroids
B. Recent contrast-enhanced CT scan
C. Excessive calcium intake
D. Urinary bladder stone

Correct answer: A. Calcified uterine fibroids.

DXA scan reveals calcified densities within the pelvis. This is confirmed to be calcified uterine fibroids on ultrasound correlation.

Unless CT scan with contrast was acquired just before DEXA scan, such an image would not be possible. DXA is a non-emergent procedure and is usually not performed on the same day as a contrast-enhanced CT scan.

Calcium tablets are sometimes visible within the bowel on a DEXA scan as small rounded densities, not a mass.

Urinary bladder stone is usually small. Such a large stone would result in renal failure, which was not mentioned in the past medical history.

56. Based on the image below, this patient has:

    A. Normal Bone density
    B. Osteopenia
    C. Osteoporosis
    D. No enough information

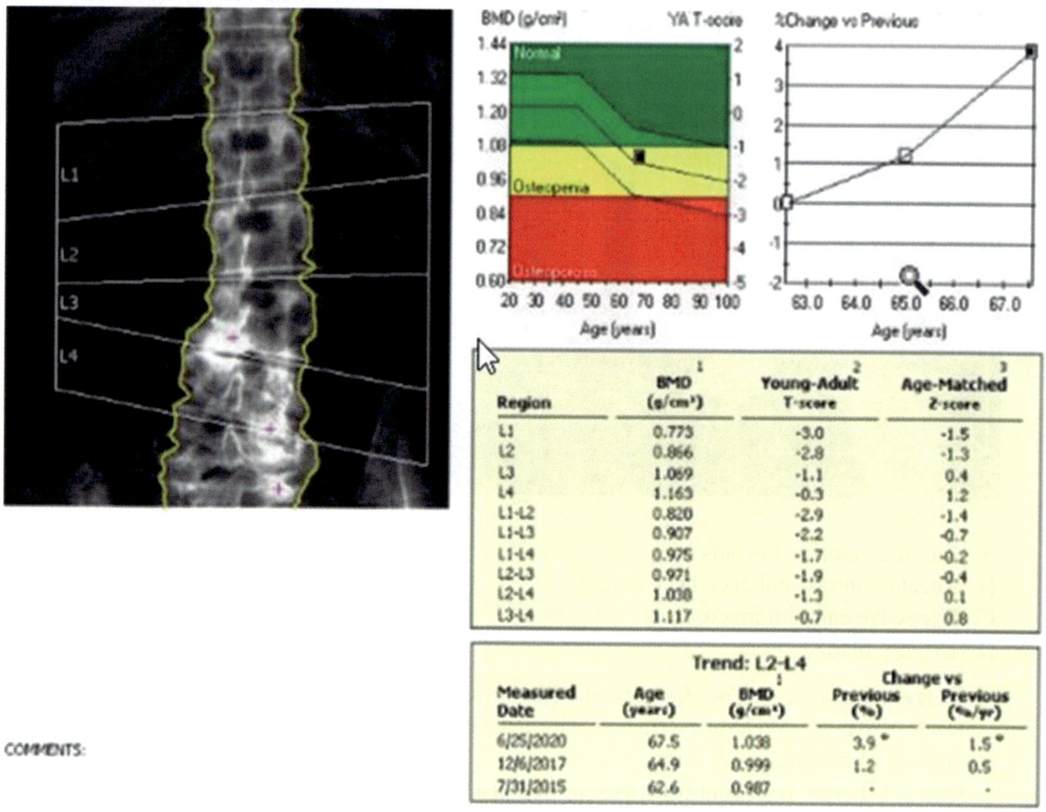

Region	BMD[1] (g/cm²)	Young-Adult[2] T-score	Age-Matched[3] Z-score
L1	0.773	-3.0	-1.5
L2	0.866	-2.8	-1.3
L3	1.069	-1.1	0.4
L4	1.163	-0.3	1.2
L1-L2	0.820	-2.9	-1.4
L1-L3	0.907	-2.2	-0.7
L1-L4	0.975	-1.7	-0.2
L2-L3	0.971	-1.9	-0.4
L2-L4	1.038	-1.3	0.1
L3-L4	1.117	-0.7	0.8

Trend: L2-L4

Measured Date	Age (years)	BMD[1] (g/cm²)	Change vs Previous (%)	Change vs Previous (%/yr)
6/25/2020	67.5	1.038	3.9 *	1.5 *
12/6/2017	64.9	0.999	1.2	0.5
7/31/2015	62.6	0.987	.	.

According to the WHO, the patient has osteoporosis because two consecutive vertebrae have a T-score of equal to or less than –2.5.

57. Patients scheduled for a DEXA study should:

    A. Wait to have it performed until 72 h after a barium study
    B. Refrain from drinking milk before a study
    C. Take their calcium or vitamin pills as usual
    D. Avoid scheduling a nuclear medicine isotope study afterwards

Answer: A

Calcium and vitamin pills may artificially increase the BMD of a vertebral body if it overlaps or is included in the region of interest. Generally, patients are asked not to take calcium or vitamin pills for 24–48 h before a DEXA study. Although there is evidence that it only minimally impacts the BMD for the L1–L4 region, it may significantly alter the BMD for a single vertebral body [1].

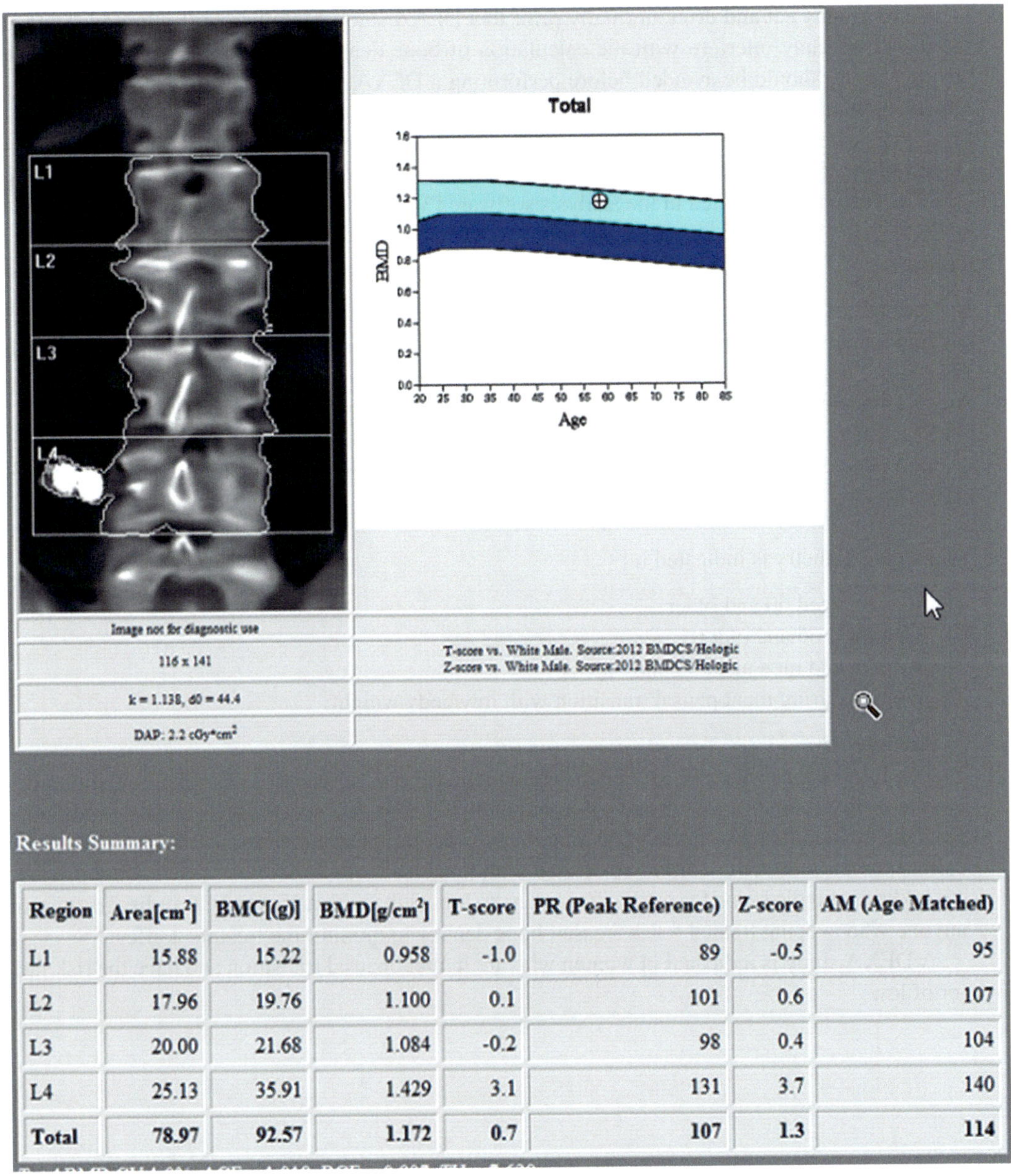

Results Summary:

Region	Area[cm²]	BMC[(g)]	BMD[g/cm²]	T-score	PR (Peak Reference)	Z-score	AM (Age Matched)
L1	15.88	15.22	0.958	-1.0	89	-0.5	95
L2	17.96	19.76	1.100	0.1	101	0.6	107
L3	20.00	21.68	1.084	-0.2	98	0.4	104
L4	25.13	35.91	1.429	3.1	131	3.7	140
Total	**78.97**	**92.57**	**1.172**	**0.7**	**107**	**1.3**	**114**

Patients may eat and drink normally prior to a DEXA study, including milk. Barium retained in the bowel may interfere with the calculation of bone density. Likewise, a nuclear medicine isotope study should be avoided before performing a DEXA study as the emitted radiation may interfere with the density determination [2].

Navel jewelry may significantly impact the calculation of bone density when it overlies a vertebral body. If such jewelry cannot be removed, the area of the jewelry should be excluded using the DEXA toolset provided in the analysis software [3].

References

1. Kendler DL, Kiebzak GM, Ambrose CG, Dinu C, Robertson S, Schmeer P, Van Pelt JL. Effect of calcium tablets on interpretation of lumbar spine DXA scans. J Clin Densitom. 2006;9(1): 97–104.
2. https://clinicalcenter.nih.gov/drd/nucmed/patientinfo.html.
3. Ott SM, Ichikawa LE, LaCroix AZ, Scholes D. Navel jewelry artifacts and intravertebral variation in spine bone densitometry in adolescents and young women. J Clin Densitom. 2009;12(1):84–8. https://doi.org/10.1016/j.jocd.2008.09.002.

58. Bone densitometry is indicated in:

A. Women aged 60 and older
B. Men aged 65 and older
C. Women and men aged 50 and older.
D. Women during menopausal transition with low body weight.

Answer: D

The International Society for Clinical Densitometry official positions for 2015 in adults recommends that women aged 65 and older and men 70 and older receive BMD testing. Additional risks that may indicate earlier testing are low body weight, prior fractures, use of high-risk medications, or a disease or condition associated with bone loss.

Additional indications for testing bone density are adults with a history of fragility fracture or anyone who is being treated for decreased bone density to monitor the treatment effect.

A DEXA study is indicated in women who are in menopausal transition and have the risk factor of low

Case 12

59. A 71-year-old female underwent bone densitometry. Images are shown below.

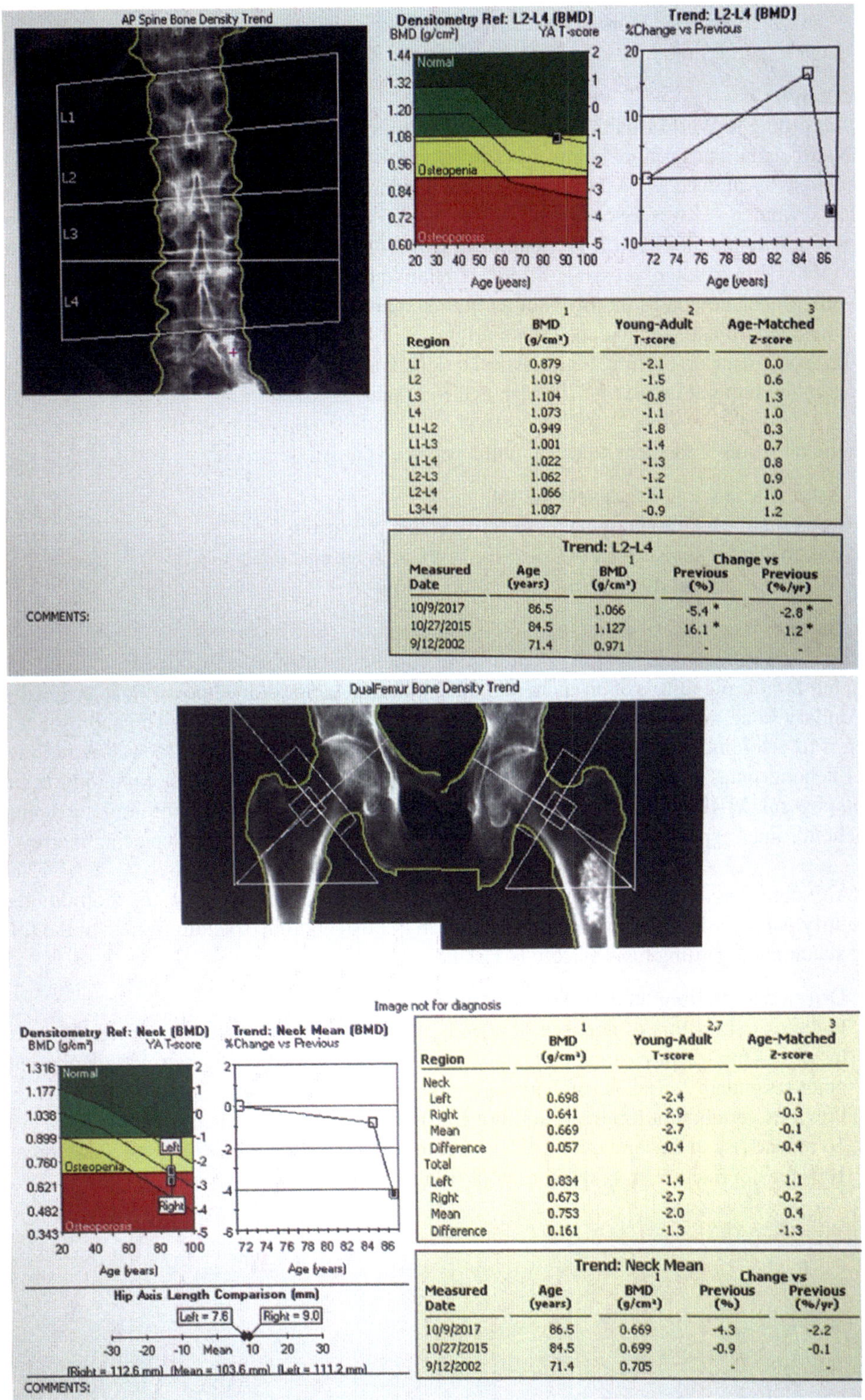

Region	BMD[1] (g/cm²)	Young-Adult[2] T-score	Age-Matched[3] Z-score
L1	0.879	-2.1	0.0
L2	1.019	-1.5	0.6
L3	1.104	-0.8	1.3
L4	1.073	-1.1	1.0
L1-L2	0.949	-1.8	0.3
L1-L3	1.001	-1.4	0.7
L1-L4	1.022	-1.3	0.8
L2-L3	1.062	-1.2	0.9
L2-L4	1.066	-1.1	1.0
L3-L4	1.087	-0.9	1.2

Trend: L2-L4

Measured Date	Age (years)	BMD[1] (g/cm²)	Change vs Previous (%)	Change vs Previous (%/yr)
10/9/2017	86.5	1.066	-5.4 *	-2.8 *
10/27/2015	84.5	1.127	16.1 *	1.2 *
9/12/2002	71.4	0.971	-	-

Region	BMD[1] (g/cm²)	Young-Adult[2,7] T-score	Age-Matched[3] Z-score
Neck			
Left	0.698	-2.4	0.1
Right	0.641	-2.9	-0.3
Mean	0.669	-2.7	-0.1
Difference	0.057	-0.4	-0.4
Total			
Left	0.834	-1.4	1.1
Right	0.673	-2.7	-0.2
Mean	0.753	-2.0	0.4
Difference	0.161	-1.3	-1.3

Trend: Neck Mean

Measured Date	Age (years)	BMD[1] (g/cm²)	Change vs Previous (%)	Change vs Previous (%/yr)
10/9/2017	86.5	0.669	-4.3	-2.2
10/27/2015	84.5	0.699	-0.9	-0.1
9/12/2002	71.4	0.705	-	-

Based on the data above, this patient has the following result in the femoral necks:

A. Normal bone densitometry in the femoral necks
B. Abnormal bone densitometry in the femoral necks consistent with osteopenia
C. Abnormal bone densitometry in the femoral necks consistent with osteoporosis
D. Low risk of femoral neck fracture, based on the Z-score

Answer: C

According to World Health Organization (WHO) criteria, osteoporosis is defined as BMD 2.5 standard deviations or more below average value for young healthy women (T-score ≤ –2.5 SD). Osteopenia is defined as BMD > 1.0 but < 2.5 standard deviations below average value for young healthy women (–1.0 > T-score > –2.5 SD). The T-score is calculated by taking the difference between patient's measured BMD and mean BMD of healthy young adults, matched for gender and ethnic group, and expressing difference relative to young adult population. The Z-score, on the other hand, compares the BMN of an individual to the mean BMD expected for healthy subject matched for age, gender, and ethnic origin.

In this particular patient, femoral necks densitometry is consistent with osteoporosis. The mean in the femoral neck is –2.7, which is lower than –2.5 SD.

60. After evaluation of the densitometry in this patient, you will:

A. Ask the technologist to repeat the study
B. Recommend additional imaging to the referring physician
C. Ask the technologist to fix one or more of the regions of interest
D. Assure the patient that her risk for fracture is low

Answer: B

This densitometry incidentally demonstrated a heterogeneously sclerotic lesion in the proximal left femur, suggestive of an enchondroma—a benign tumor of hyaline cartilage originating in medullary bone. Metaphyseal or metadiaphyseal location is very typical. The appropriate next step is to recommend further evaluation with a radiograph of the pelvis and lower extremities.

Enchondromas usually present as a geographic central lesion. The most common location is metaphyseal. **MRI** is useful in evaluating soft tissue extension and for confirming the diagnosis. Enchondromas appear as well-circumscribed somewhat lobulated masses replacing marrow.

61. FRAX score is assigned based on clinical risk factors, bone mineral density, and mortality data to quantify patient's 10-year probability of hip or major osteoporotic fracture. Which of the following statement regarding FRAX score is TRUE:

A. Driven mainly from the BMD of the lumbar spine
B. Includes risk factors of age, sex, low body mass index, prior fracture after age 50, parental history of hip fracture, smoking history, past or current use of corticosteroids, alcohol intake, other secondary causes of osteoporosis, and rheumatoid arthritis
C. Previous vertebral or hip fracture is not considered a predictor of fracture risk
D. To reduce risk of fractures, recommend treating patients with FRAX 10-year risk scores of ≥ 10% for hip fracture or ≥30% for major osteoporotic fracture

Answer: B

The FRAX tool has been developed to evaluate the fracture risk of patients. It is based on individual patient models that integrate the clinical risk factors with the bone mineral density (BMD) of the femoral neck and may be modified by the lumbar spine trabecular score. The clinical factors that may affect the score include age, sex, low body mass index, prior fracture after age 50, parental history of hip fracture, smoking history, past or current use of corticosteroids, alcohol intake, other secondary causes of osteoporosis, and rheumatoid arthritis. The previous vertebral or hip fracture is considered the most important predictor of fracture risk.

FRAX score is a useful tool to aid clinical decisions regarding the use of pharmacologic therapies in patients with low bone mass. To reduce fracture risk, recommend treating patients with FRAX 10-year risk scores of ≥3% for hip fracture or ≥20% for major osteoporotic fracture.

*StatDx. https://my.statdx.com/.

Pulmonary System

Medhat Sam Gabriel, Bital Savir-Baruch, Robert H. Wagner, Bruch J. Barron, Davide Bova, Aamna Hassan, Stephanie Rodriguez-Santiago, and Yekaterina Kucerova

Case 1

1. A 28-year-old female with history of Tetralogy of Fallot, status post repair in childhood, presented to ED with chest pain. She underwent ventilation/perfusion lung scan. Chest radiograph was obtained the same day. Images are shown below.

M. S. Gabriel (✉) · B. Savir-Baruch · R. H. Wagner · D. Bova
Division of Nuclear Medicine, Department of Radiology, Loyola University Medical Center, Maywood, IL, USA
e-mail: sgabriel@lumc.edu; Bital.savir-Baruch@lumc.edu; rwagner@lumc.edu; dbova@lumc.edu

B. J. Barron
Division of Nuclear Medicine and Molecular Imaging, Department of Radiology and Imaging Sciences, Atlanta, GA, USA
e-mail: bjbarro@emory.edu

A. Hassan
Department of Nuclear Medicine, Shaukat Khanum Memorial Cancer Hospital and Research Center, Lahore, Punjab, Pakistan

S. Rodriguez-Santiago
Department of Nuclear Medicine, C.W. Bill Young VA Medical Center, Bay Pines, FL, USA
e-mail: stephanie.rodriguezsantiago@va.gov

Y. Kucerova
Division of Nuclear Medicine, Department of Radiology, Loyola University Medical Center, Maywood, IL, USA

Department of Radiology, Cook County Healthy, Chicago, IL, USA
e-mail: yekaterina.kucerova@cookcountyhealth.org

M. S. Gabriel, B. Savir-Baruch (eds.), *RadTool Nuclear Medicine MCQs*,
https://doi.org/10.1007/978-3-030-69281-0_7

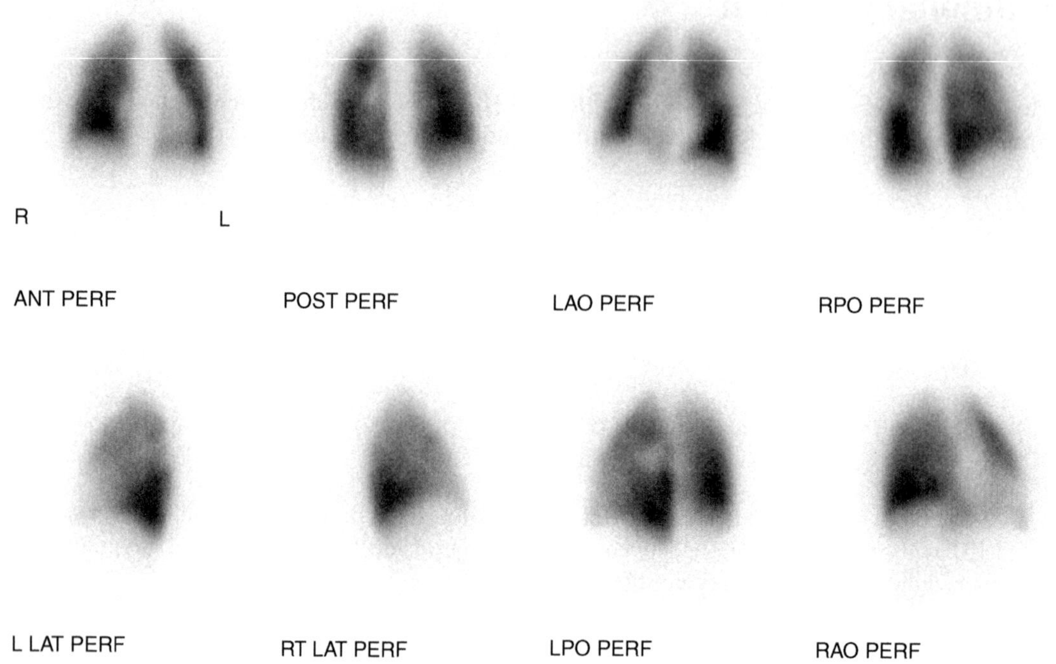

Image 1

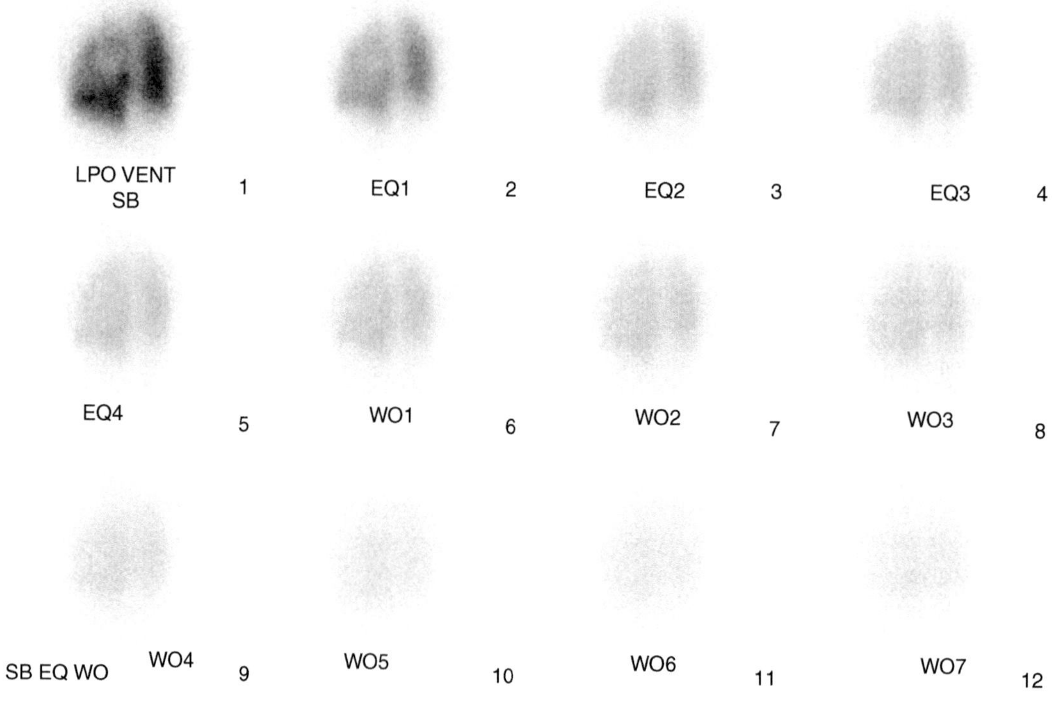

Image 2

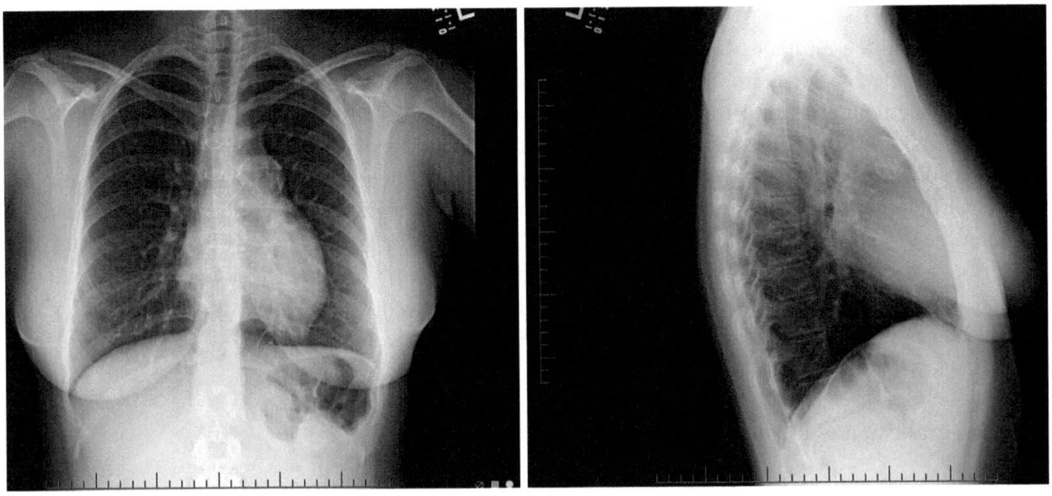

Image 3

Which of the following is correct:

A. Tc-99m-MAA is used in the above scan to obtain ventilation images
B. Tc-99m-MAA preparations contain particles ranging in size from 0.5 to 1 mm
C. In practice, 200,000–500,000 MAA particles are administered
D. After Tc-99m-MAA administration, there is an approximate obstruction of 2–5% of pulmonary vessels.

Answer: C

Tc-99m macroaggregated albumin (Tc-99m-MAA) is used to perform perfusion images (image 1), not ventilation images (image 2). Tc-99m-MAA localizes by capillary blockade. The areas of decreased or absent perfusion will also demonstrate decrease or absent particles and appear "cold" on the scan.

In general, less than 1 in 1000 (<0.1%) of the capillaries are blocked. In normal individuals, 95% of the particles are removed from the circulation on the first pass through the pulmonary capillary bed. About 5% of particles are very small measuring less than 5 μm (micrometer) and pass through the capillary system.

Tc-99m-MAA preparations contain particles ranging in size from 5 to 100 μm with the majority between 10 and 30 μm. Red blood cells have a diameter of about 7 μm, and the capillaries are up to 10 μm. This is the rationale for having the MMA particles larger than 10 μm.

Although the study can be done with 100,000 particles, preferably the particle number is higher between 200,000 and 500,000.

2. In which category of patients, the maximum number of MAA particles may be given (200,000–500,000)?

A. Patients with Right-to-Left Shunt
B. Neonates
C. Patients with pulmonary hypertension
D. Patients with chronic kidney disease

Answer: D

Reducing the number of MMA particles is warranted in neonates/children and certain conditions in adults, such as pulmonary hypertension, right-to-left cardiac shunt, and pregnancy. Many nuclear medicine laboratories reduce the number of particles to half for patients with pulmonary hypertension. Nowadays, because most of the doses come from central radiopharmacies as individual doses, the easiest way to reduce the particle is to decrease the dose. This might require longer imaging to achieve good quality studies. In cases of right-to-left cardiac shunt, the particles enter the systemic circulation and may deposit in the brain or the kidneys. Still, studies showed no serious effects in these patients. The recommended number of particles given to neonates is 10,000 and between 50,000 and 150,000 for small children based on age and weight. Although free Tc-99m may be excreted by the kidney, MAA particles do not. Therefore, the study is safe in patients with chronic renal disease.

3. Upon evaluation of the images above, you conclude:

A. High probability for pulmonary embolism
B. Intermediate probability for pulmonary embolism
C. Low probability for pulmonary embolism
D. Nondiagnostic study for pulmonary embolism

Answer: C

Perfusion lung images (image 1) demonstrate decreased perfusion in the left upper lung (better seen on LPO, LAO, and RPO projections). LPO projection demonstrates a "stripe" sign, which is not characteristic of PE pattern. Ventilation images (image 2) performed in the LPO projection demonstrates a matching ventilation defect on early single breath images.

The term stripe sign refers to a peripheral thin line or stripe of activity, which indicates a small area of perfused lung tissue surrounding the area of decreased perfusion. Pulmonary emphysema is the most common cause of this phenomenon. Perfusion defects presenting with a stripe sign are very unlikely to represent pulmonary emboli based on the fact that it is very rare for a non-pleural-based lesion to result from an embolus. These defects should be interpreted as a low or even a very low probability for pulmonary embolism in the absence of other perfusion abnormalities on the scan.

Case 2

4. A 52-year-old male presented with chest pain. There was suspicion of pulmonary embolism. A ventilation/perfusion lung scan was performed. Images are shown below.

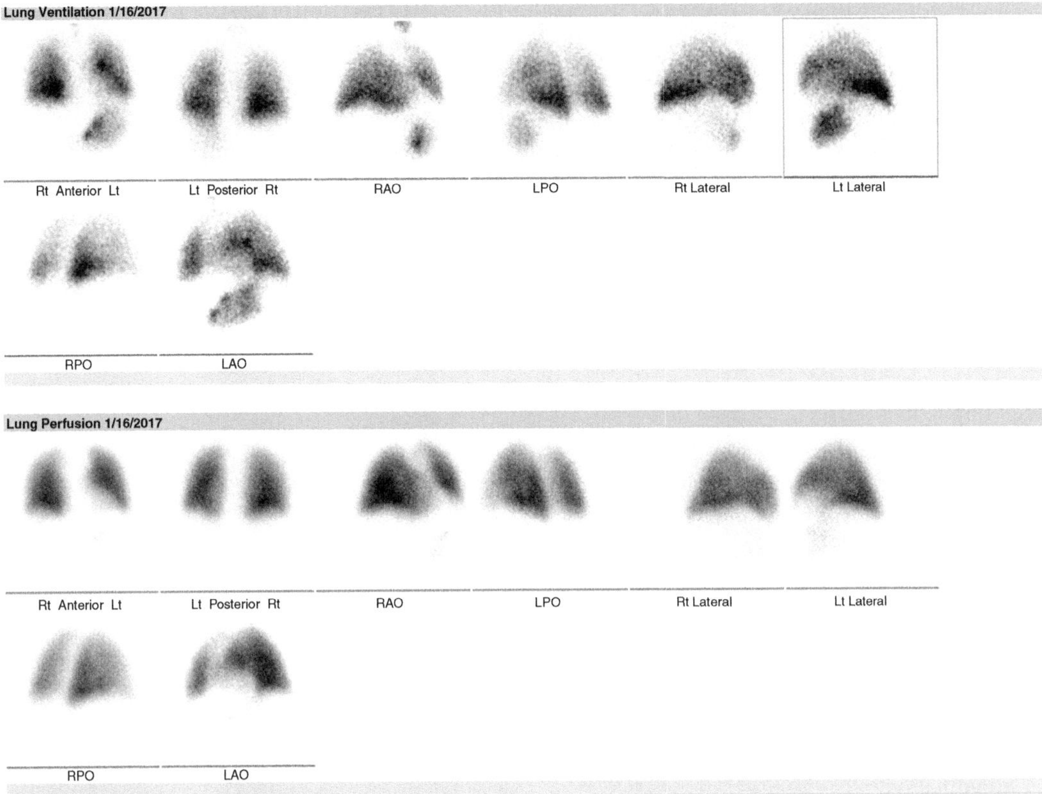

Image 1

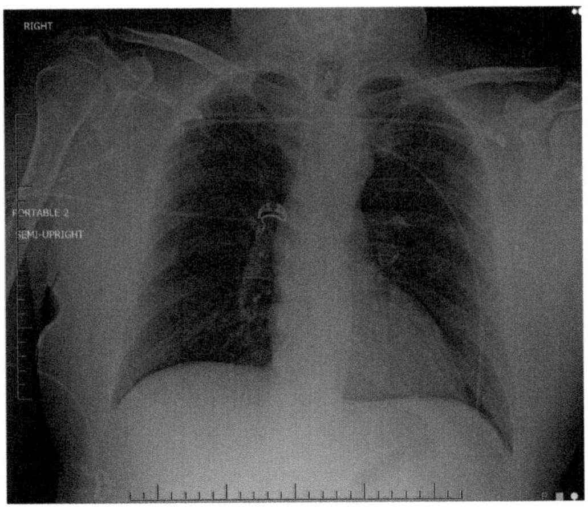

Image 2

Which of the following is correct regarding the radiotracer used for ventilation images?

A. Radioactive aerosol delivered through nebulizer system
B. Exhaled radioactive material disposed by using an activated charcoal trap until it decays to background
C. Ventilation imaging is performed in a room with positive pressure in case of accidental leakage from the closed system
D. The acquisition consists of dynamic ventilation images in multiple projections

Answer: A

Two radiopharmaceuticals are available for ventilation imaging in the United States, a radioactive gas (Xenon-133) and an aerosol (Tc-99m diethylenetriamine pentaacetic acid "DTPA").

The ventilation images shown were obtained after administration of the Tc-99m DTPA aerosol. DTPA ventilation images are generally obtained in eight static projections, similar to perfusion images. The study does require a special room with negative pressure and does not allow for dynamic images. Xenon-133 on the other hand requires a negative pressure room but allows for single breath, equilibrium, and washout views.

5. Upon evaluation of ventilation images, you conclude:

A. Normal ventilation
B. Diaphragmatic injury with tracer accumulation in the upper abdomen
C. Tracer deposition within a fatty liver
D. Splenomegaly

Answer: A

Tc-99m DTPA ventilation images obtained in multiple projections demonstrate normal radiotracer distribution within bilateral lungs. The activity seen below the diaphragm is within the stomach and is secondary to tracer ingestion during the administration of DTPA aerosol. This is a common finding and does not require further investigation. Swallowed Tc-99m DTPA in the mouth, esophagus, and stomach can sometimes shine through on the subsequent perfusion examination.

Xe-133 is fat-soluble and it may deposit in the liver, which may be difficult to separate from the right lung base.

6. Which of the following is true regarding the above perfusion and ventilation images:

A. Ventilation images were obtained first with the activity of the tracer about 30 mCi
B. Ventilation images were obtained first with the activity of the tracer about 3 mCi
C. Perfusion images were obtained first with the activity of the tracer about 30 mCi
D. Perfusion images were obtained first with the activity of the tracer about 3 mCi

Answer: A

Ventilation images were performed in multiple projections after administration of 30 mCi of 99mTc-DTPA aerosol, with approximately 1 mCi delivered to the lungs. Perfusion images were obtained in the same projections after intravenous administration of 3 mCi of 99mTc-MAA.

99mTc-DTPA is administered as an aerosol through the nebulizer system. Usually, about 30 mCi is placed in the system nebulizer, with only 1 or 2 mCi actually delivered to the patient, followed by approximately 3 mCi of Tc-99m-MAA.

References

*Ziessman HA, O'Malley JP, Thrall JH. Nuclear medicine: the requisites.
*Mettler FA, Guiberteau MJ. Essentials of nuclear medicine imaging.

Case 3

7. A 56-year-old female with a history of liver cirrhosis, recurrent ascites, and hepatic hydrothorax is being evaluated for dyspnea. The patient states that her breathing has been getting worse for about 2.5 years, since her diagnosis of cirrhosis. She is getting only partial relief with thoracentesis. Chest radiograph and perfusion lung scan were obtained.

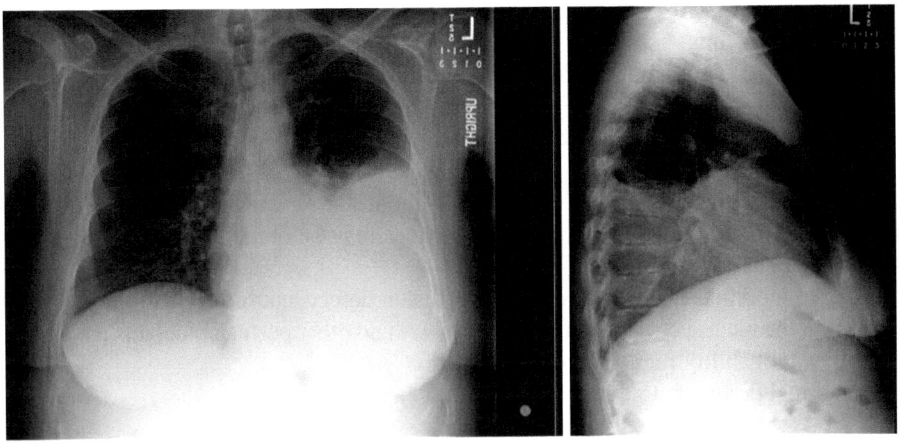

Image 1

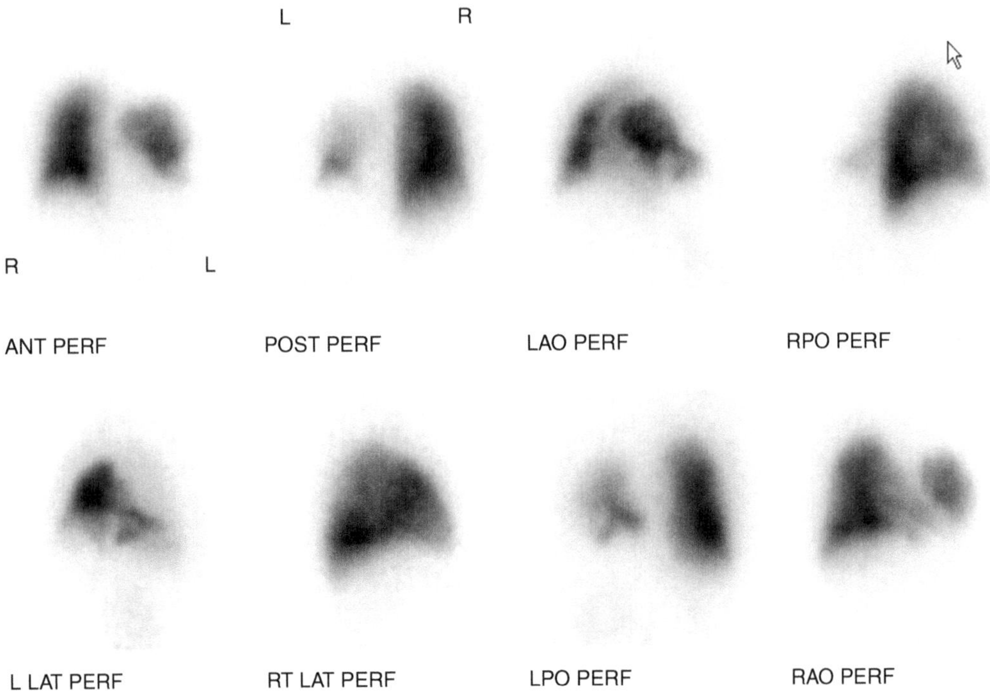

Image 2

Upon evaluation of the images, you conclude:

A. Perfusion defects in the left lung likely secondary to pulmonary embolism
B. Bilateral perfusion defects, greater on the left, consistent with pulmonary embolism
C. Nondiagnostic study.
D. Additional perfusion images are needed.

Answer: D

Chest radiograph (image 1) demonstrates a moderate left pleural effusion which limits evaluation of the left lower lung.

Tc-99m-MAA perfusion images obtained in multiple projections demonstrate nonsegmental perfusion defects in the left lung. The pattern of tracer distribution is not characteristic of acute pulmonary embolism. Given the fact that perfusion images are typically obtained in supine position, the pattern of tracer distribution could be merely secondary to the left-sided pleural effusion, rather than true perfusion defects. In a supine position, the pleural fluid might layer out and attenuate photons. Additional upright perfusion images may be obtained in an attempt to overcome this problem.

According to modified PIOPED II criteria, perfusion defects caused by a solitary large (one-third or more of the pleural cavity) pleural effusion with no other perfusion defects are associated with a very low probability for pulmonary embolism.

There is normal perfusion to the right lung.

8. Additional upright perfusion images were obtained.

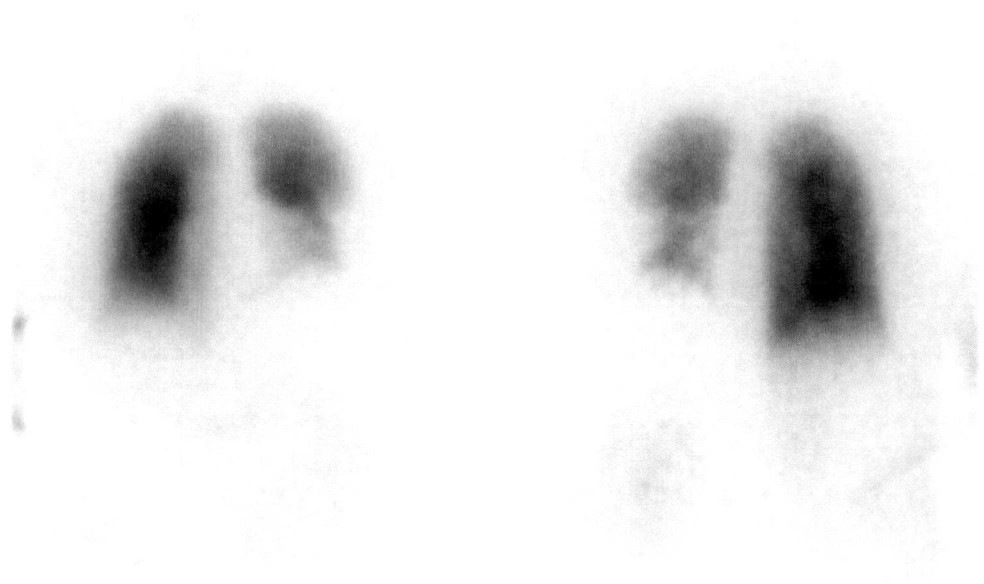

Image 1

Upon evaluation of images, you conclude:

A. High probability for pulmonary embolism
B. Intermediate probability for pulmonary embolism
C. Indeterminate for pulmonary embolism
D. No evidence for pulmonary embolism

Answer: D

Upright perfusion images demonstrate decreased "counts" in the left lower lung, rather than diffusely. This results from changing the position to upright, shifting the nonloculated pleural fluid to the lower lung field.

As mentioned before, defects caused by a solitary large effusion with no other perfusion defects are unlikely to result from pulmonary embolism.

The pulmonary clots may resolve rapidly, particularly in young patients. Clots in the elderly may require more time to resolve and sometimes they never resolve. Defects that persist for 3–6 months are unlikely to ever clear. These are called "non-resolved" pulmonary emboli. Frequently although incorrectly, they are referred to as "chronic" emboli. They may result in pulmonary hypertension.

9. Additional supine images were obtained in this patient.

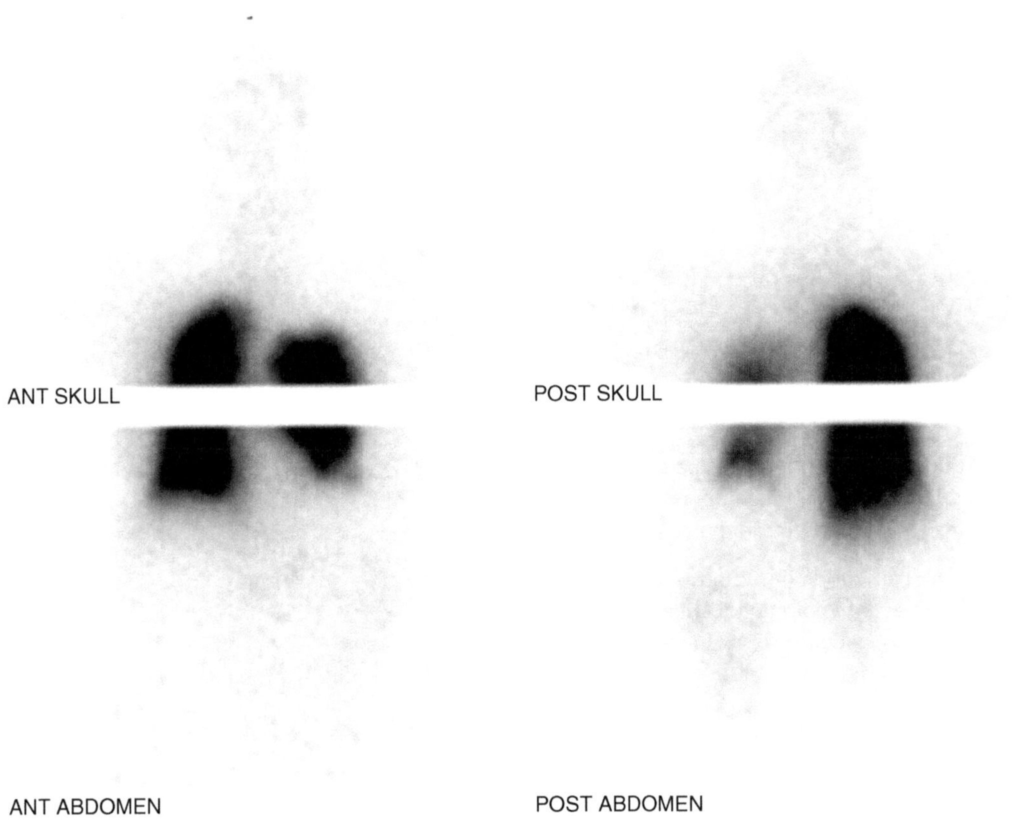

Image 1

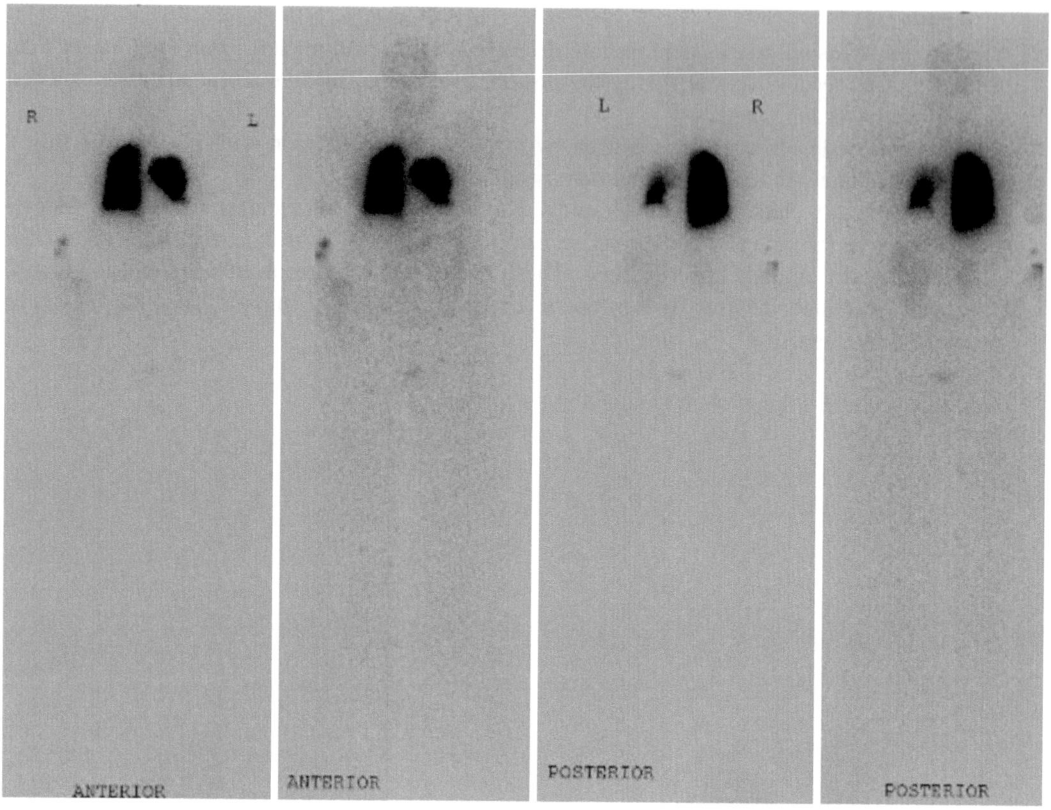

Image 2

Additional calculations were performed:
(total body counts – lung counts)/(total body counts × 100%) = 20%
Upon evaluation of the study, you conclude:

A. The distribution of radiotracer is within normal limits
B. The tracer preparation technique was poor
C. There is evidence for right-to-left shunt
D. The interpretation of the study should be revised to a "high probability for pulmonary embolism"

Answer: C

Additional whole-body scan and spot images of the brain and the kidneys demonstrate abnormal Tc-99m-MAA accumulation in the capillary bed of kidneys, brain, spleen, and soft tissues, which is most suggestive of a right-to-left shunt. Right-to-left shunt is simply shunting of blood through cardiovascular or pulmonary systems, bypassing the pulmonary capillary circulation. It may result in venous to arterial emboli, causing ischemia to multiple organs such as the brain and the bowel. The presence of a right-to-left shunt can be easily recognized on posterior images by the immediate presence of renal activity. Then it may be confirmed by the demonstration of activity in the brain.

A right-to-left shunt must be differentiated from extrapulmonary uptake of free Tc-99m pertechnetate, which may be seen in the thyroid, salivary glands, stomach, and choroid plexus. Free Tc-99m does not cross the blood-brain barrier to enter the brain. Therefore, cerebral uptake typically results from a right-to-left cardiac shunt.

The percentage, and subsequently, the severity of the shunt can be calculated using the following formula:

Right-to-left shunt % = (total body counts – lung counts)/(total body counts × 100). This is only used when a shunt is suspected and not as a part of a routine quality control procedure. The presence or absence of a shunt should not change the interpretation of the study regarding the likelihood of pulmonary embolism.

10. Which of the following condition is most likely associated with scan findings in this patient:

    A. Left-to-right intracardiac shunt
    B. Hepatopulmonary syndrome
    C. Pulmonary embolism
    D. Mucous plug

    Answer: B

    In this patient with a past medical history significant for cirrhosis, impaired hepatic function, and progressively worsening dyspnea, hepatopulmonary syndrome should be first suspected as an etiology of right-to-left shunt.

    Left-to-right intracardiac shunt is usually evaluated with first-pass technique (first-pass radionuclide angiography). Uncomplicated pulmonary embolism or mucous plug is not associated with a shunt.

References

*Gandhi SJ, Babu S, Subramanyam P, Shanmuga Sundaram P. Tc-99m macro aggregated albumin scintigraphy—indications other than pulmonary embolism: a pictorial essay.

Behnia F, Hippe DSS, Saad Bermo M, Elojeimy S, Vesselle H, Tc-99m MAA mapping prior to Y-90 liver radioembolization; Factors that may affect pulmonary shunt fraction.

*Ziessman HA, O'Malley JP, Thrall JH. Nuclear medicine: the requisites.

*Mettler FA, Guiberteau MJ. Essentials of nuclear medicine imaging.

11. After reviewing the perfusion lung scan (image 1), you ask for a spot view of the head (image 2) to confirm your suspicion of:

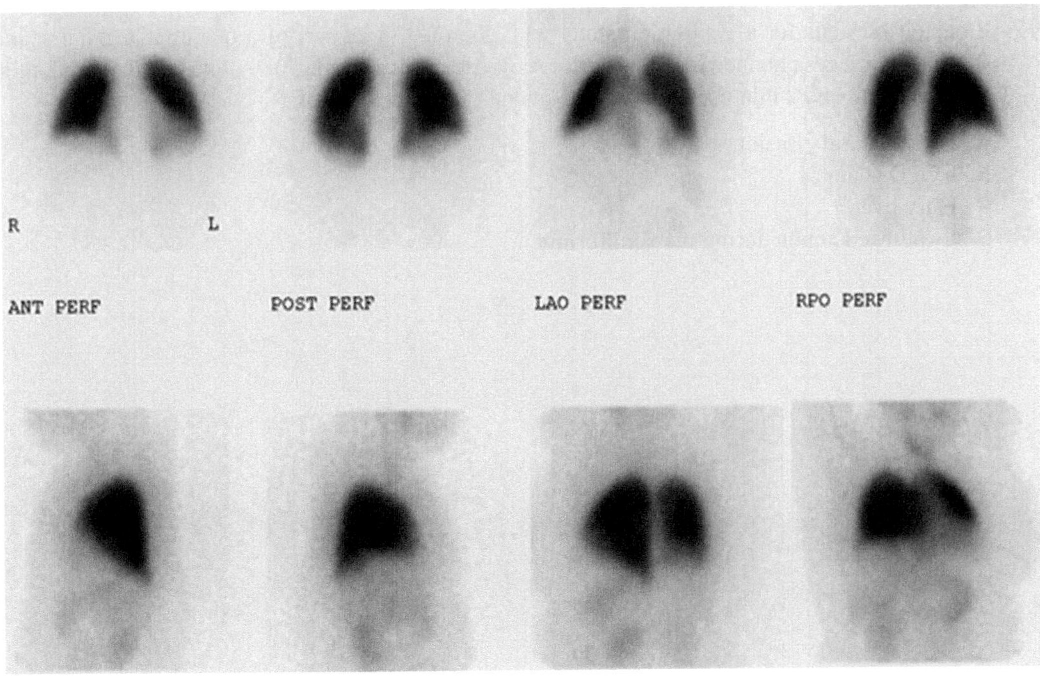

Image 1

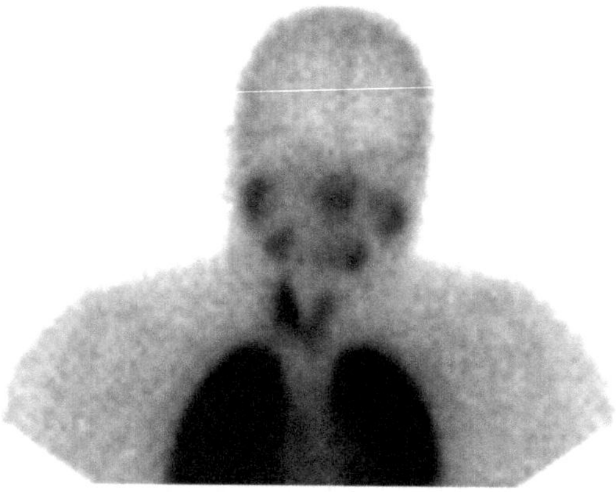

Image 2

A. Tc-99m dissociation from the Tc-99m-MAA
B. A right-to-left shunt
C. A left-to-right shunt
D. Thyroid malignancy

Answer: A

The study demonstrates progressive degradation of the Tc99m-MAA with activity in the stomach, kidneys, thyroid, and salivary glands. This pattern may be seen from “old” MAA preparation near its expiration. Fresh Tc-99m-MAA usually does not dissociate easily. Salivary gland uptake of free Tc99m pertechnetate is a clue to the dissociation. The thyroid uptake is asymmetric with more activity on the right lobe. Although Tc-99m pertechnetate may be used to assess thyroid nodules, the diagnosis of thyroid cancer is not determined by the presence of “cold” nodules only because the majority of these nodules are benign.

If a right-to-left shunt were present, Tc-99m-MAA would be present in the brain. Left-to-right shunt cannot be evaluated by a perfusion lung scan.

12. A patient presents for a ventilation study with Xenon-133 as part of a quantitative lung scan. Ventilation images obtained in the posterior projection illustrate the 30 s single breath hold, 2 min of equilibrium and 2 min of wash out. The images are suggestive of:

A. Right pleural effusion
B. COPD changes
C. Fatty liver
D. Swallowed xenon during the equilibrium

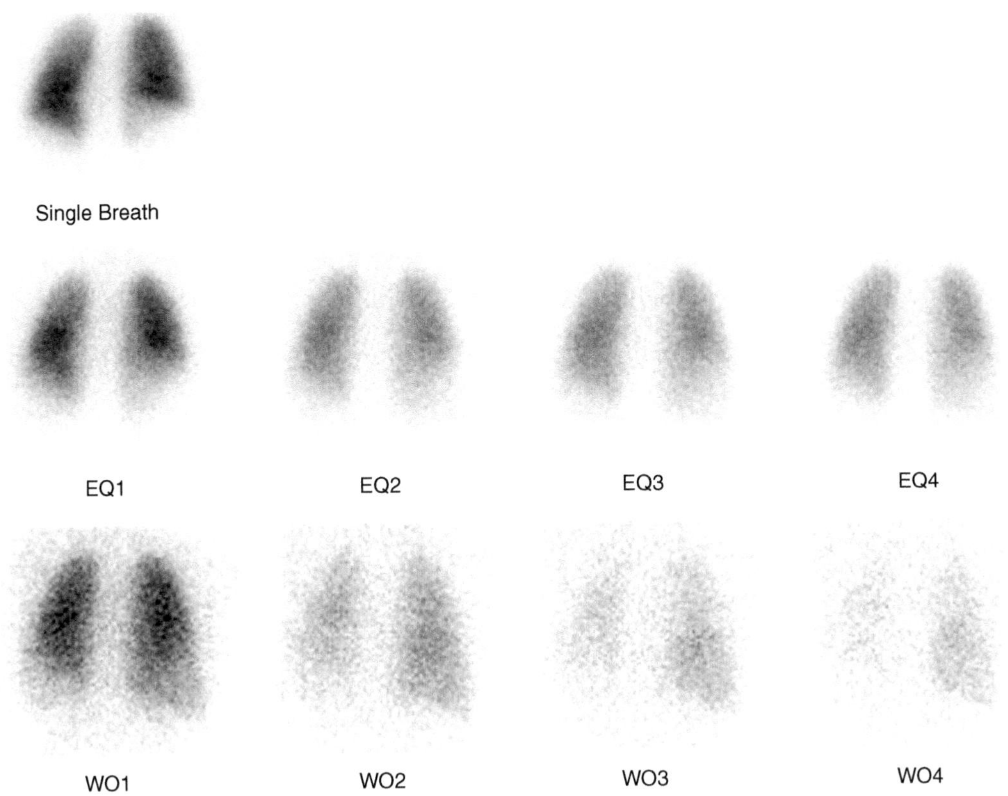

Answer: C

The liver can be seen during the late equilibrium and wash out phases, typically in the cases of fatty liver. There is some xenon that is absorbed into the bloodstream and circulates through the body. Since xenon is lipophilic, it will occasionally be seen in the liver, especially when it is a fatty liver. The finding is best seen on the wash out images when there is little activity remaining in the lungs.

The photopenic area at the inferior right lung, best seen on a single breath, represents the liver. For correlation and illustration, a coronal slice from a recent CT scan was horizontally reversed to match the lung scan. No significant pleural effusion was present and the area in question gradually fills in with xenon. COPD changes are most commonly seen involving the apices of the lungs. It would be difficult to swallow that much xenon from the equilibrium images. In addition, the shape of the area of activity resembles the liver and it is on the right side of the body.

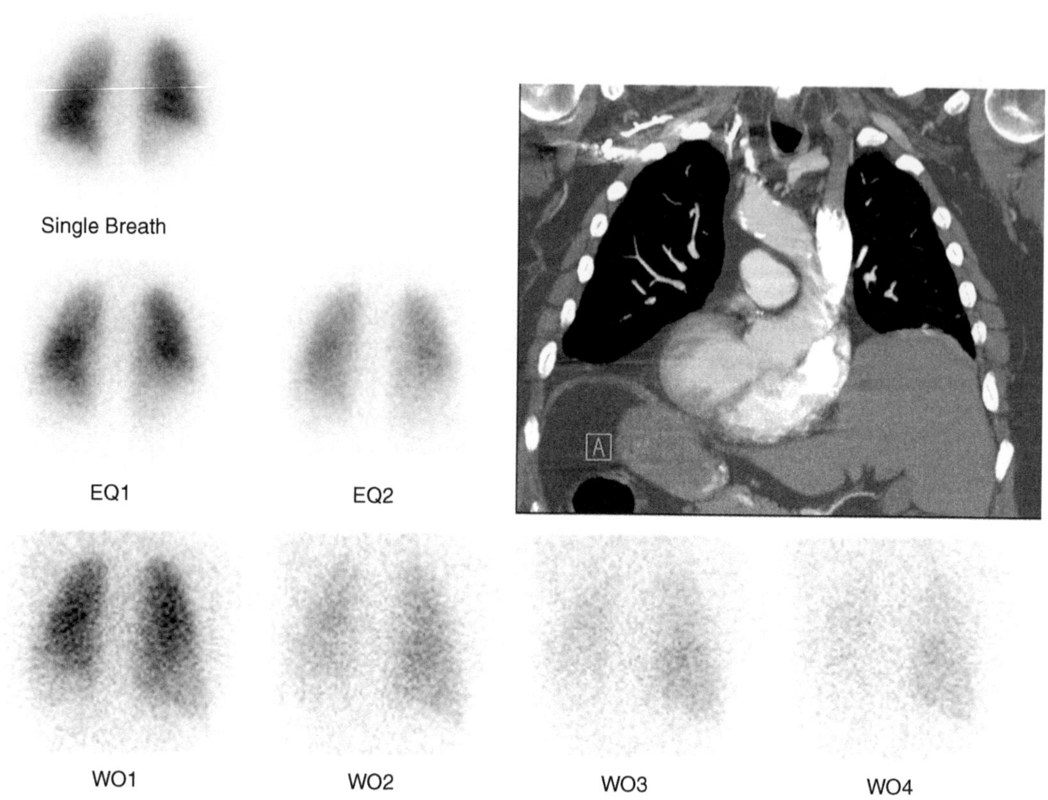

13. This patient underwent a V/Q scan to evaluate chronic cough and shortness of breath. What diagnosis is most likely?

    A. Non-resolved "chronic" pulmonary emboli
    B. Bronchiectasis
    C. Lobar pneumonia
    D. Main pulmonary artery occlusion

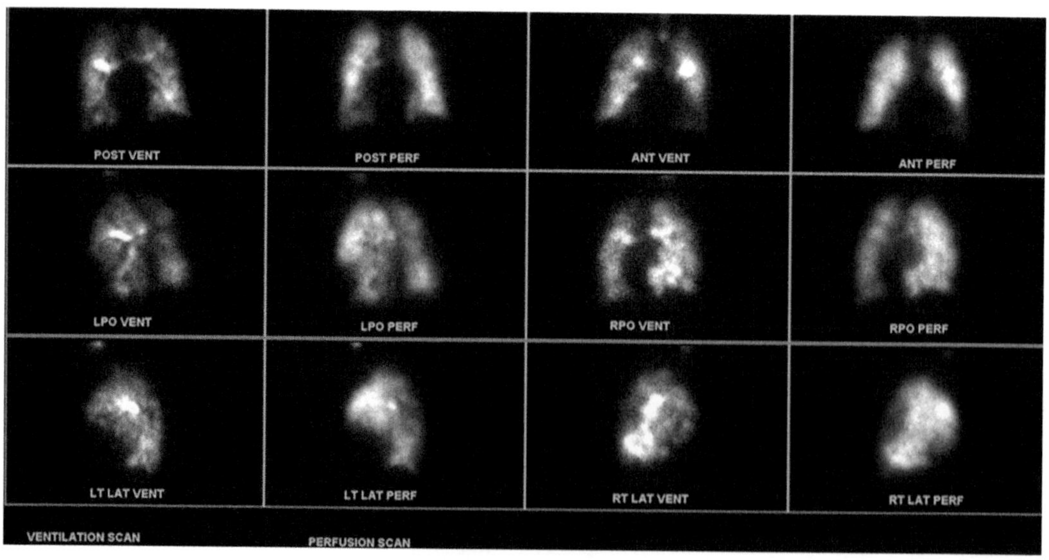

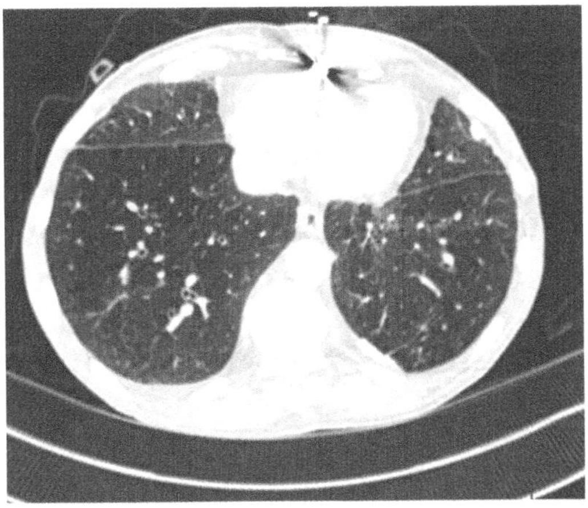

Answer: B

Note the prominence of bronchioles on the CT. In an extreme condition known as lymphangioleiomyomatosis (LAM), the muscle cells that line the airways multiply abnormally leading to what appears like bronchiectasis. A xenon ventilation study would show diffuse but heterogeneous trapping of tracer. This condition is seen mostly in women of child-bearing age.

Non-resolved, frequently referred to as "chronic" pulmonary embolism is distinctive by wedge shape defects. Lobar pneumonia may result in an area of decreased perfusion, similar in shape to the affected region. Pulmonary artery occlusion may result in non-visualization of an entire lung while, less commonly seen, partial occlusion is usually seen as diffuse decrease perfusion of one lung.

14. This patient is a 70-year-old man who presented with a hip fracture 2 days earlier. He subsequently developed shortness of breath. No fat globules were noted within his urine specimen. What is his diagnosis?

    A. Asthma
    B. Acute pulmonary emboli
    C. Chronic pulmonary emboli
    D. Fat emboli

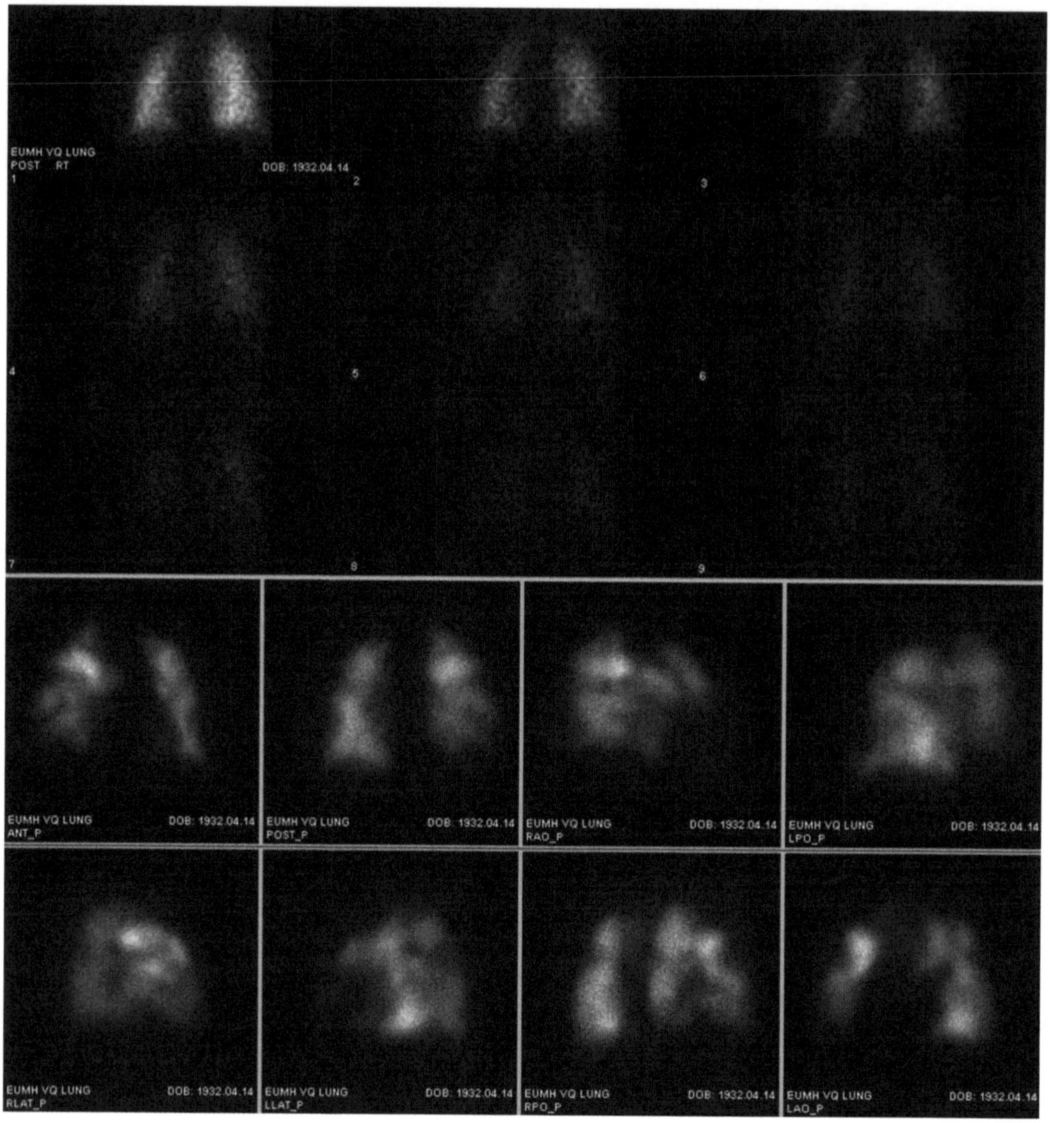

Answer: B

The typical imaging finding for fat emboli is a saw-toothed pattern at the periphery of the lung due to the small fat particles. Other causes of peripheral saw-toothed defects are air emboli, amniotic fluid emboli, tumor emboli, and septic emboli.

Wedge-shaped segmental mismatched perfusion defects are more commonly seen in acute pulmonary emboli. A simple assay is to look for fat globules in the urine, which are fairly common in fat emboli. Chronic or non-resolved emboli cannot be excluded, particularly no prior studies are available. However, in this patient with acute onset of symptoms, acute pulmonary embolism is more likely. After reading the study as a high probability for acute pulmonary embolism, the patient was treated with anti-coagulation and symptoms recovered.

Gastrointestinal System

Medhat Sam Gabriel, Bital Savir-Baruch, Robert H. Wagner, Bruch J. Barron, Davide Bova, Aamna Hassan, Stephanie Rodriguez-Santiago, and Yekaterina Kucerova

Case 1 A patient is referred for a hepatobiliary study to rule out biliary dyskinesia given persistent postprandial right upper quadrant pain and nausea. A prior abdominal US was significant for a small polypoid lesion measuring less than 4 mm (Fig. 1a).

1. What is the best interpretation for the findings seen in the initial 60 min of images post radiotracer injection (Fig. 1b)?
   (a) Cystic duct obstruction
   (b) Common bile duct obstruction
   (c) Impaired hepatic function
   (d) Enterogastric reflux

M. S. Gabriel (✉) · B. Savir-Baruch · R. H. Wagner · D. Bova
Division of Nuclear Medicine, Department of Radiology, Loyola University Medical Center, Maywood, IL, USA
e-mail: sgabriel@lumc.edu; Bital.savir-Baruch@lumc.edu; rwagner@lumc.edu; dbova@lumc.edu

B. J. Barron
Division of Nuclear Medicine and Molecular Imaging, Department of Radiology and Imaging Sciences, Atlanta, GA, USA
e-mail: bjbarro@emory.edu

A. Hassan
Department of Nuclear Medicine, Shaukat Khanum Memorial Cancer Hospital and Research Center, Lahore, Punjab, Pakistan

S. Rodriguez-Santiago
Department of Nuclear Medicine, C.W. Bill Young VA Medical Center, Bay Pines, FL, USA
e-mail: stephanie.rodriguezsantiago@va.gov

Y. Kucerova
Division of Nuclear Medicine, Department of Radiology, Loyola University Medical Center, Maywood, IL, USA

Department of Radiology, Cook County Healthy, Chicago, IL, USA
e-mail: yekaterina.kucerova@cookcountyhealth.org

M. S. Gabriel, B. Savir-Baruch (eds.), *RadTool Nuclear Medicine MCQs*,
https://doi.org/10.1007/978-3-030-69281-0_8

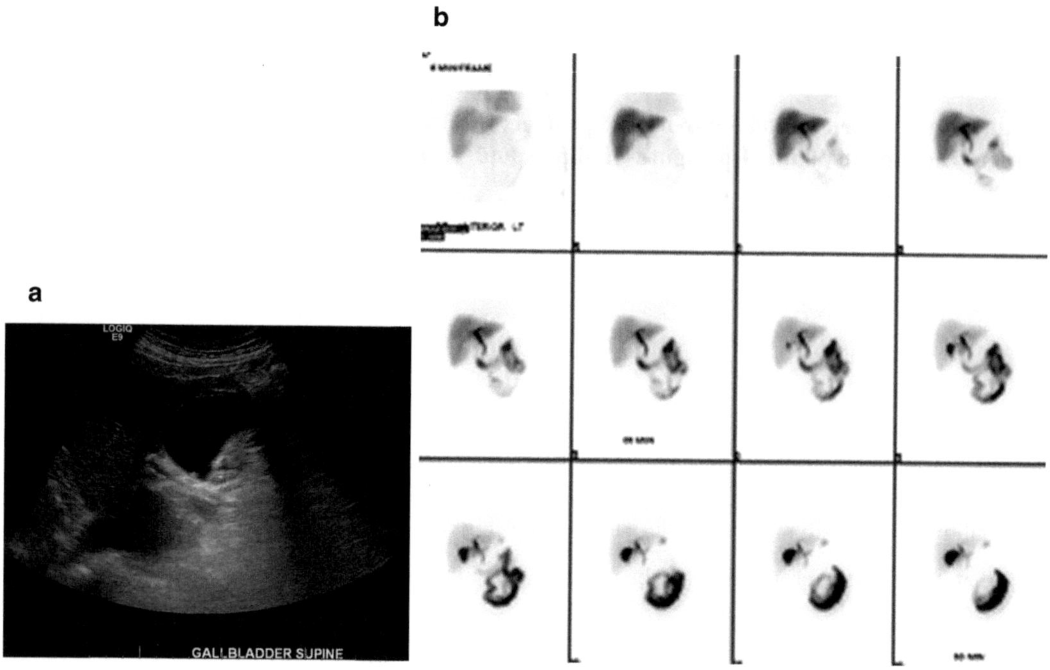

Fig. 1 (**a**, **b**)

Answer: d

The initial images following radiotracer administration demonstrate adequate hepatic uptake and activity in the gallbladder and the small bowel, excluding cystic or common bile duct obstruction. There is normal biliary to bowel transit time. However, there is enterogastric reflux is noted.

2. Based on the findings shown in Fig. 1b, what is a possible explanation for the patient's symptomatology?

(a) Biliary duodenogastric reflux
(b) Biliary dyskinesia
(c) Chronic cholecystitis
(d) Acute cholecystitis

Answer: a

Enterogastric reflux is not an uncommon finding in hepatobiliary scintigraphy. It can be seen during the initial passive phase of the study or following the CCK infusion or fatty meal challenge. A significant and progressive biliary duodenogastric reflux evidenced by progressive increased retrograde tracer activity into the stomach could be the cause of the patient's pain.

3. The patient was given 8 ounces of Lactose-Free Ensure Plus and images were acquired for 60 min. The gallbladder ejection fraction was calculated at 45%. Your conclusion in the study is:

(a) Normal since the EF is greater than 33%
(b) Normal since the EF is greater than 38%
(c) Equivocal since the EF is between 30% and 50%
(d) Abnormal since the EF is lower than 50%

Answer: a

There are several publications supporting the 60-min Sincalide (0.02 μg/kg) infusion as the preferred method for the determination of the gallbladder ejection fraction due to its lower percentage variability. A lower range of 38% has been proposed as normal with this methodology. Borderline abnormal studies should be read with caution. Currently, many centers have adopted the use of a fatty challenge for the gallbladder ejection fraction calculation. The recommended time for acquisition is also 60 min of imaging following the ingestion of the fatty meal. The published lower range of normal is 33% if Lactose-Free Ensure Plus is used.

4. What is the minimum fat content required in a meal to induce adequate gallbladder contractility?

   (a) 10 g
   (b) 15 g
   (c) 20 g
   (d) 25 g

   Answer: a
   The normal stimulus of cholecystokinin release from the proximal small bowel that will signal receptors in the gallbladder wall to contract is elicited by the fat and protein content in the ingested meal. However, there should be at least 10 g of fat in the meal to provoke this process. If a fatty meal challenge is used for the determination of gallbladder ejection fraction, it is very important to instruct the patient to ingest the entire volume. If the patient is unable to ingest the entire volume, this should be clearly stated in the report, particularly if the calculated ejection fraction is below normal.

5. What are the recommended dose and time of infusion for Sincalide when used to determine the gallbladder ejection fraction during a hepatobiliary study?

   (a) 0.01 μg/kg infused over 30 min
   (b) 0.02 μg/kg infused over 30 min
   (c) 0.01 μg/kg infused over 60 min
   (d) 0.02 μg/kg infused over 60 min

   Answer: d
   Comparison studies have been done evaluating the reproducibility of 15, 30, and 60-min infusions of 0.01–0.02 μg/kg of Sincalide. The 60-min infusion at a rate of 0.02 μg/kg of Sincalide showed the least variability.

References

1. Ziessman HA, Hepatobiliary Scintigraphy in 2014. J Nucl Med Technol. 2014;42:249–59.
2. ACR-SNM-SPR Practice Guideline for the Performance of Gastrointestinal Scintigraphy. Revised 2010.
3. Dillehay G et al. Appropriate use criteria for hepatobiliary scintigraphy in abdominal pain. J Nucl Med. 2017;195,602-be n 6/5/17.
4. Ziessman HA, Cholecystokinin Cholescintigraphy: Methodology and normal values using a lactose-free fatty-meal food supplement. J Nucl Med. 2003;44:1263–6.
5. Tulchinsky M, SNM practice guideline for hepatobiliary scintigraphy 4.0*. J Nucl Med Technol. 2010;38:4.

Case 2 A hepatobiliary study for a patient in the MICU has been approved to rule out obstruction. The patient is status post cholecystectomy, and choledochoduodenostomy in an attempted Whipple procedure (Fig. 2).

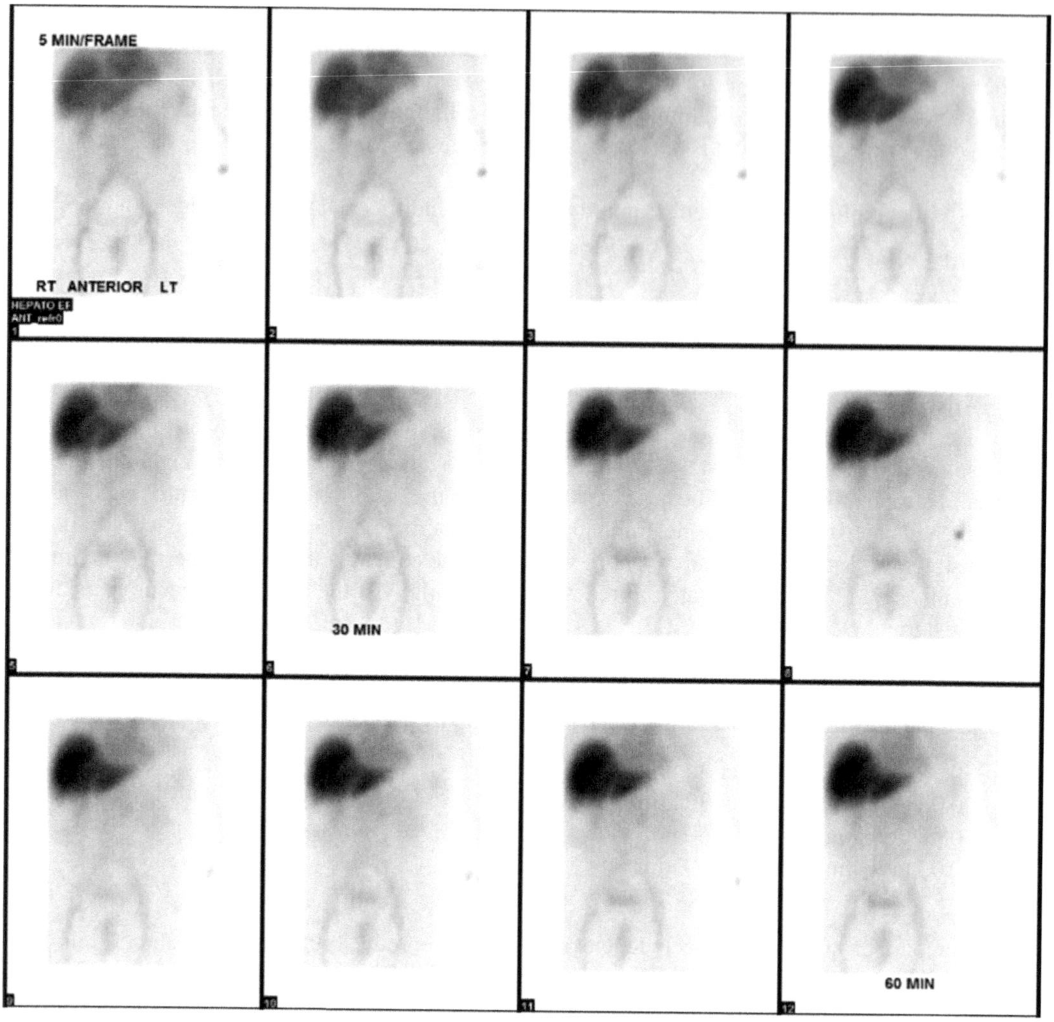

Fig. 2 Initial 60-min images

6. While reviewing the patient's chart for possible contraindications, you notice that his bilirubin level is high. What is the standard recommended dose for patients with normal bilirubin levels?
   (a) 3–5 mCi
   (b) 5–10 mCi
   (c) 10–15 mCi
   (d) 15–20 mCi

   Answer: a

   Patients with hepatic dysfunction and elevated bilirubin levels usually require delayed images, which possibly include 24-h static views, to ensure adequate assessment. The standard recommended dose is 3–5 mCi of Mebrofenin or Disofenin for patients with normal bilirubin levels. Higher doses should be considered for patients with severe hyperbilirubinemia. Mebrofenin is recommended for this group of patients given its greater hepatic extraction (approximately 98%), compared to 88% for Disofenin. Delayed images, particularly at 24 h postinjection, can be of very poor quality in patients with hyperbilirubinemia, limiting evaluation if a low dose is used. As a matter of fact, this patient's bilirubin level was 22 mg/dL. A higher radiotracer dose is suggested to ensure sufficient circulating activity in these patients, who would likely require delayed imaging. A maximum of 10 mCi has been recommended if the bilirubin levels are greater than 10 mg/dL.

7. What is the expected normal time at which blood pool clearance occurs during hepatobiliary scintigraphy?

   (a) 1–2 min
   (b) 5–10 min
   (c) 15–20 min
   (d) 25–30 min

   Answer: b
   In patients with preserved hepatic function, normal prompt blood pool clearance of hepatobiliary radiotracers should occur by 10 min. In patients with impaired liver function, the time for blood pool clearance is usually longer, correlating to the degree of liver dysfunction.

8. The delayed static image performed at 4-h postinjection (Fig. 3) is consistent with?

   (a) Acute cholecystitis
   (b) Patent common bile duct
   (c) High-grade obstruction
   (d) Small biliary leak

   Answer: b
   The delayed image demonstrates a small amount of bowel activity. This confirms patent transit through the biliary anastomosis. The genitourinary activity is expected given radiotracer renal excretion increased in patients with hepatic dysfunction. These patients require more time to evaluate for cystic duct obstruction, often 24 h.

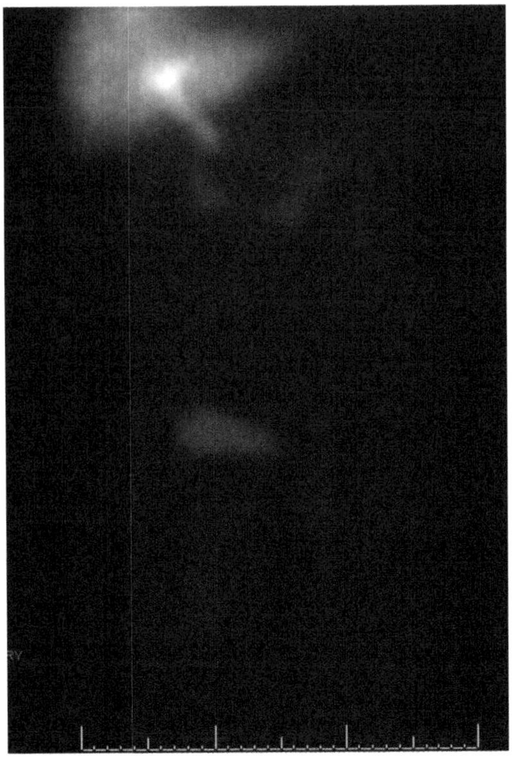

Fig. 3 Static 4-h image

9. When considering patients with bilirubin levels above 15 mg/dL,
   (a) A hepatobiliary study can be helpful; however a higher radiotracer dose should be considered.
   (b) A hepatobiliary study should not be performed since the study yield would be low
   (c) Hyperbilirubinemia above 15 mg/mL is considered a contraindication for hepatobiliary scintigraphy
   (d) Disofenin should be considered over Mebrofenin given its improved hepatic extraction

   Answer: a

   Hyperbilirubinemia is not a contraindication to hepatobiliary scintigraphy. The newer radiotracers Disofenin and Mebrofenin demonstrate high hepatic uptake. Studies with preserved diagnostic quality have been reported in patients with elevated bilirubin levels of up to 30 mg/dL. In cases of significant hyperbilirubinemia, a higher radiotracer dose should be given. The use of Mebrofenin is favored given its higher hepatic uptake.

References

1. Ziessman HA, Hepatobiliary scintigraphy in 2014. J Nucl Med Technol. 2014;42:249–59.
2. ACR-SNM-SPR Practice Guideline for the Performance of Gastrointestinal Scintigraphy. Revised 2010.
3. Dillehay G et al. Appropriate Use criteria for hepatobiliary scintigraphy in abdominal pain. J Nucl Med. 2017;195,602-be n 6/5/17.
4. Ziessman HA, Cholecystokinin cholescintigraphy: methodology and normal values using a lactose-free fatty-meal food supplement. J Nucl Med. 2003;44:1263–6.
5. Tulchinsky M, SNM practice guideline for hepatobiliary scintigraphy 4.0*. J Nucl Med Technol. 2010;38:4.

Case 3

10. During the interview prior to her scheduled HIDA study, a 43-year-old female patient mentioned that her last solid meal was approximately 2 days prior to the examination date. She has been drinking only apple juice for the past 2 days due to severe nausea. What is the best option for this patient?
    (a) Cancel the study and reschedule once the patient can tolerate solid food.
    (b) Premedicate the patient with 0.04 mg/kg of morphine and inject the radiotracer 30 min later.
    (c) Premedicate the patient with a 60 min infusion of 0.02 μg/Kg of Sincalide and inject the radiotracer 30 min later
    (d) Inject the radiotracer without premedication since it was ordered as outpatient

    Answer: c

    The patient's gallbladder is probably filled with viscous bile given the prolonged fasting and should be stimulated to contract and empty prior to the study. The radiotracer should not be injected without proper patient preparation to avoid false-positive results. There are two options to perform the study the same day. The simplest one is to instruct the patient to eat a solid meal, if possible and putting into consideration it contains at least 10 g of fat. The second option is to premedicate the patient with sincalide. The radiotracer can then be injected 15–30 min after the sincalide infusion is completed.

    Morphine augmentation is used to stimulate sphincter of Oddi contraction and, to a lesser extent, relax the gallbladder in an asymptomatic patient if the gallbladder does not fill by 30–60 min into the study.

11. Why do we instruct patients to fast for at least 3–4 h prior to a HIDA study?

(a) To prevent gallbladder relaxation
(b) To prevent gallbladder contraction
(c) To prevent Sphincter of Oddi contraction
(d) To prevent nausea or vomit if CCK is to be used

Answer: b

Fasting for at least 3–4 h, although ideally for 6 h, is crucial to ensure adequate relaxation of an empty gallbladder. If the patient eats shortly prior to the radiotracer injection, the gallbladder might still be contracting during the acquisition of the images, as a normal response resulting in non-visualization of the gallbladder, i.e., false-positive scan.

Fasting for more than 24 h, receiving total parenteral nutrition or being on a clear liquid diet results in inadequate stimulus for gallbladder contraction for a prolonged period of time. If a patient is injected without previous adequate emptying of the gallbladder, there is a high possibility that the study will demonstrate false-positive findings with non-visualization of the gallbladder.

12. The patient was pretreated with a CCK infusion prior to radiotracer administration. The images shown below demonstrate the biliary transit during the initial 60 min. Which of the following statements is true?

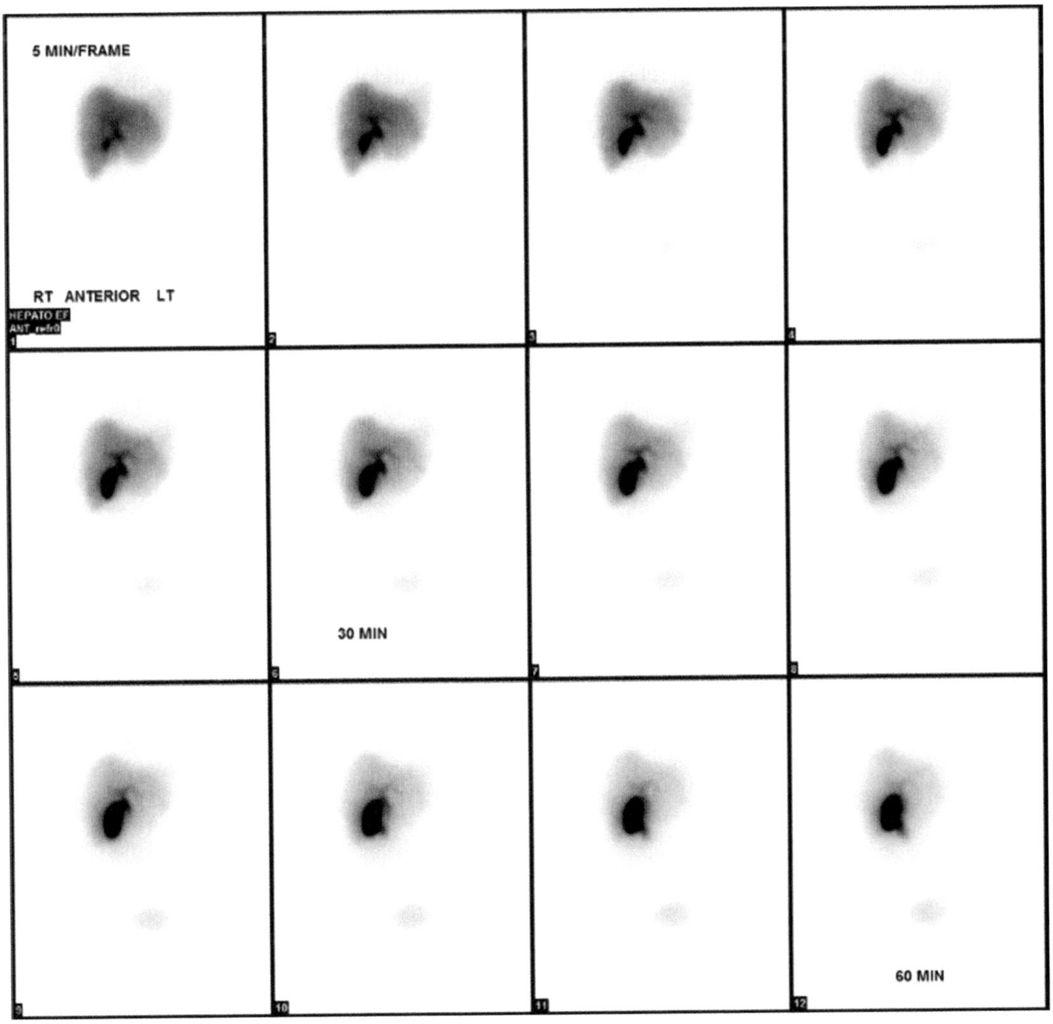

(a) A fatty meal challenge cannot be performed to determine the gallbladder ejection fraction given premedication with CCK
(b) CCK should not be given again during the same study to determine the gallbladder ejection fraction
(c) The patient needs to wait for at least 2 days before attempting to calculate the ejection fraction.
(d) CCK or a fatty meal may be given to evaluate the ejection fraction on the same day without a delay

Answer: d

Pretreatment of a patient with CCK to ensure adequate emptying of the gallbladder in preparation for a hepatobiliary study is not a contraindication for the subsequent determination of the gallbladder ejection fraction with either an additional infusion of CCK or with a fatty meal/supplement challenge immediately after the first part of the study.

References

1. Ziessman HA, Hepatobiliary scintigraphy in 2014. J Nucl Med Technol. 2014;42:249–59.
2. ACR-SNM-SPR practice guideline for the performance of gastrointestinal scintigraphy. Revised 2010.
3. Dillehay G et al. Appropriate use criteria for hepatobiliary scintigraphy in abdominal pain. J Nucl Med. 2017;195,602-be n 6/5/17.
4. Ziessman HA, Cholecystokinin cholescintigraphy: methodology and normal values using a lactose-free fatty-meal food supplement. J Nucl Med. 2003;44:1263–6.
5. Tulchinsky M, SNM practice guideline for hepatobiliary scintigraphy 4.0*. J Nucl Med Technol. 2010;38:4.

13. A 52-year-old male with history of COPD, s/p bilateral lung transplant, underwent gastric emptying study with solid meal. The images are shown below.

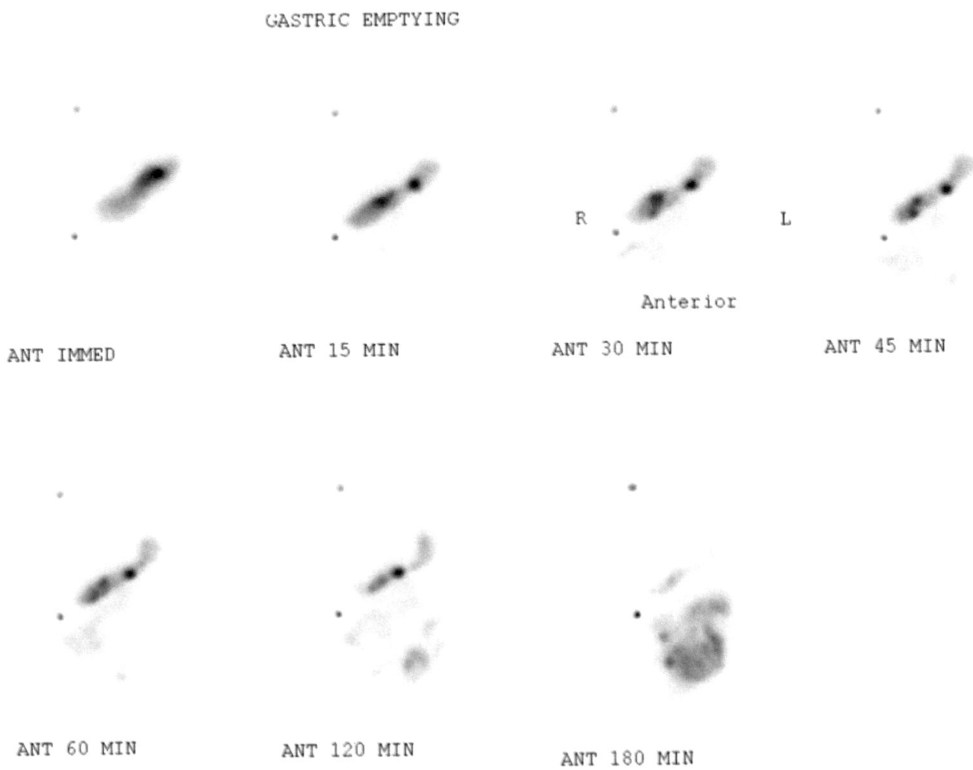

Image 1

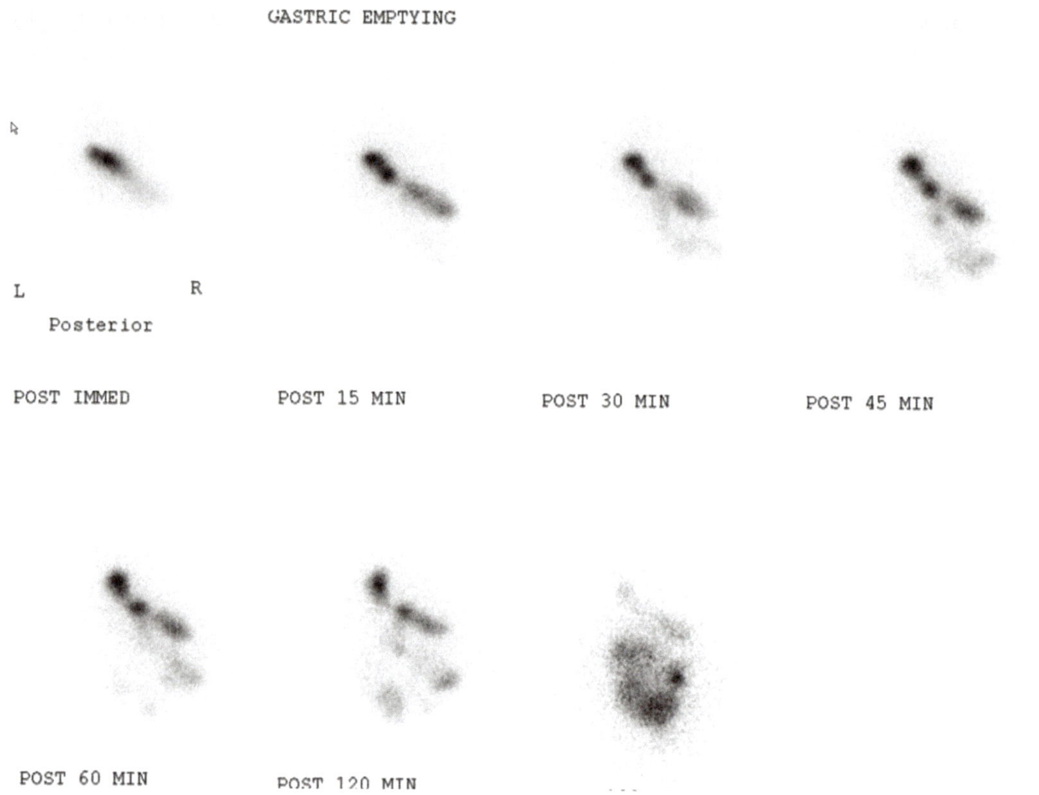

Image 2

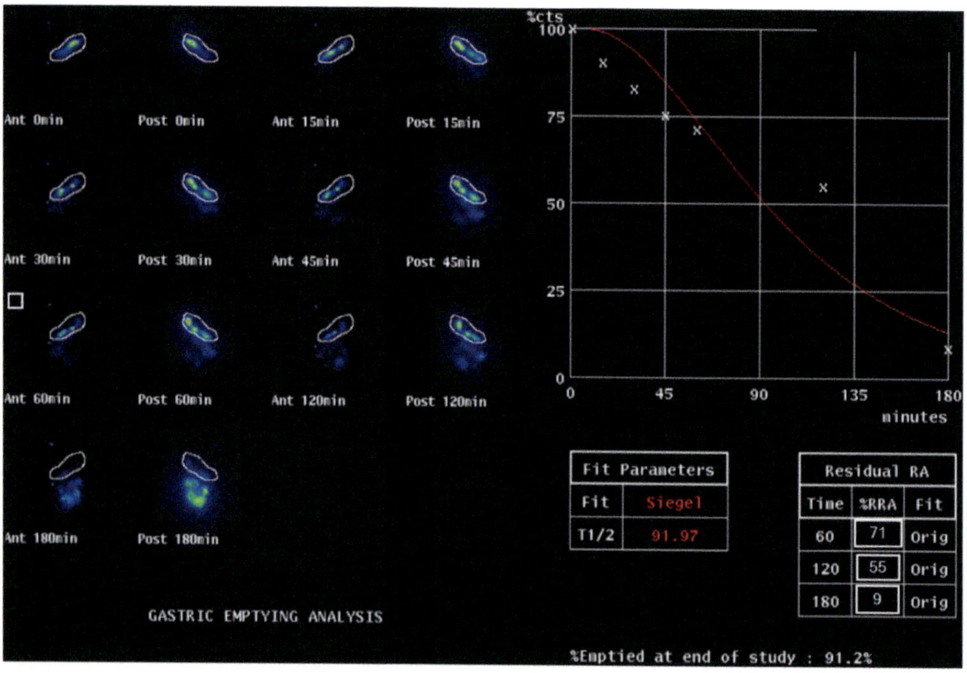

Image 3

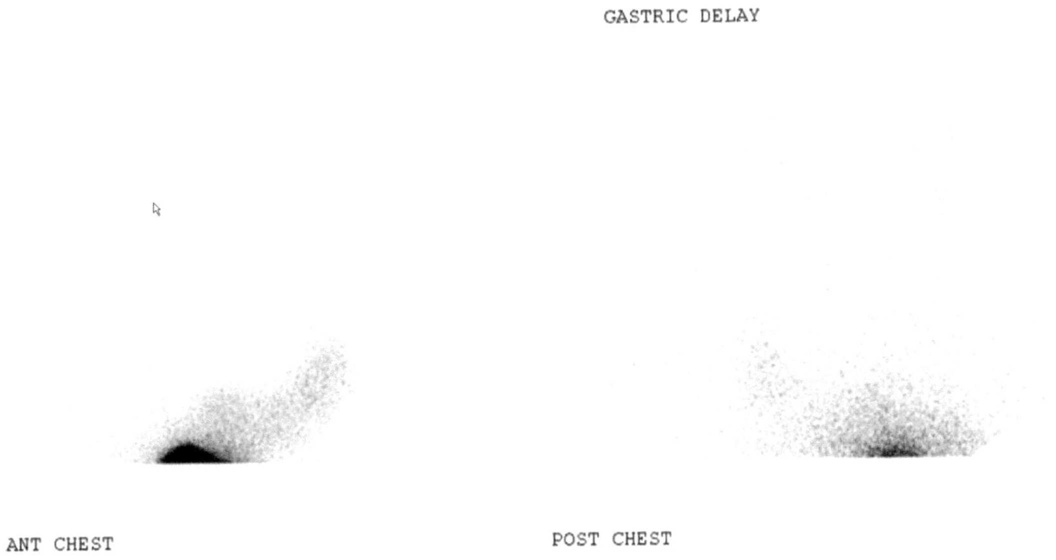

Image 4

Upon evaluation of the study, you conclude that the patient most likely has:

A. Delayed gastric emptying study
B. Aspiration of the radiotracer to the lungs
C. Accelerated gastric emptying study
D. Normal gastric emptying study

Answer: D

Gastric emptying study with solid meal was obtained in intervals over the abdomen for 3 h. It demonstrated emptying into the small bowel by 15 min (image 1 and 2).

By the end of the study (3 h imaging), the majority of tracer emptied to the small bowel with only 9% of radiotracer retained in the stomach.

Computer analysis demonstrated:

71% of gastric contents remain in the stomach by 60 min. (Normal >30 and < 90%).

55% of gastric contents remain in the stomach by 120 min. (Normal <60%).

9% of gastric contents remain in the stomach by 180 min. (Normal < 30%).

The findings are consistent with normal gastric emptying.

Delayed gastric emptying study of solids will demonstrate values above upper limits of normal, i.e., more than 10% remaining in the stomach after 4 h.

Accelerated transit of solids is defined as gastric emptying of > 70% by 60 min.

Delayed anterior and posterior views of the chest (image 4), obtained 5 h after the radioactive meal ingestion demonstrate no activity in the lungs to suggest aspiration.

References

- *Ziessman HA, O'Malley JP, Thrall JH, Nuclear medicine: the requisites.
- *Mettler FA, Guiberteau MJ, Essentials of nuclear medicine imaging.
- *StatDx. https://my.statdx.com/.

14. A 74-year-old female underwent gastric emptying study with solid food. Images are shown below.

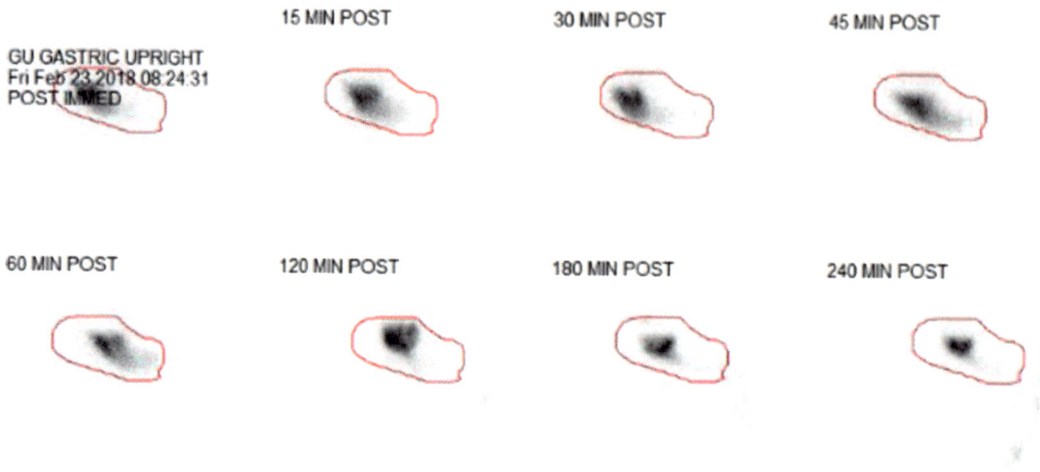

Image 1

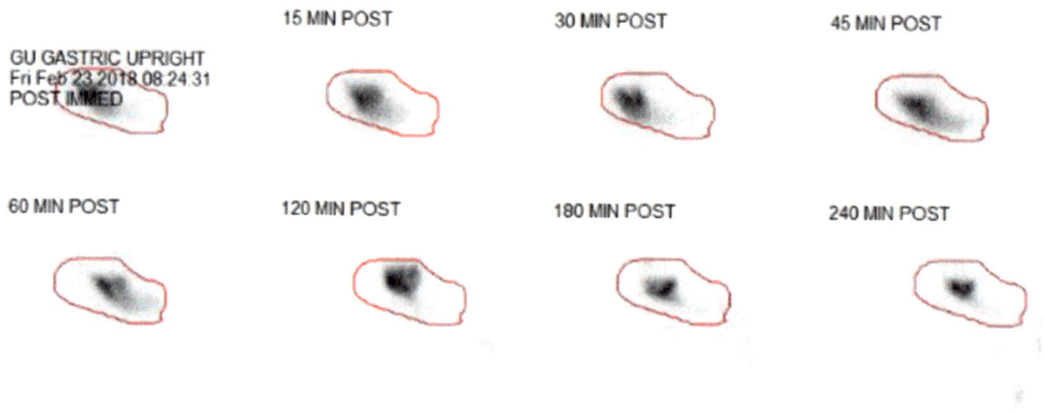

Image 2

Marker	Time	Total	Retention
Start	0 min	101669 Counts	100 %
1	61 min	83950 Counts	83%
2	121 min	92494 Counts	91%
3	179 min	66579 Counts	65%
4	240 min	54749 Counts	54%
Manual	240 min	54749 Counts	54%

Image 3

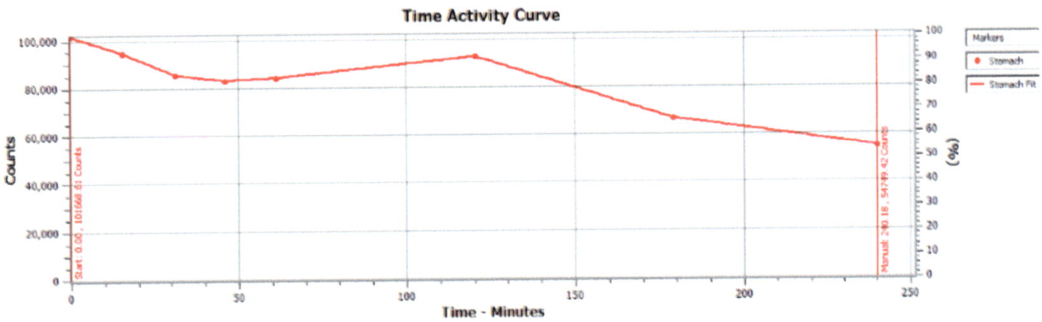

Image 4

Which of the following clinical conditions will likely be associated with imaging findings:

A. Remote history of gastrectomy
B. Zollinger Ellison syndrome
C. Pancreatic insufficiency
D. Diabetes Mellitus

Answer: D

Gastric emptying study was obtained in intervals over the abdomen for 4 h. It demonstrates emptying into the small bowel by 15 min.

By the end of 4 h imaging, the majority of tracer remained within the stomach with retention of 54%. The findings are consistent with delayed gastric emptying.

Among the clinical conditions listed above, diabetes is the most appropriate choice. Other clinical conditions, such as remote gastrectomy, Zollinger Ellison syndrome, and pancreatic insufficiency will likely result in accelerated emptying.

Chronic gastric stasis may be secondary to anatomical or functional abnormalities. Anatomical causes include gastric ulcer, surgery, vagotomy, pyloric hypertrophy, postradiotherapy, and tumors. Functional causes of chronic delayed gastric emptying are diabetic gastroenteropathy, which is the most common cause, systemic lupus erythematosus, amyloidosis, hypothyroidism, and gastroesophageal reflux. Many commonly used therapeutic drugs may also delay gastric emptying as well as opiates, alcohol, and nicotine.

15. An 86-year-old male presented with bright red blood per rectum. Upper endoscopy and colonoscopy were unremarkable. Red blood cell tagged nuclear medicine study was performed. Images are shown below.

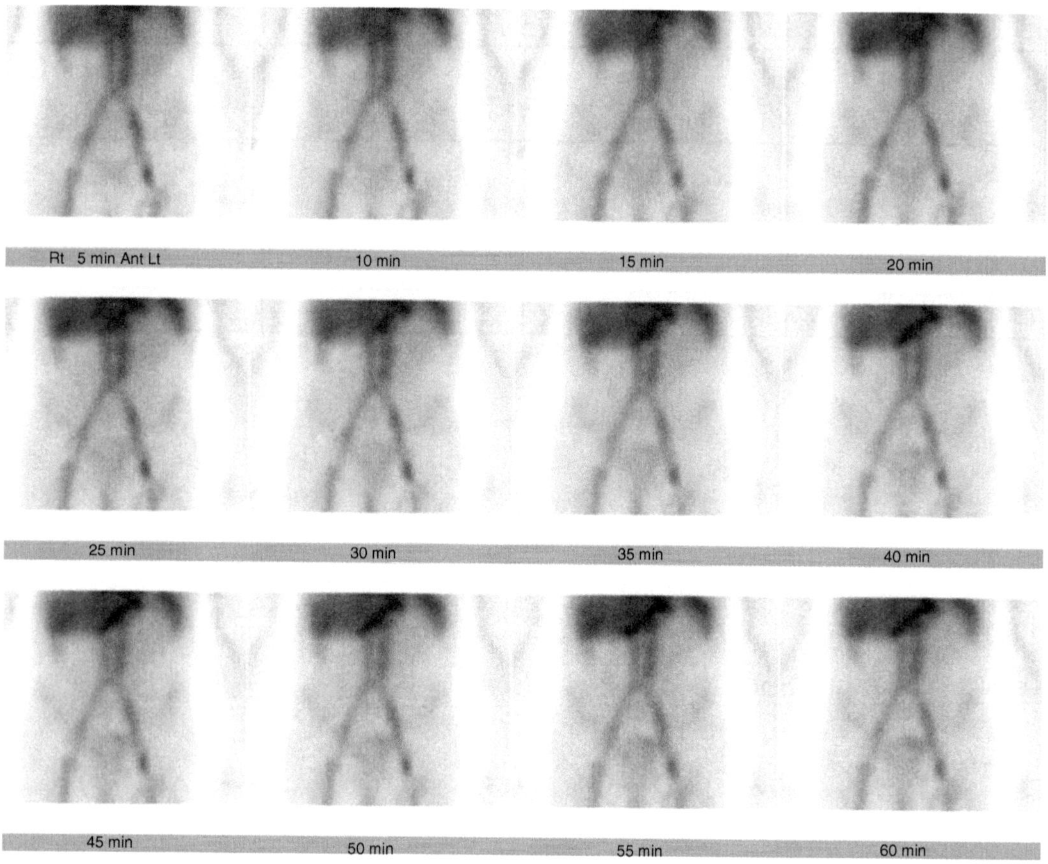

Image 1

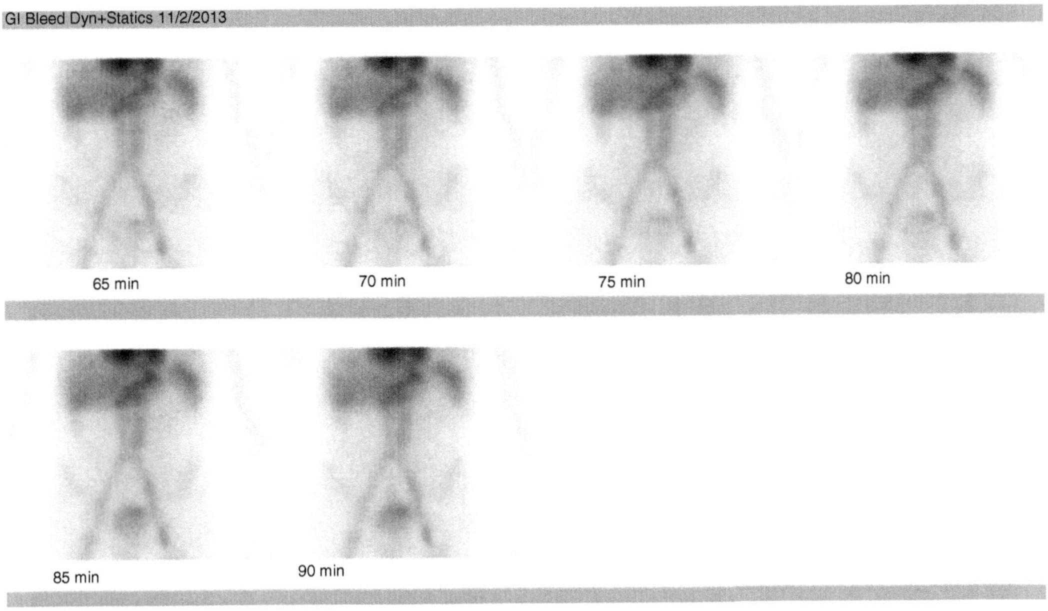

Image 2

Upon evaluation of the study, you conclude:

A. Bleeding within the stomach
B. Bleeding in the splenic flexure of the large bowel
C. Bleeding within the rectum
D. No evidence of gastrointestinal bleed

Answer: D

Red blood cell tagged nuclear medicine study was obtained in anterior projection over the abdomen for the period of 90 min. Images demonstrate normal tracer activity in the liver, spleen, aorta, and iliac arteries. Additional activity is seen in the mid-upper abdomen that does not move with time and likely corresponding to free technetium in the stomach. Later, activity in the bladder is noted. The study demonstrates no evidence of active GI bleeding.

16. A gastric emptying study is provided on a patient with a history of abdominal pain after eating. The patient has had multiple abdominal surgeries secondary to gunshot wounds.

    The time-activity curve demonstrates increasing activity near the end of the study. This is most likely secondary to:

    A. Formation of a postsurgical enterogastric fistula with reflux of intestinal contents back into the stomach.
    B. The presence of Mo99 breakthrough in the Tc-99 m eluate used to prepare the meal with subsequent decay and formation of additional Tc-99 m.
    C. De-labeling of the Tc-99 m sulfur colloid creating free Tc-99 m excretion into the stomach.
    D. The drawn region of interest includes a loop of bowel.

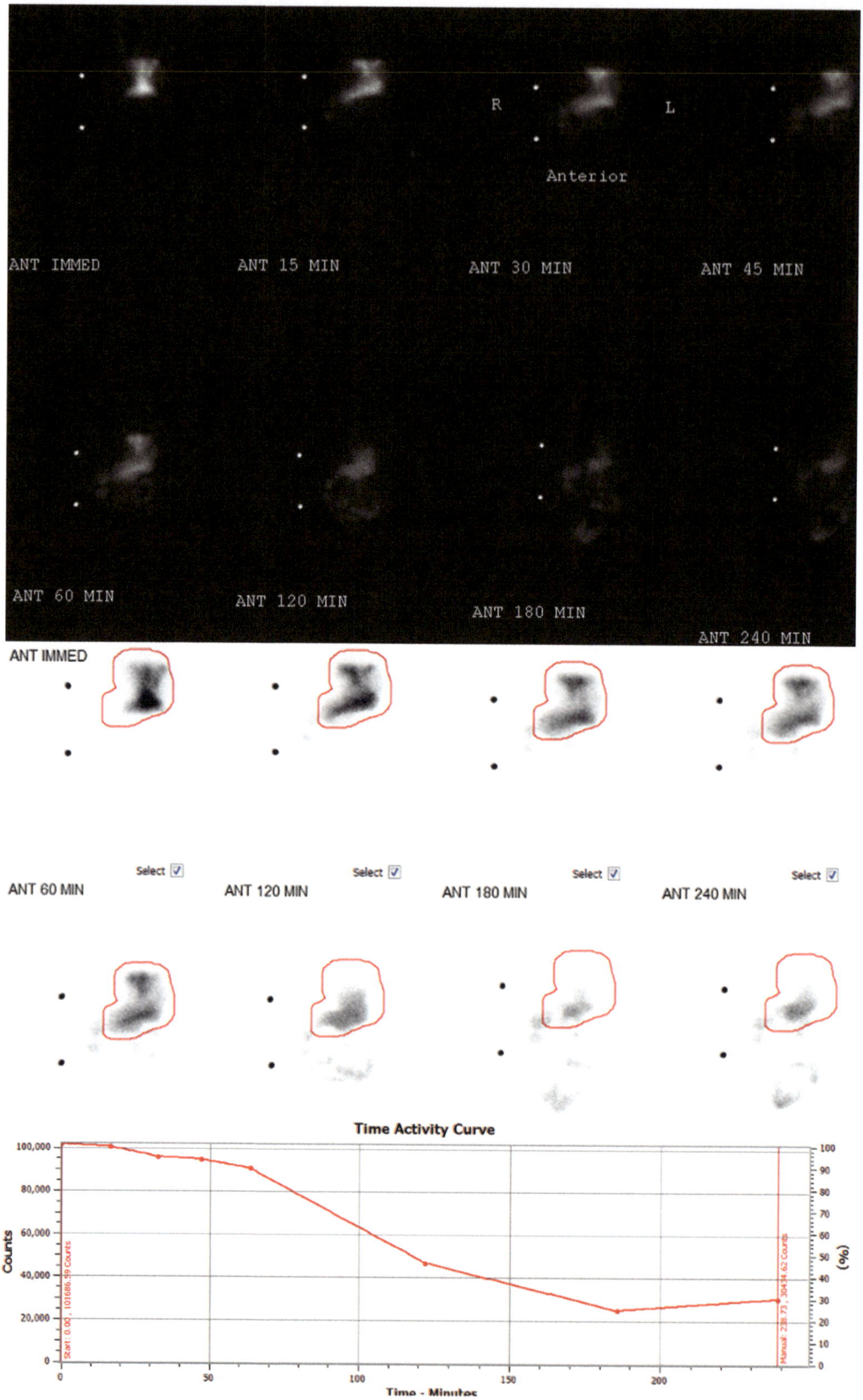
R
L
Anterior
ANT IMMED
ANT 15 MIN
ANT 30 MIN
ANT 45 MIN
ANT 60 MIN
ANT 120 MIN
ANT 180 MIN
ANT 240 MIN
ANT IMMED
Select
ANT 60 MIN
ANT 120 MIN
ANT 180 MIN
ANT 240 MIN
Time Activity Curve
Counts
(%)
Time - Minutes

Answer: D

The region of interest includes a loop of small bowel. This can best be seen in the 180 min image. Despite this, the contents of the stomach do not fall below 10% at 4 h and are consistent with delayed emptying.

If an enterogastric fistula were present, emptying would occur faster from the stomach into the small bowel.

For sufficient Tc-99 m generation to occur from Mo99 breakthrough, a serious violation of radiochemical purity would be present.

Tc-99m sulfur colloid which is used to prepare the meal is highly stable and does not dissociate in the stomach.

17. A 63-year-old male presents with a history of GI bleeding of suspected diverticular origin. The most commonly used radiopharmaceutical is:

    A. Tc-99 m heat-damaged red cells
    B. Histamine H2 blockers followed by Tc-99 m pertechnetate
    C. In-111 labeled platelets
    D. RBCs labeled with Tc-99 m in vitro

Answer: D

Tc-99 m labeled RBCs is the most commonly used radiopharmaceutical for the detection of GI bleeding. A high-efficiency method (in vitro technique) can provide up to 98% labeling with very little free pertechnetate. Free pertechnetate will confuse the results of the study since it is normally excreted by the gastric mucosa. This may give the impression of a GI bleed originating in the stomach.

Tc-99 m pertechnetate is used when a Meckel diverticulum is suspected. Most commonly seen in children, the Meckel diverticulum contains gastric mucosa that will accumulate the Tc-99 m pertechnetate. H2 blockers are often used to improve the quality of the scan by causing greater retention of Tc-99 m pertechnetate in the gastric mucosa.

Heat-damaged red cells have been studied for GI bleeding, but are most commonly used for the detection of splenules. In-111 platelets are used to determine the location of platelet sequestration in idiopathic thrombocytopenia purpura.

18. This 55-year-old patient presents with abdominal pain and concern for cholecystitis. You perform the study and review the images. Your findings suggest:

    A. Acute cholecystitis
    B. Duodenal diverticulum
    C. Bile reflux
    D. Vicarious excretion

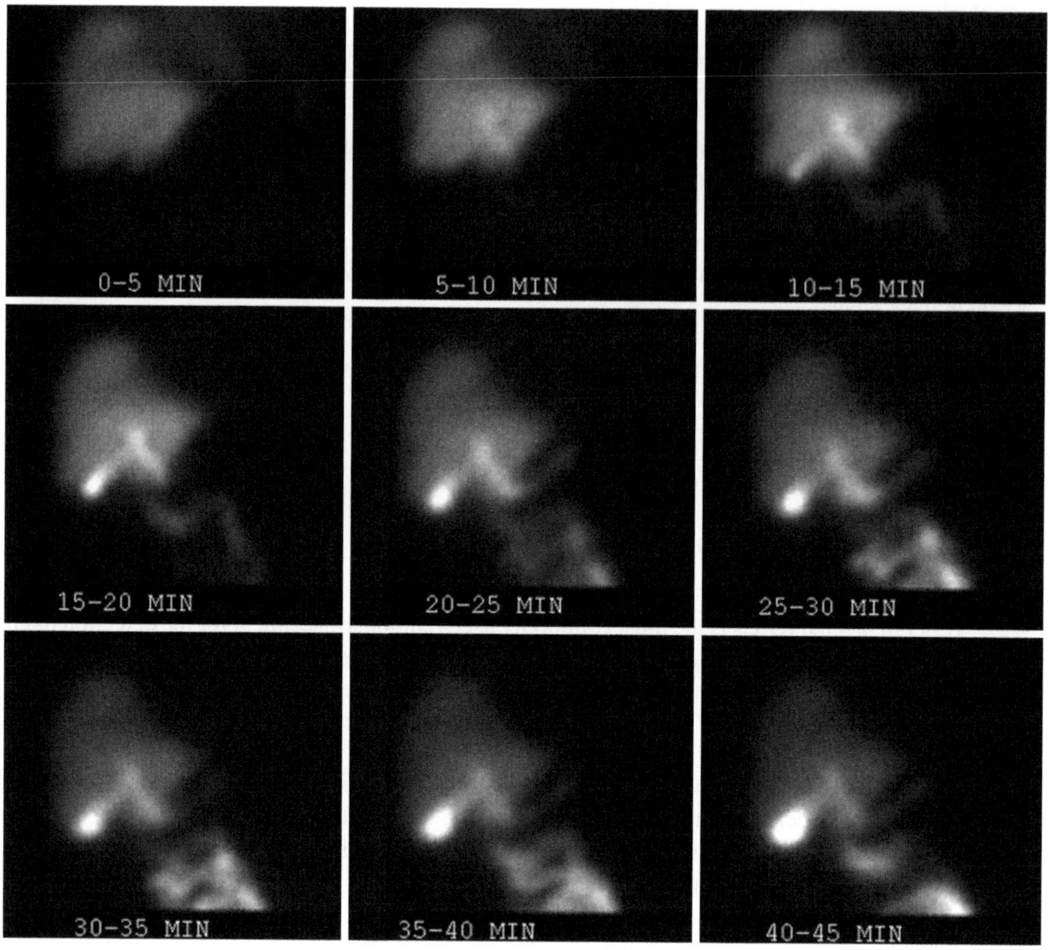

Answer: C

Bile reflux into the stomach can be one of the causes for abdominal pain. Sometimes, it is severe. As this is upper abdominal pain, it may be confused with biliary disease or even heart disease. On the hepatobiliary study, activity can be seen to appear on the 20–25 min image and persists through the remainder of the study.

The gallbladder visualizes promptly and activity is seen in the small bowel consistent with a patent cystic duct and common bile duct, respectively. A duodenal diverticulum is most frequently seen along the medial second part of the duodenum or the superior aspect of the third part of the duodenum. The gallbladder clearly fills, which speaks against the presence of acute cholecystitis. Vicarious excretion occurs in the presence of severe hepatocellular dysfunction when tracer is noted to be significantly excreted by the renal system.

19. This 71-year-old female presents with a history of intermittent right upper quadrant pain. The ultrasound demonstrates no cholelithiasis. The clinician suspects cholecystitis. You perform the study and after seeing the results after 1 h, you decide to give the patient a fatty meal. Additional imaging is performed and the gallbladder ejection fraction is calculated at 20%.
The study is most consistent with:

A. Acute cholecystitis
B. Chronic cholecystitis
C. Biliary dyskinesia
D. Hepatocellular dysfunction

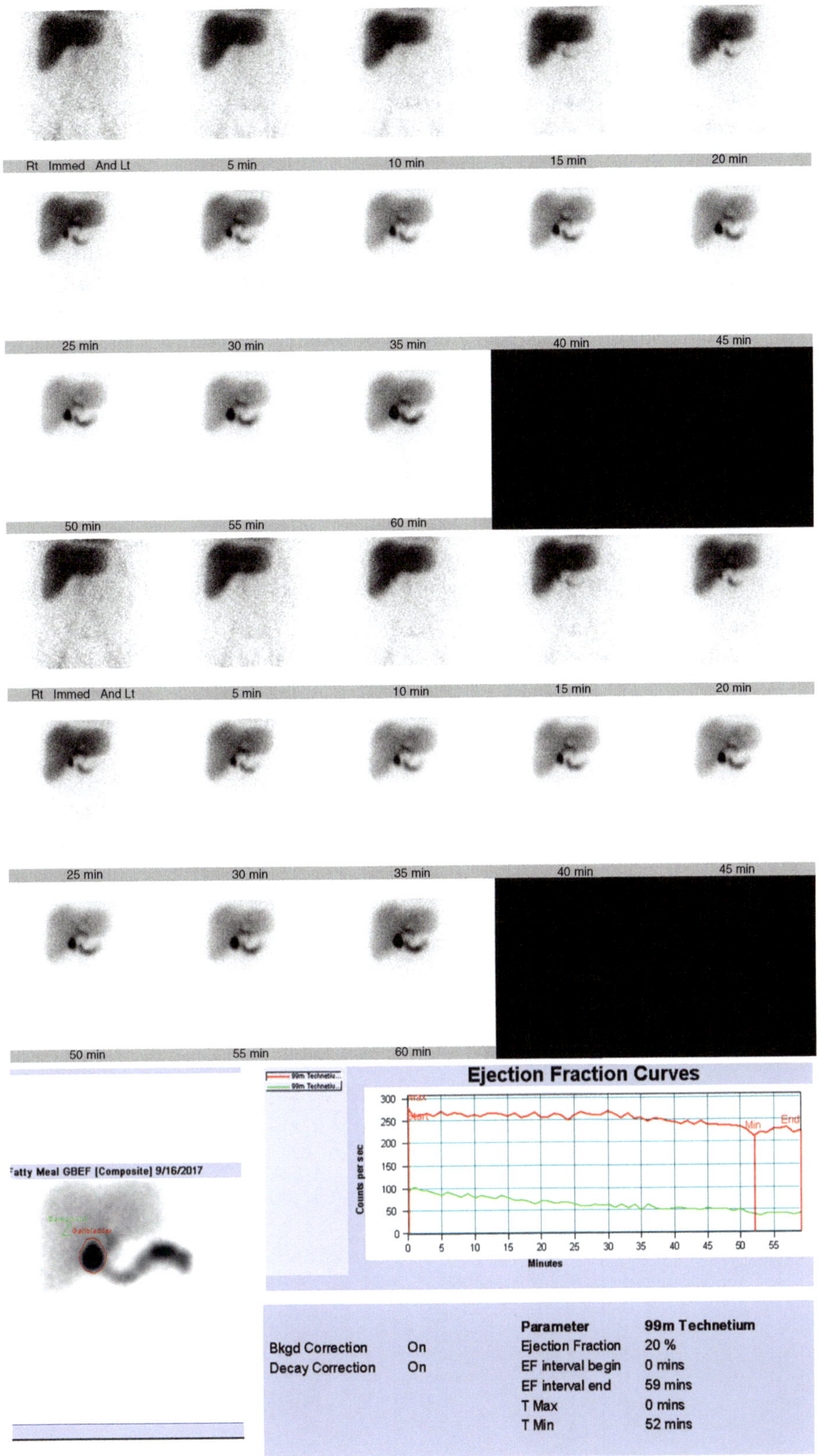
Rt Immed And Lt
5 min
10 min
15 min
20 min
25 min
30 min
35 min
40 min
45 min
50 min
55 min
60 min
Rt Immed And Lt
5 min
10 min
15 min
20 min
25 min
30 min
35 min
40 min
45 min
50 min
55 min
60 min
'atty Meal GBEF [Composite] 9/16/2017
Gallbladder
99m Technetiu...
99m Technetiu...
Ejection Fraction Curves
Max
Start
Min
End
Counts per sec
300
250
200
150
100
50
0
0
5
10
15
20
25
30
35
40
45
50
55
Minutes
Bkgd Correction On
Decay Correction On
Parameter 99m Technetium
Ejection Fraction 20 %
EF interval begin 0 mins
EF interval end 59 mins
T Max 0 mins
T Min 52 mins

Answer: C

Biliary dyskinesia is a disorder where the common and cystic ducts are patent and the gallbladder fills normally. The gallbladder however does not contract normally in response to either Sincalide or fatty meal stimulation. With Sincalide, the normal gallbladder ejection fraction should be 38% or greater. For a fatty meal, the gallbladder ejection fraction should be 33% or greater.

In acute cholecystitis, the gallbladder does not fill even after morphine augmentation. In chronic cholecystitis, there is often cholelithiasis. The gallbladder may fill after the first 60 min of the study or after morphine augmentation. In hepatocellular dysfunction, there is persistent blood pool activity as the hepatocytes are unable to extract the tracer from the circulation.

20. This 40-year-old female presents with a history of bloating, abdominal pain, nausea, vomiting, and diarrhea. An outside physician told her that she had gastroparesis. You perform a gastric emptying study with the standard meal of egg whites cooked with sulfur colloid, two pieces of bread, a packet of jelly, and a half glass of water. At 60 min, 21% of the gastric contents remain. At 120 min, less than 10% remains. You ended the study concluding:

    A. The outside physician is correct
    B. The study is normal
    C. The study is consistent with accelerated emptying
    D. The study is technically inaccurate and needs to be repeated

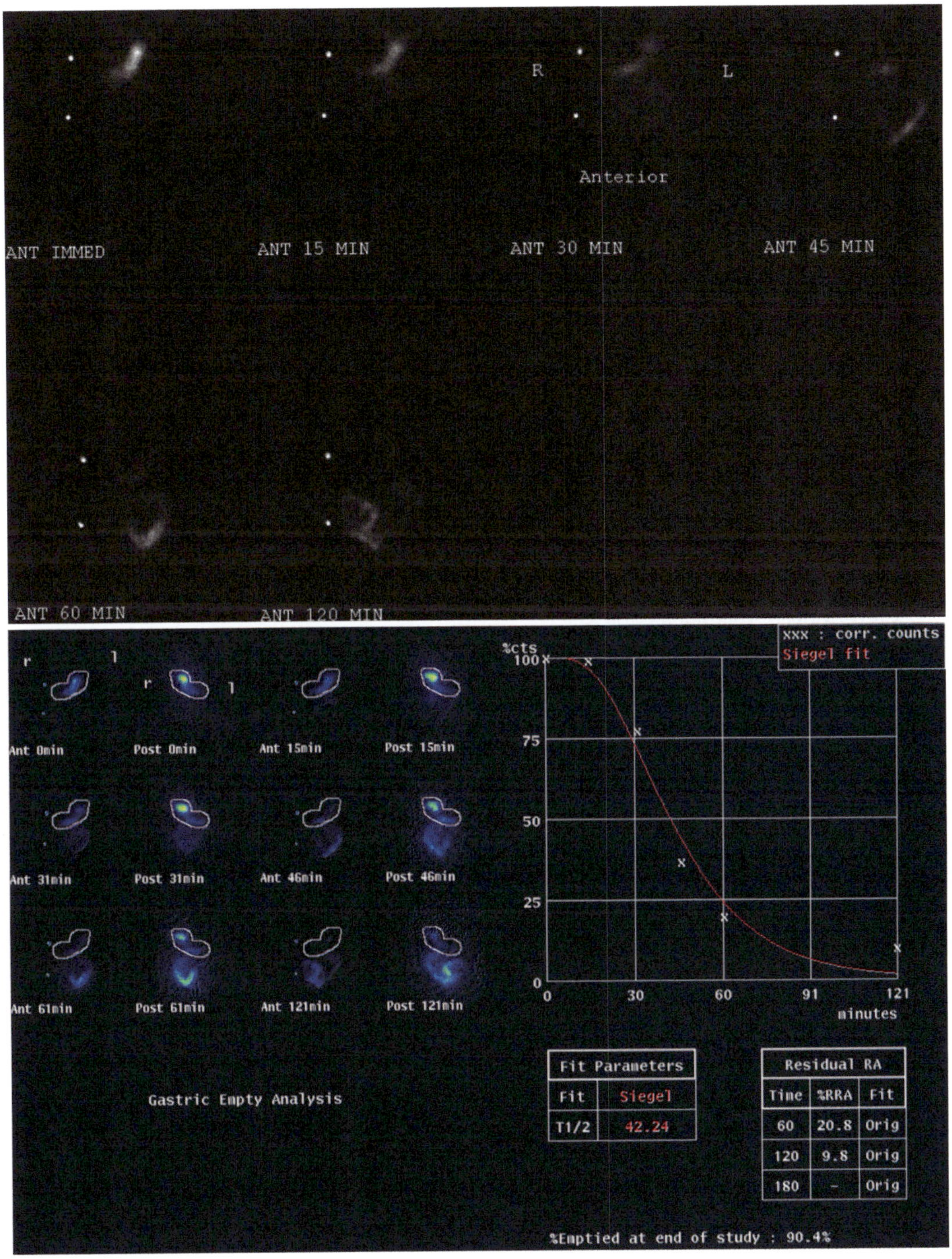
R
L
Anterior
ANT IMMED
ANT 15 MIN
ANT 30 MIN
ANT 45 MIN
ANT 60 MIN
ANT 120 MIN
xxx : corr. counts
Siegel fit
%cts
100
75
50
25
0
0
30
60
91
121
minutes
Ant 0min
Post 0min
Ant 15min
Post 15min
Ant 31min
Post 31min
Ant 46min
Post 46min
Ant 61min
Post 61min
Ant 121min
Post 121min
Gastric Empty Analysis
Fit Parameters
Fit Siegel
T1/2 42.24
Residual RA
Time %RRA Fit
60 20.8 Orig
120 9.8 Orig
180 - Orig
%Emptied at end of study : 90.4%

Answer: C

Patients who have dumping syndrome may have symptoms of bloating, abdominal pain, nausea, and vomiting. Patients with gastroparesis may also complain of bloating, abdominal pain, nausea, and vomiting. Thus, by symptoms alone, it may be difficult to differentiate one entity from the other. Since the lines of treatment are significantly different, it is important to distinguish between them. In a normal person, between 30% and 90% of the contents of the meal remain in the stomach at 1 h with less than 10% remaining at 4 h. In patients with accelerated gastric emptying, less than 30% remains at 60 min. It is important to confirm the region of interest is correct since placing it over a smaller portion of the stomach may make the study look like dumping. Placing the region of interest over the stomach and part of the small bowel will decrease the apparent emptying and may make the study appear like gastroparesis. However, the above study was done accurately with no technical errors.

21. This 78-year-old female presented to the emergency room with a history of abdominal pain and eight episodes of bloody stool. She has a prior history of atrial fibrillation and anticoagulation with Coumadin, hypertension, diabetes, and gastritis. On admission to the hospital, she is hypotensive with a hemoglobin level of 6.4, decreased from 9.1 on presentation to the emergency room. A colonoscopy was performed which demonstrated bleeding from a diverticulum in the distal ascending colon. Epinephrine was injected locally and clips were placed to stop the bleeding.

    Twenty-four hours later, the patient passed additional bright red blood. This GI bleeding study was performed. Your impression is:

    A. Likely de-labeling of tracer with accumulation of free Tc-99 m in the stomach and migration into the duodenum.
    B. Bleeding has recurred in the distal ascending colon
    C. Bleeding is present from adenocarcinoma in the pancreatic duct
    D. There is no active bleeding and the activity seen on the scan only reflects clots from earlier bleeding.

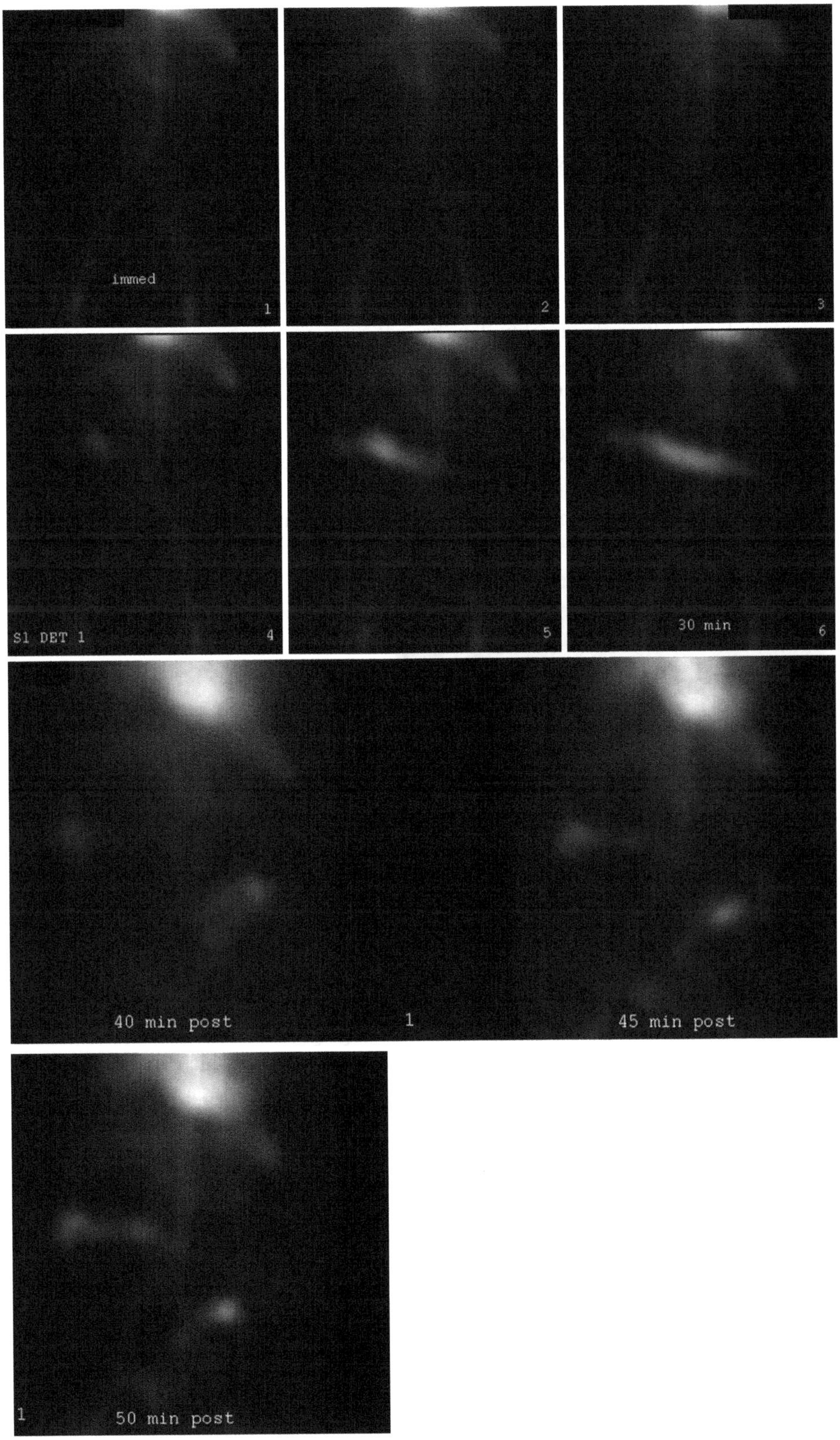
immed
1
2
3
S1 DET 1
4
5
30 min
6
40 min post
1
45 min post
1
50 min post

25. This 41-year-old female presented with right upper quadrant pain. Linear uptake is noted in the right lobe of the liver separating the upper right lobe from the lower right lobe. This finding is secondary to:

    A. Small angle scatter artifact
    B. Congenital biliary tract abnormality
    C. Gilbert's syndrome
    D. Superior vena cava obstruction

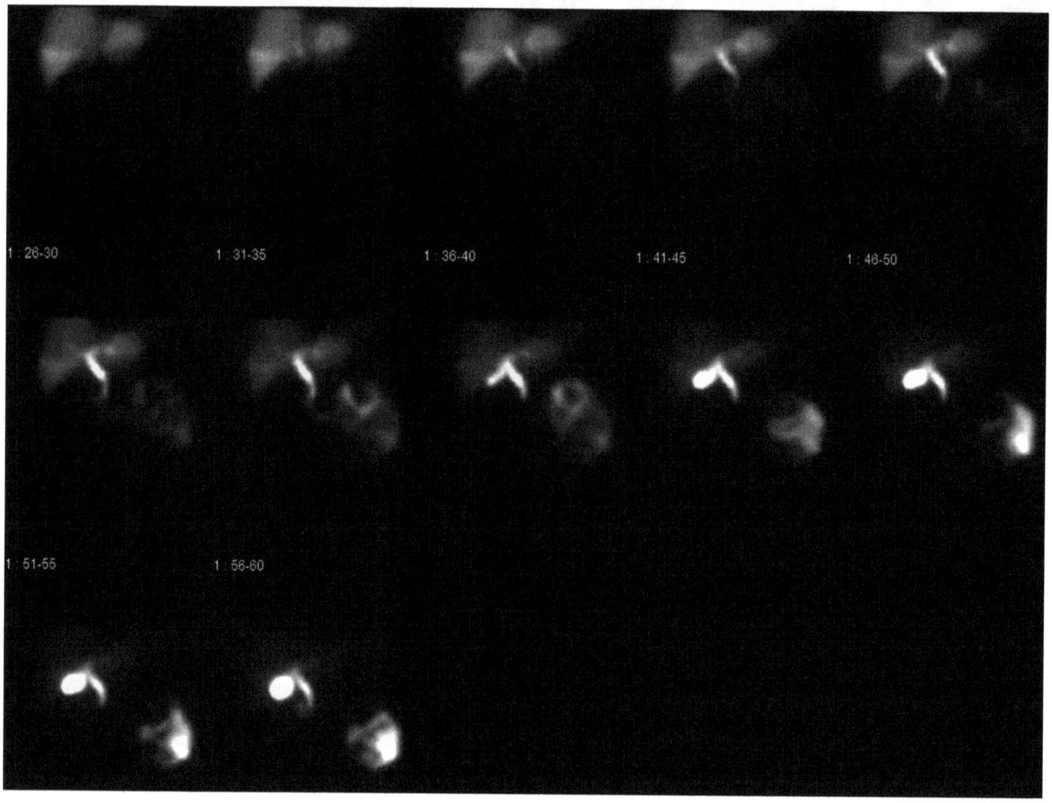

Answer: A

Small angle scatter artifact, which is also known as the "hot stripe sign" is caused by the overlying soft tissue of the breast. Careful examination of the image will show decreased activity in the upper portion of the right lobe of the liver with some adjacent decreased activity of the medial left lobe and lateral left lobe corresponding to breast shadow.

Gilbert's syndrome is a disorder of conjugation of bilirubin. There is poor uptake of tracer by the liver in these cases. Obstruction of the superior vena cava may lead to recanalization of the umbilical vein and retrograde flow. This, however is usually best seen on a liver-spleen scan with Tc-99 m sulfur colloid as a focal area of uptake anteriorly.

26. This neonate presents with situs inversus, a horizontal appearing liver and questionable asplenia. A CT was unable to definitively determine the presence of a spleen. A liver-spleen scan is provided in the anterior, posterior, and lateral projections.

    What additional scintigraphic study would be helpful in confirming the presence or absence of a spleen?

    A. A Ga67 study
    B. A labeled WBC study
    C. A HIDA study
    D. Heat damaged red cells

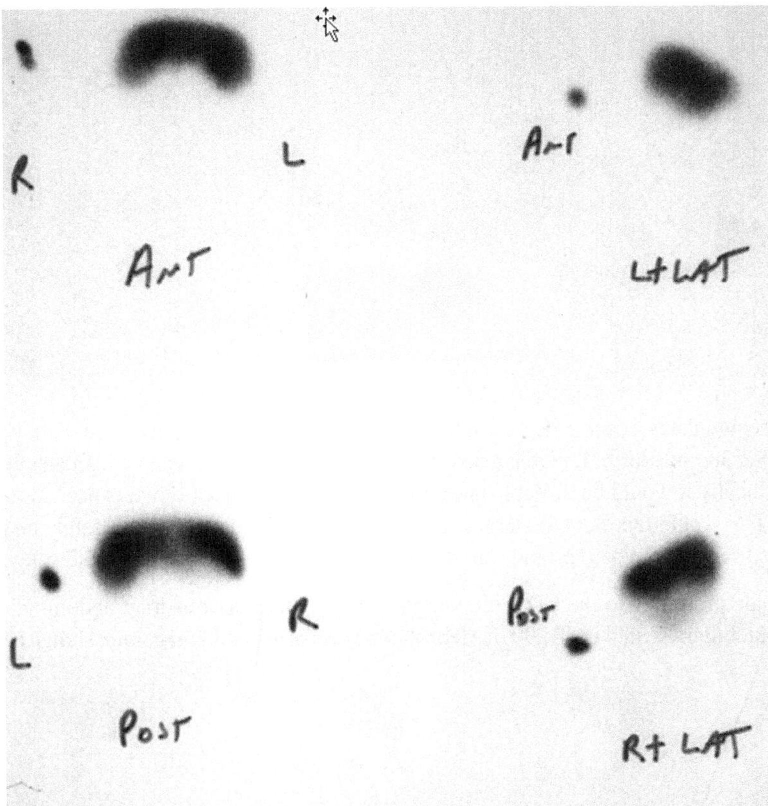

Answer: C

A HIDA study would be most helpful. When viewing this study, you should remember that the organs are reversed. On the first image (a liver-spleen study performed with Tc-99 m sulfur colloid) demonstrates what appears to be the right lobe on the left side of the patient, but there is no distinction between the liver and spleen. Two days later, a HIDA study (below) was performed for comparison. Since the sulfur colloid is collected by the liver and spleen while the HIDA agent is collected by the liver only, the difference in the image appearance is due to the presence of a spleen. Note the presence of progressive visualization of the renal collecting systems on the HIDA study.

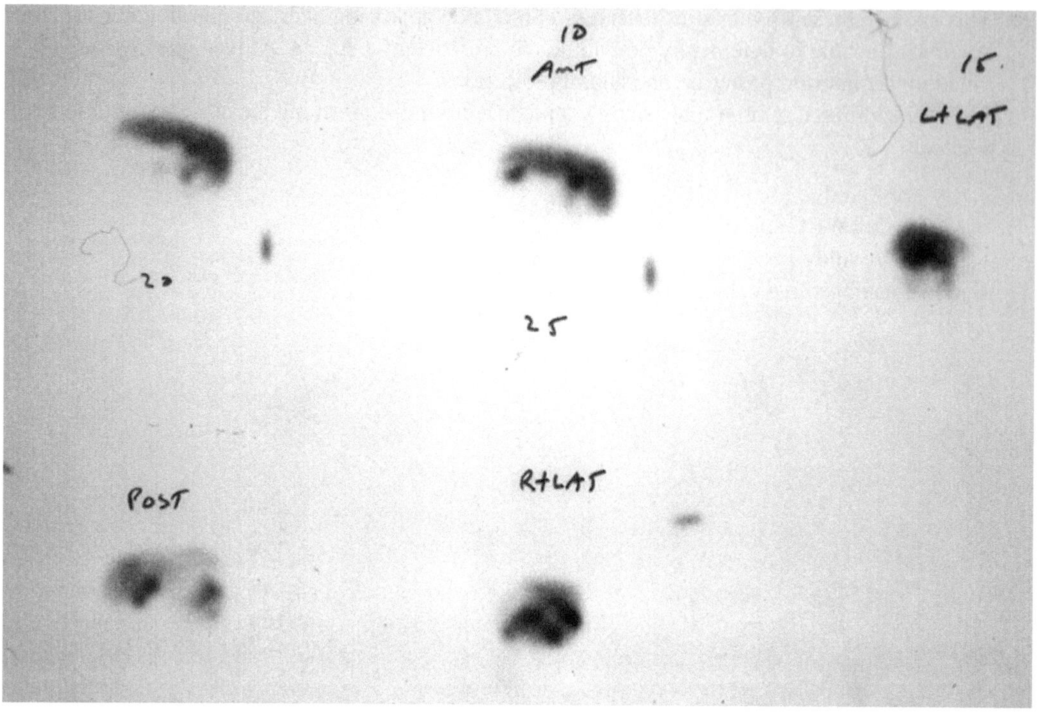

Gallium often accumulates to some degree in the spleen and would not be the optimal choice for separation of the liver and spleen. Although labeled WBC's would show the spleen with greater uptake than the liver, technically it would be difficult since approximately 50 cm^3 of blood is needed to perform the labeling. This would represent too large a fraction of the blood of a neonate to safely remove. A heat-damaged red cell study would provide similar images as the Tc-99 m sulfur colloid study.

Case 5 31-year-old male presented to the ED with complaints of right upper quadrant abdominal pain since last evening. Hepatobiliary scan with Tc-99 m Mebrofenin was performed. Images are shown below.

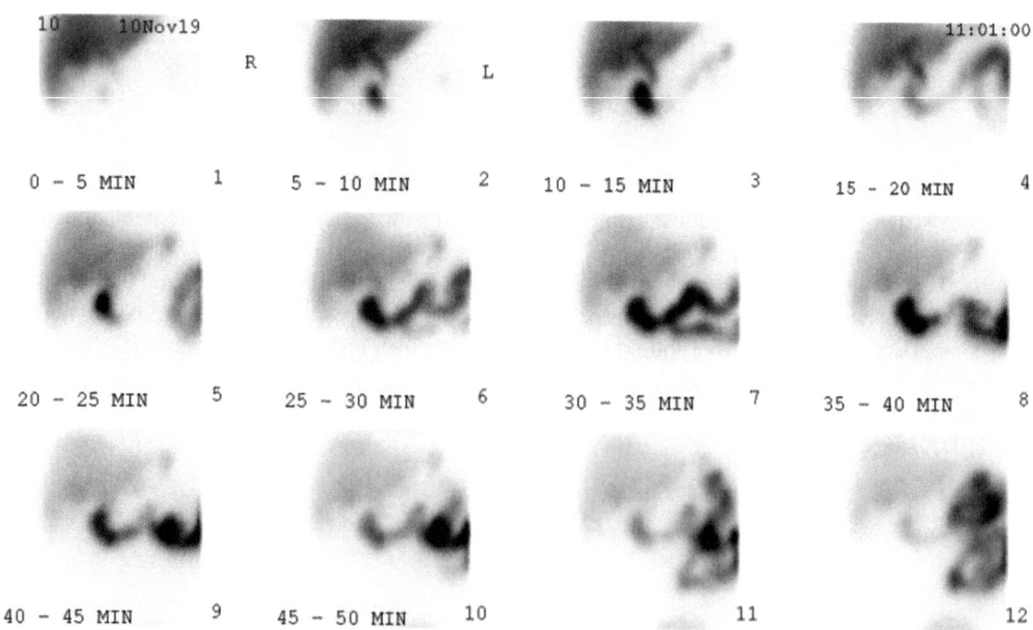

27. Which of the following is the next best step:

   A. Rescan the patient at 24 h
   B. Inject Morphine and additional radiotracer
   C. Inject Cholecystokinin
   D. Send the patient home

   Answer: B

   Tc-99 m Mebrofenin hepatobiliary scan demonstrates good extraction of tracer from the circulation, consistent with good hepatocellular function. There is a visualization of the tracer within the small bowel by the 5–10 min image. This is confirmed on subsequent images and consistent with patient common bile duct.

   As the study progresses, the gallbladder was not visualized. Instead, a curvilinear area of slightly increased radiotracer uptake is noted in the hepatic parenchyma adjacent to the expected location of the gallbladder fossa. This can be referred to as the "rim sign," which is often seen in cases of severe acute cholecystitis.

   Morphine, an opiate drug, is commonly used to aid in the diagnosis of acute cholecystitis. Given intravenously at a low dose of 0.04 mg/kg, morphine sulfate causes the sphincter of Oddi to constrict, which in turn increases the intrabiliary pressure. A booster dose of Tc-99 m Mebrofenin should be given because most of the tracer would be already in the bowel. The standard protocol is to image for an additional 30 min. If the cystic duct is patent, activity should be seen in the gallbladder, but if it is obstructed, obviously the gallbladder will not be visualized.

   Non-filling of the gallbladder by 60 min after Tc-99 m Mebrofenin injection is abnormal, though not diagnostic of acute cholecystitis. It adds confirmation to the clinical suspicion of acute cholecystitis. Non-visualization of the gallbladder might occur in patients without acute cholecystitis (false-positive). This is particularly seen in extensively sick patients or in those who do not fast. Also, prolonged fasting may result in false-positive results.

   Cholecystokinin, CCK, is a polypeptide hormone released by mucosal cells of the proximal small bowel in response to fat and protein. Exogenous administration of CCK will cause the gallbladder to contract and the sphincter of Oddi to relax. This would be an inappropriate step in this patient with presumed cystic duct obstruction.

28. The patient underwent an additional 30 min of imaging after the Morphine augmentation and a booster dose of Mebrofenin. Images are shown below.

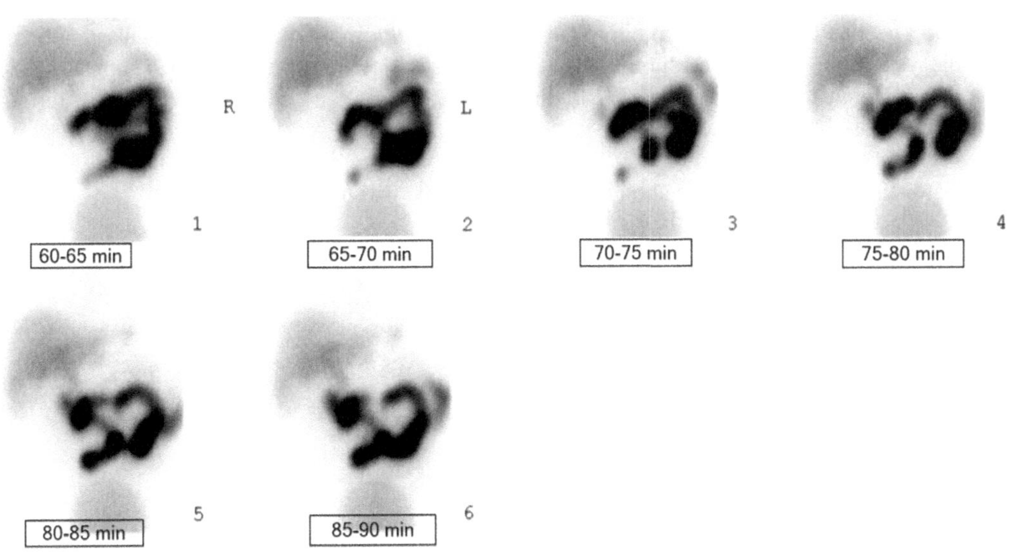

Upon evaluation of images, you conclude that the study is consistent with:

A. Acute cholecystitis
B. Partial biliary obstruction
C. Biliary leak
D. Patent cystic duct

Answer: A

Additional post-Morphine images demonstrate progressively increasing radiotracer accumulation within the small bowel but still non-visualization of the gallbladder. A picture is typically seen with cystic duct obstruction, supporting the clinical suspicion of acute cholecystitis.

The most common cause for partial biliary obstruction is choledocholithiasis. With partial obstruction, prompt hepatic uptake and secretion into biliary ducts occur. Delayed transit into the intestinal tract (delayed biliary to bowel transit time) is commonly observed in these situations. In our case, there is normal biliary to bowel transit and normal clearance from biliary ducts.

There is no scintigraphic evidence of radiotracer extravasation outside the biliary tree to suggest biliary leak.

29. During the initial 60 min of a HIDA study, there is clear evidence of gallbladder activity but no bowel activity. The patient is asymptomatic. Practically, what is the next step?

A. The study is abnormal, and no additional images are needed.
B. A lactose-free fatty supplement may be given, but CCK administration is contraindicated.
C. CCK administration may be given, but fatty meat is contraindicated.
D. A lactose-free fatty supplement or CCK may be used to evaluate the GBEF.

Answer: C

Delayed biliary to bowel transit can be seen in normal asymptomatic patients with subsequently proven normal gallbladder contractility. Different groups continue to debate regarding the need of scintigraphic evidence of biliary to bowel transit prior to gallbladder ejection fraction calculation with CCK or a fatty meal. However, a gallbladder ejection fraction calculation utilizing a fatty laden meal or a CCK infusion can still be performed if there is no concern for a high-grade obstruction, particularly in asymptomatic patients with normal gallbladder on ultrasound.

Case 6 This 58-year-old male patient with a history of diabetes mellitus was admitted with leukocytosis and complaints of abdominal pain and nausea. Abdominal sonography performed the day of admission revealed cholelithiasis, minimal gallbladder wall thickening, and a very small amount of pericholecystic fluid. A positive Murphy sign was reported. Later on the day, a follow-up CT of the abdomen and pelvis demonstrated similar findings. Two days later, a HIDA was ordered for further evaluation of possible acute cholecystitis.

30. What is the most likely explanation for the scintigraphic findings shown in Fig. 4?

A. Partial obstruction of the common bile duct.
B. Partial obstruction of the cystic duct.
C. High-grade obstruction.
D. Complicated acute cholecystitis.

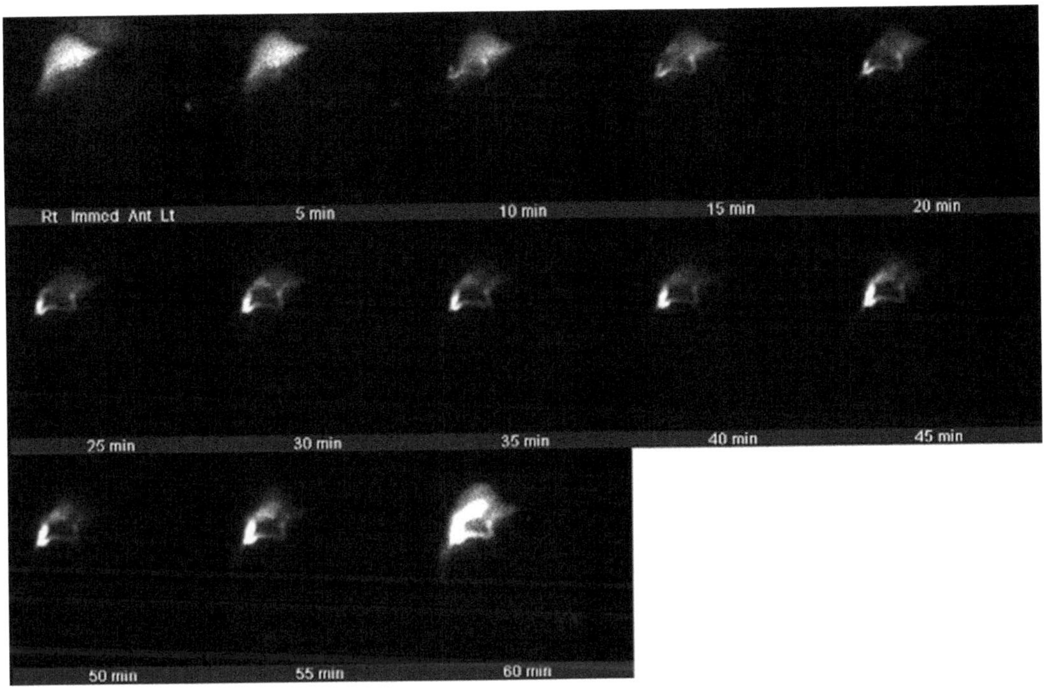

Fig. 4 Dynamic HIDA Scan

Answer: D

The hepatobiliary scan shown in Fig. 4 initially demonstrates prompt and predominantly uniform hepatic uptake soon after injection. At approximately 10 min postinjection, there is unexpected extrahepatic transit of activity from the margin of the gallbladder fossa, into the perihepatic space. The activity continues to accumulate and moves upward until it covers the surface of the liver. By the end of the initial 60 min of imaging, there is evidence of some extension of activity freely into the peritoneum. Neither biliary to bowel transit nor filling of the gallbladder is evidenced by the end of the study. In this patient without previous history of cholecystectomy, the findings are most consistent with the sequelae of complicated acute cholecystitis and a ruptured gallbladder.

31. What is the best next step in the management of this patient?

    A. Inject 2 mg of morphine and image for additional 30 min
    B. Perform delayed images at 4 h
    C. Cholecystectomy with drainage placement
    D. Initiation of IV antibiotics and observation

Answer: C

The severity of the findings described above requires prompt surgical management. Further delayed imaging is not necessary or appropriate. Administration of morphine for further imaging to confirm the cystic duct patency is contraindicated in this scenario given the level of inflammation and degree of obstruction. Antibiotics treatment may only be considered if surgery is contraindicated.

32. What is the accuracy of hepatobiliary scintigraphy for the diagnosis of acute calculous cholecystitis?

    A. About 80%–90% sensitivity and specificity
    B. About 95% sensitivity and 70% specificity
    C. About 70% sensitivity and 90% specificity
    D. About 95% sensitivity and 90% specificity

    Answer: D

    Systematic literature reviews of the diagnostic accuracy of HIDA scintigraphy and abdominal sonography for the detection of acute cholecystitis determined sensitivity and specificity of 96% and 90%, respectively, for cholescintigraphy.

33. What is the accuracy of the abdominal ultrasound for the diagnosis of acute cholecystitis?

    A. About 60% sensitivity and specificity
    B. About 70% sensitivity and specificity
    C. About 80% sensitivity and specificity
    D. About 90% sensitivity and specificity

    Answer: B

    The abdominal sonography has an 81% sensitivity and 83% specificity [2].

34. If clinically suitable, what would be the best technique to recommend for confirmation of suspected complicated gallbladder disease, including gallbladder perforation?

    A. Abdominal sonography
    B. Contrast-enhanced CT
    C. ERCP
    D. MRI

    Answer: B

    The recommended image algorithm for acute upper abdominal pain depends on the patient's presentation and suspected diagnosis. Cholescintigraphy usually follows abdominal sonography, particularly in patients with inconclusive sonographic findings for acute cholecystitis. The most recently published criteria for the use of hepatobiliary scintigraphy in patients presenting with acute abdominal pain is defined as *appropriate* (Score: 7). Although ultrasound is a useful tool in the initial evaluation of abdominal pain, if complications are suspected confirmation with CT and MRI is frequently needed. If gallbladder perforation is suspected based on the scintigraphic findings and no other cross-sectional imaging studies are available for comparison, further evaluation with a contrast-enhanced CT of the abdomen should be recommended. CT findings that can help confirm a complication include; a collapsed gallbladder with increased pericholecystic fluid, the presence of an abscess, and an extraluminal gallstone [1]. ERCP and/or MRI/MRCP could be performed with the additional benefit of potential intervention. However, these modalities are more time-consuming and should be considered carefully in unstable patients.

References

1. Owens JO, et al., Imaging of cholecystitis. Am J Roentgenol. 2011;196(4):W367–W374.
2. Gary D et al. Appropriate use criteria for hepatobiliary scintigraphy in abdominal pain. Soc Nucl Med Molecul Imag. DOI: https://doi.org/10.2967/jnumed.117.195602.
3. Ziessman HA, Hepatobiliary scintigraphy in 2014. J Nucl Med Technol. 2014;42:249–59.

Genitourinary System

Medhat Sam Gabriel, Bital Savir-Baruch, Robert H. Wagner, Bruch J. Barron, Davide Bova, Aamna Hassan, Stephanie Rodriguez-Santiago, and Yekaterina Kucerova

Case 1

1. A 16-year-old girl underwent Tc-99m MAG3 renal scintigraphy for evaluation of the left hydronephrosis. Images are shown below.

M. S. Gabriel (✉) · B. Savir-Baruch · R. H. Wagner · D. Bova
Division of Nuclear Medicine, Department of Radiology, Loyola University Medical Center, Maywood, IL, USA
e-mail: sgabriel@lumc.edu; Bital.savir-Baruch@lumc.edu; rwagner@lumc.edu; dbova@lumc.edu

B. J. Barron
Division of Nuclear Medicine and Molecular Imaging, Department of Radiology and Imaging Sciences, Atlanta, GA, USA
e-mail: bjbarro@emory.edu

A. Hassan
Department of Nuclear Medicine, Shaukat Khanum Memorial Cancer Hospital and Research Center, Lahore, Punjab, Pakistan

S. Rodriguez-Santiago
Department of Nuclear Medicine, C.W. Bill Young VA Medical Center, Bay Pines, FL, USA
e-mail: stephanie.rodriguezsantiago@va.gov

Y. Kucerova
Division of Nuclear Medicine, Department of Radiology, Loyola University Medical Center, Maywood, IL, USA

Department of Radiology, Cook County Healthy, Chicago, IL, USA
e-mail: yekaterina.kucerova@cookcountyhealth.org

M. S. Gabriel, B. Savir-Baruch (eds.), *RadTool Nuclear Medicine MCQs*,
https://doi.org/10.1007/978-3-030-69281-0_9

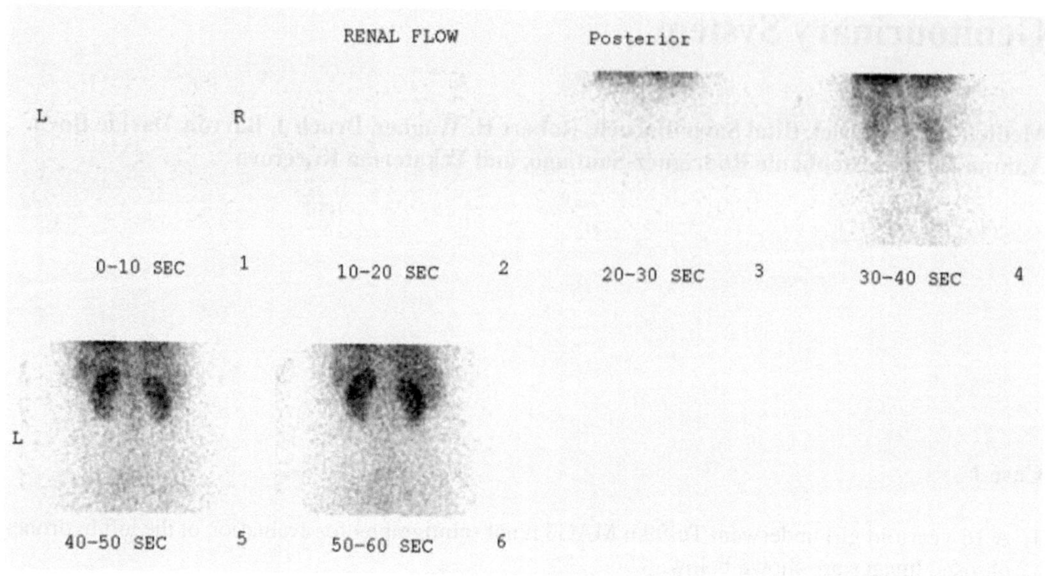

Image 1

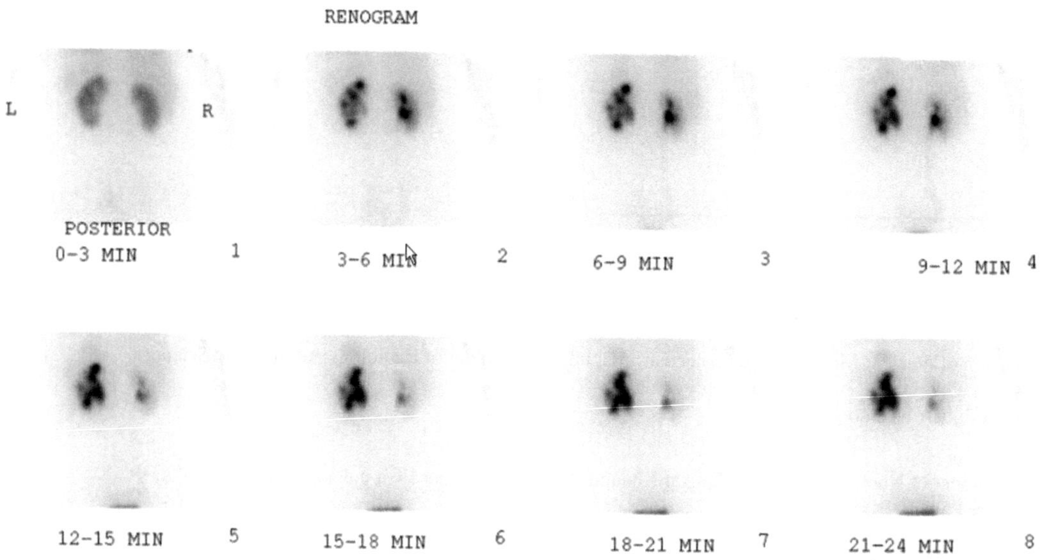

Image 2

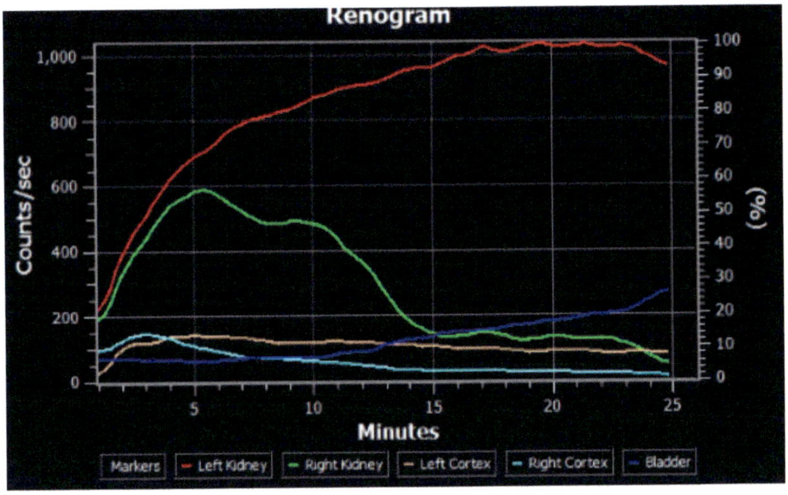

Image 3

Upon evaluation of images, you conclude:

A. Normal left kidney
B. Abnormal tracer uptake by right renal cortex; Lasix should be administered
C. No significant tracer washout from the left collecting system; Lasix should be given
D. Check patient's creatinine level; if it is two times the upper level of normal, reduce the Lasix dose by a factor of 2

Answer: C

Tc-99m MAG3 renal scintigraphy dynamic images of the left kidney demonstrate prompt tracer uptake by the cortex with some delayed tracer motion from the renal cortex towards the renal pelvis. There is significant tracer retention within known dilated renal pelvis with no significant tracer excretion towards the urinary bladder. In order to evaluate for mechanical obstruction, Lasix should be administered. The dose should be calculated based on the creatinine level. The usual adult dose is 40 mg intravenously. A larger dose, up to 80 mg, may be required in patients with impaired renal function. Upon increasing urine flow after administration of a diuretic, a nonmechanical obstruction (sometimes is referred to as functional obstruction) may be overcome by increasing pressure in the renal pelvis and thus allowing urine to flow from the collecting system to the ureter and the bladder. Diuresis will not overcome an anatomic obstruction such as an impacted stone.

2. After the initial set of images (21 min), the patient was advised to use the bathroom and evacuate the urinary bladder. Then, Lasix was administered. Images are shown below:

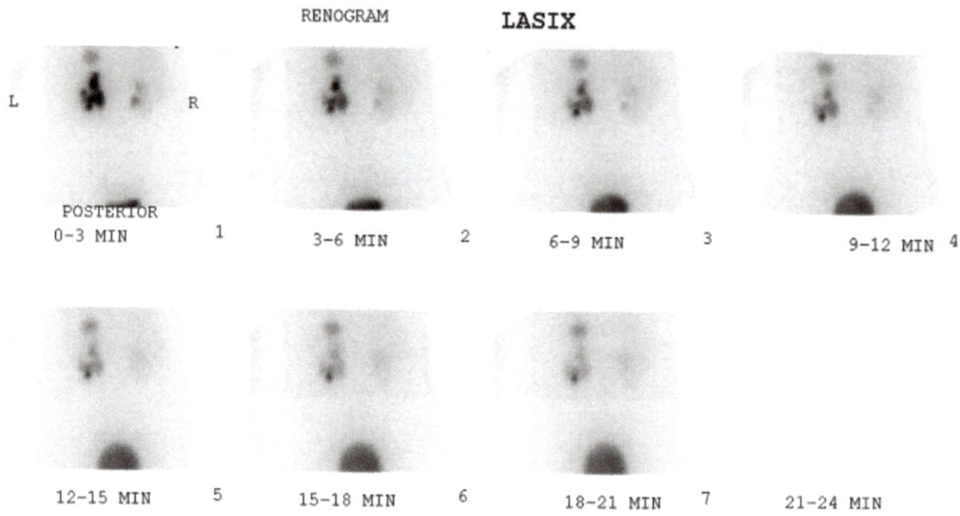

Image 1

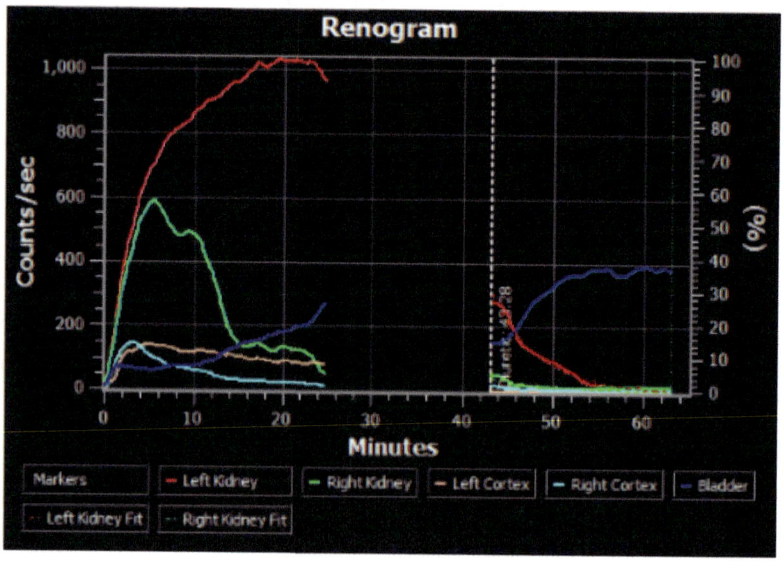

Image 2

Upon evaluation of the study, you conclude:

A. Normal renal scintigraphy study
B. Patient should be on constant diuretic therapy, given the significant response to Lasix
C. Mechanical obstruction of the left kidney
D. Dilated left collecting system with but evidence of mechanical obstruction

Answer: D

Tc-99m MAG3 renal scintigraphy (image 2) demonstrates significant emptying of the left kidney collecting system post-void suggesting motion of the tracer from the renal pelvis towards the bladder secondary to gravity (counts difference between initial 24 min and beginning of Lasix administration at 43 min). More than 50% of the tracer washed out prior to the administration of Lasix.

Post-Lasix images demonstrate further tracer washout. The findings are suggestive of a dilated left collecting system but without evidence of mechanical obstruction. There is no reason to put the patient on Lasix therapy, given the fact, that the left renal collecting system emptied well by means of gravity.

Right kidney demonstrates normal function.

Case 2

3. A 47-year-old male with history of nephrolithiasis underwent renal scintigraphy with Tc-99m MAG-3. Images are shown below.

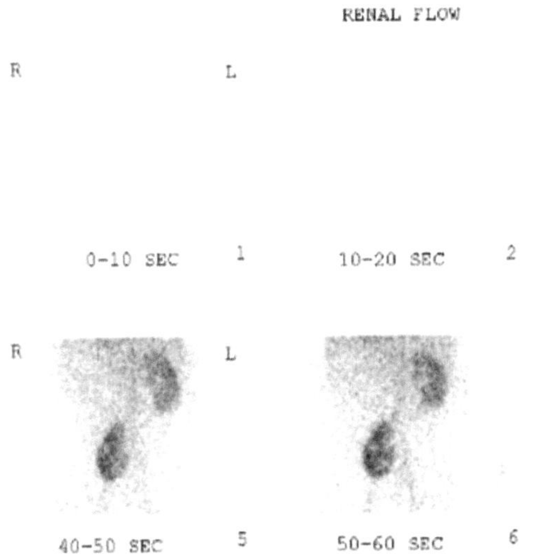

Image 1

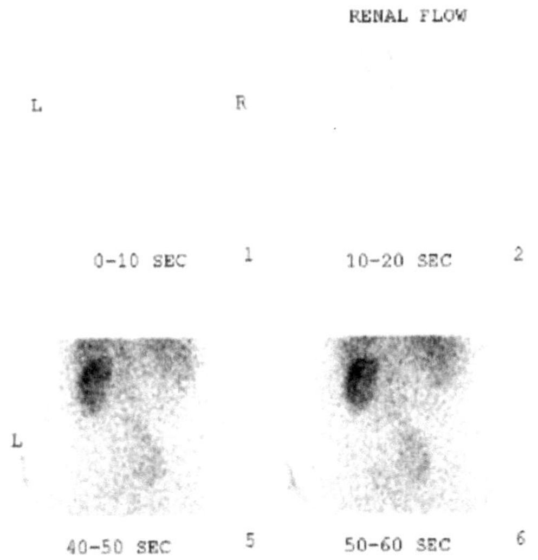

Image 2

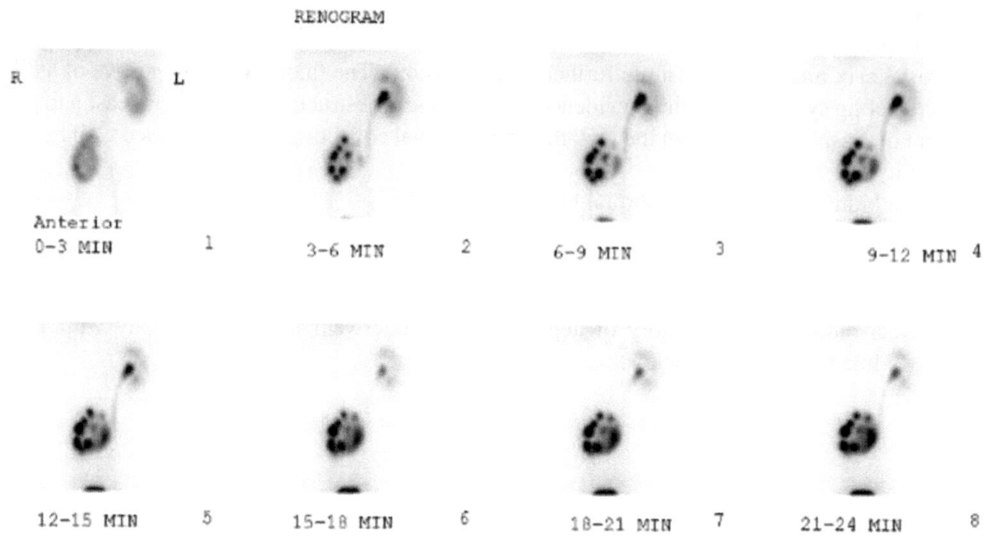

Image 3

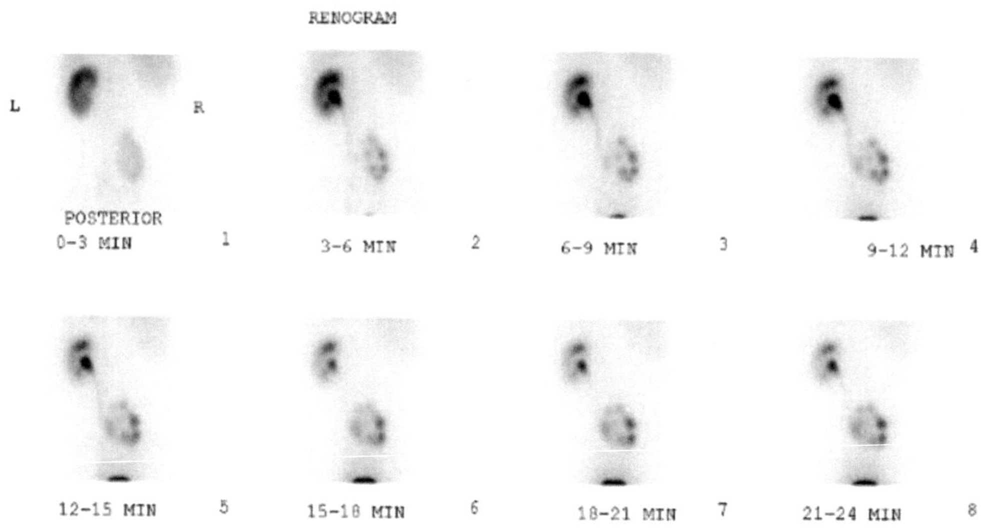

Image 4

The above images demonstrate:

A. Tracer uptake in the pelvis, concerning malignancy
B. Tracer uptake in the pelvis, likely within the urinary bladder
C. Ectopic right pelvic kidney
D. Horseshoe kidney

Answer: C

Because Tc-99m MAG3 is protein-bound and not filtered, it is exclusively cleared from the kidney by tubular secretion. Therefore, initial tracer visualization on renogram would be consistent with renal parenchyma (not urinary bladder or cancer).

Above Tc-99m MAG3 flow images (image 1, anterior projection and image 2, posterior projection) and functional images (image 3, anterior projection and image 4, posterior projection) demonstrate initial flow and subsequent functional tracer distribution in normal left kidney. An additional area of tracer accumulation in the pelvis is consistent with the ectopic right pelvic kidney. A pelvic kidney is a normal native kidney located in the pelvis, instead of the abdomen. This occurs when a kidney does not ascend from its original location in the pelvis to its final location during fetal development. Similar appearance will be characteristic for a transplanted kidney, which is usually placed in the iliac fossa.

The horseshoe kidney is the most common renal fusion anomaly, which consists of two renal masses connected at their lower poles by a parenchymal or fibrous isthmus.

4. Tc-99m MAG3 renal dynamic imaging was performed for 24 min using dual-detector acquisition of both anterior and posterior projections. In order to determine differential renal function of the right pelvic kidney, which of the following calculations should be performed:

A. Right kidney counts = $\sqrt{\text{anterior counts Right kidney} \times \text{posterior counts Right kidney}}$

B. Right kidney counts = $\sqrt{\text{anterior counts Right kidney} \times \text{anterior counts Left kidney}}$

C. Right kidney counts = $\sqrt{\text{anterior counts Right Kidney} + \text{posterior counts Right Kidney}}$

D. Right kidney counts = $\sqrt{\text{anterior counts Right Kidney} + \text{anterior counts Left kidney}}$

Answer: A

Differential renal function (or split function) is the relative renal function of one kidney versus the other. The calculation is typically driven from the counts at 2–3 min post-injection, i.e., in the cortical tracer uptake phase, which takes place after the blood flow phase and before the beginning of excretion into the collecting system. Region of interest (ROI) drawn around kidneys and actual counts in each kidney are expressed as a fraction of 100% total function. Normally, the relative contribution for each kidney lies between 45% and 55%, although values between 42% and 58% are still considered within normal limits by many scholars. This value does not indicate whether the overall or global renal function is normal or decreased. It merely reflects the function of one kidney compared to the other.

The native kidneys are posterior organs, and the calculation of differential renal function in normal individuals is driven from counts obtained in posterior projection for each kidney. However, in this patient, the ectopic pelvic kidney will be closer to an anterior detector. In order to obtain accurate measurements for both kidneys, anterior and posterior views are acquired, typically simultaneously using a dual-headed gamma camera. Then, the geometric mean is calculated, which is a weighted average of the anterior and posterior counts in each kidney. The following formula is used:

Right kidney counts = $\sqrt{\text{anterior counts Right kidney} \times \text{posterior counts Right kidney}}$

A similar calculation is used for the left kidney. After the geometric mean for each kidney is determined, the calculation of differential function is performed.

5. Time-activity curves from anterior and posterior renogram were also generated.

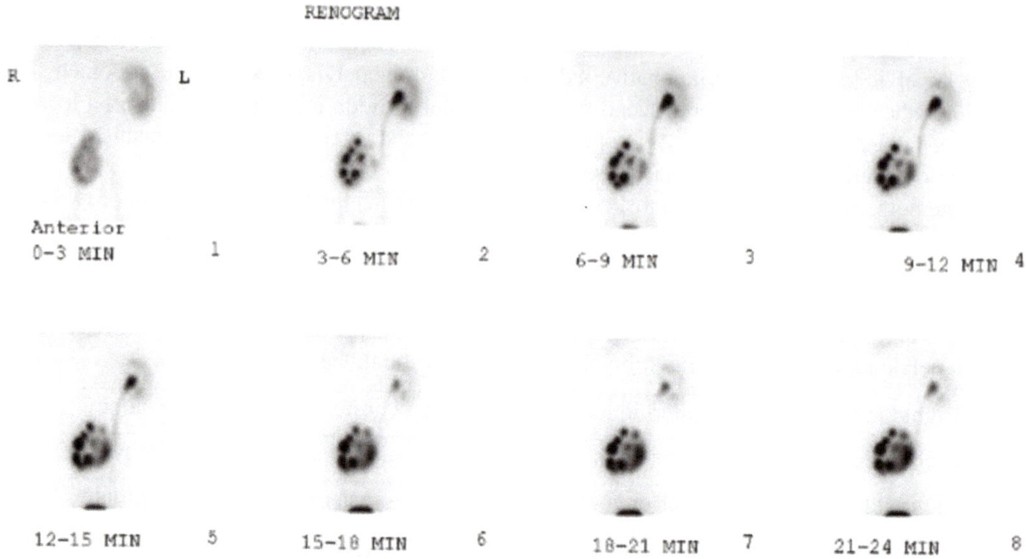

Image 1

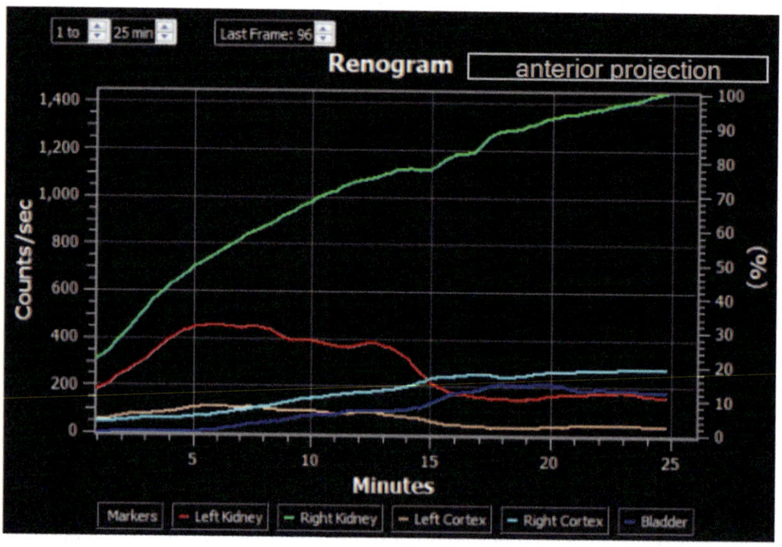

Image 2

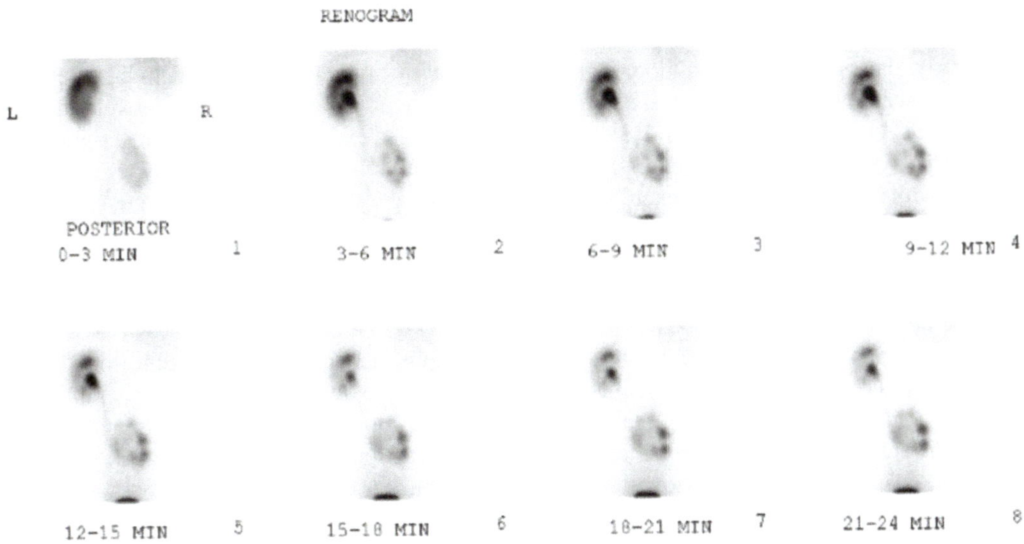

Image 3

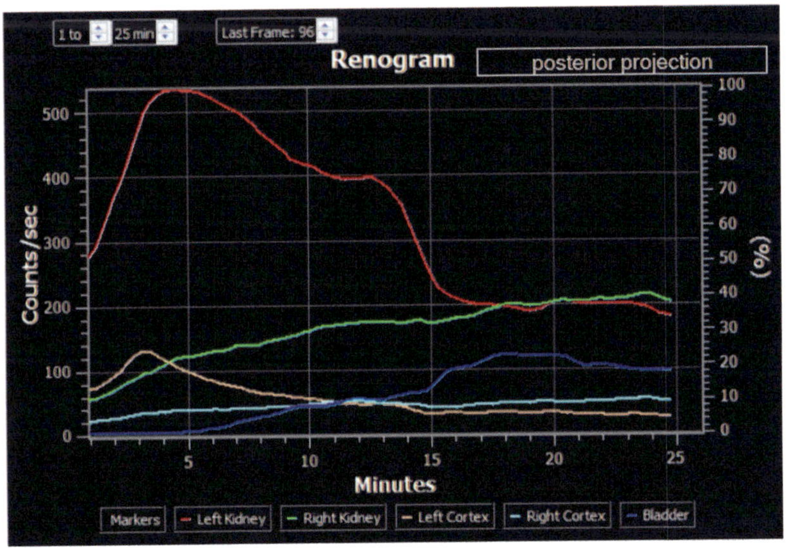

Image 4

Upon evaluation of the above images, you conclude:

A. Mechanical obstruction of the right kidney
B. Dilated left collecting system
C. Lasix should be administered
D. There is a significant difference in anterior and posterior counts for each kidney, which makes the study nondiagnostic

Answer: C

Left kidney has more counts on the posterior projection (closer to the posterior detector). Similarly, the right kidney has more counts on the anterior projection. The left kidney demonstrates essential normal tracer uptake, secretion, and washout towards the collecting system. The right kidney demonstrates slightly decreased tracer uptake and excretion into the dilated right renal collecting system. There is a progressive tracer accumulation in the right renal calyxes and renal pelvis, with no significant tracer excretion towards to urinary bladder. Time-activity curves (TACs) confirm these findings (image 2, anterior projection and image 4, posterior projection). There is a diminished initial slope on the TAC of the right kidney, progressively increasing cortical activity over time with no significant tracer excretion into the collecting system.

The addition of a diuretic such as furosemide (Lasix) to the protocol will allow to assess for mechanical obstruction. Furosemide will help to distinguish mechanical obstruction from "functional" obstruction (nonobstructive collecting system dilatation attributed to vesicoureteral reflux, urinary tract infection, congenital malformations, previous obstruction, or a noncompliant bladder). The author does not prefer the term functional obstruction but rather uses dilated collecting system to describe the delayed excretion without mechanical obstruction.

6. The patient received an intravenous injection of 40 mg of furosemide and additional dynamic images were obtained for 24 min. Post-Lasix images are shown below.

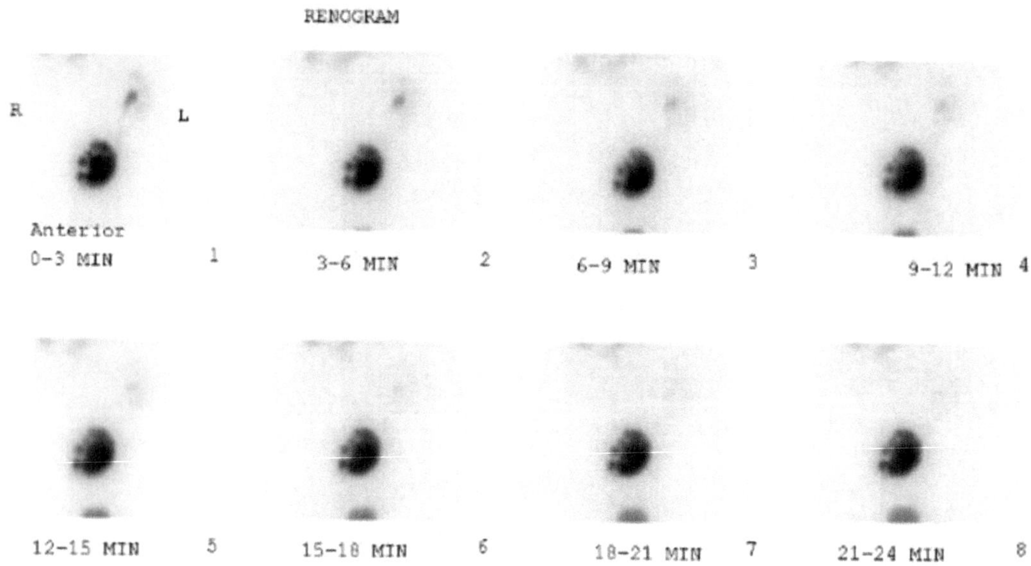

Image 1

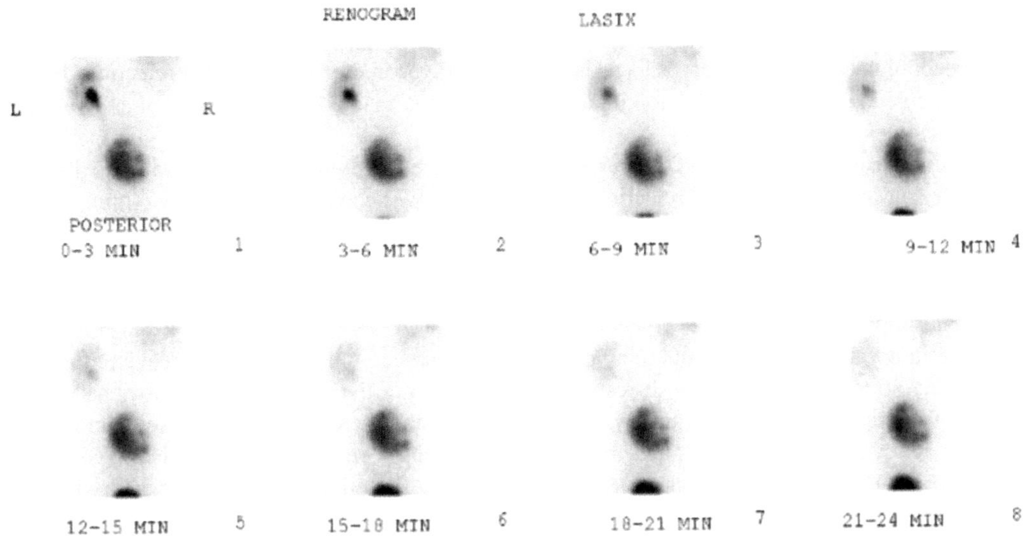

Image 2

Which of the following is the most likely explanation of the above findings?

A. Dilated right renal collecting system, possibly secondary to vesicoureteral reflux
B. Left obstructing renal stone
C. Left ureteropelvic junction (UPJ) obstruction
D. Right ureteropelvic junction (UPJ) obstruction

Answer: D

Post-Lasix images of the right kidney demonstrate persistent tracer retention within the dilated right renal calyxes and right pelvis. No appreciable amount of tracer was excreted towards the bladder by the end of the study. The tracer activity continues to accumulate in the pelvicalyceal system with no response to the diuretic challenge. The findings are consistent with mechanical obstruction. Among the above conditions, right ureteropelvic junction (UPJ) obstruction is the most logical explanation.

Dilated right renal collecting system secondary to vesicoureteral reflux will likely present with nonobstructive hydronephrosis and/or hydroureter, which will demonstrate a good response to the diuretic challenge.

Post-Lasix images of the left kidney demonstrate continued tracer washout from the left renal pelvis towards the bladder with the majority of the tracer is excreted by the end of the study. This appropriate response to the diuretic would exclude the presence of mechanical obstruction.

- Lessne ML, Shah NP, Stinchon JF, Cross ME, CNMT, Elizabeth Oates M. Optimal scintigraphic evaluation of a hydronephrotic horseshoe kidney. J Nucl Med Technol. 2007;35(4):255–8.
- Ziessman HA, O'Malley JP, Thrall JH, Nuclear medicine: the requisites.
- Mettler FA, Guiberteau MJ. Essentials of nuclear medicine imaging.

7. Regarding posttransplant Vasomotor Nephropathy, which of the following is true?

A. Usually occurs a few weeks to a few months postoperatively
B. Most commonly seen in renal transplant from a deceased donor (cadaveric kidney)
C. Tc-99m MAG3 renal scintigraphy usually demonstrate markedly diminished blood flow
D. It requires surgical intervention

Answer: B

Renal transplantation is a well-established procedure. Complications of renal transplantation are generally acute tubular necrosis (ATN), also called vasomotor nephropathy, rejection (hyperacute, acute, chronic), anti-rejection medication (cyclosporin) toxicity, and surgical and vascular complications.

Acute rejection is a cell-mediated process characterized by lymphocytic infiltration. Although it can occur at any time, acute rejection does not typically occur until 10–14 days (minimum of 5–7 days) after transplantation and is most common in the first 3 months. It is common for the patient to be febrile and the allograft to be swollen and painful. Typically, the patient becomes desensitized to the allograft over time. It is rarely seen after 1 year in a patient taking appropriate immunosuppressive therapy.

Vasomotor nephropathy (acute tubular necrosis) is most commonly seen in cadaveric kidneys, occurring up to 50% of the time. It results from ischemia in the renal transplant after harvesting and before transplantation. Frequently, it is self-limited and the condition will resolve in the following few weeks. It is to be noted that, good perfusion on renal scintigraphy is linked with a favorable outcome. Significant decrease in perfusion, on the other hand, is associated with prolonged recovery, if any.

Case 3

8. 42-year-old male with the past medical history of end-stage renal disease underwent renal transplantation from a deceased donor. Postoperative period complicated by anuria, hyperkalemia, and progressive elevation of serum creatinine. The patient underwent Tc-99m MAG3 renal scintigraphy. Images are shown below.

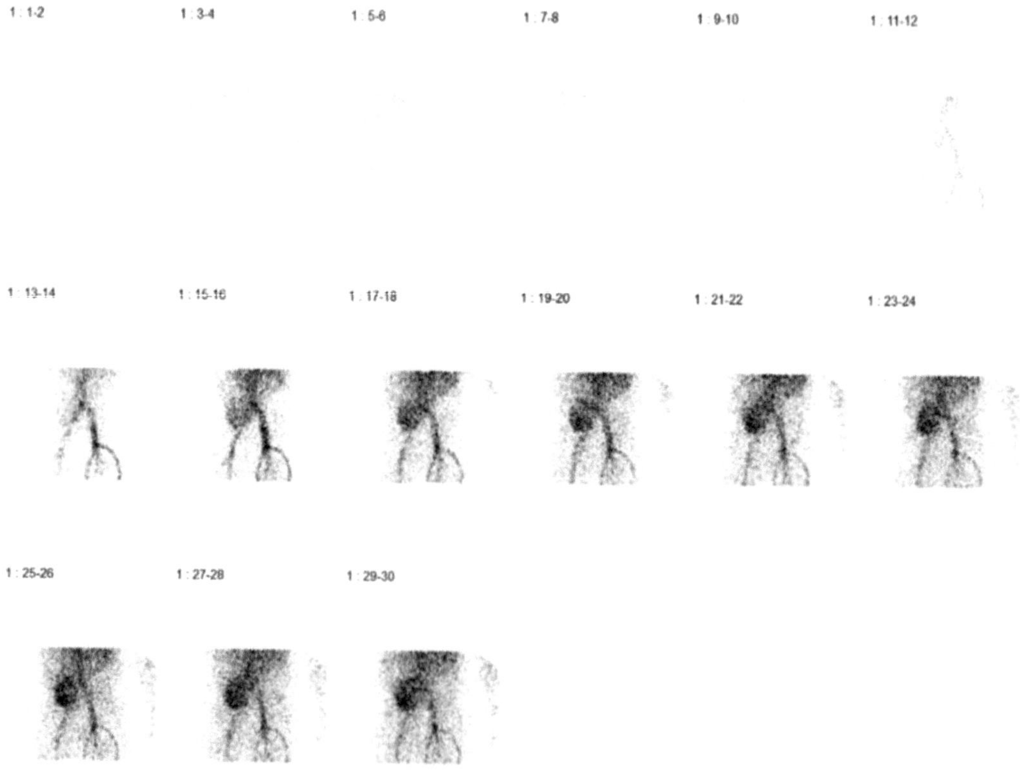

Image 1

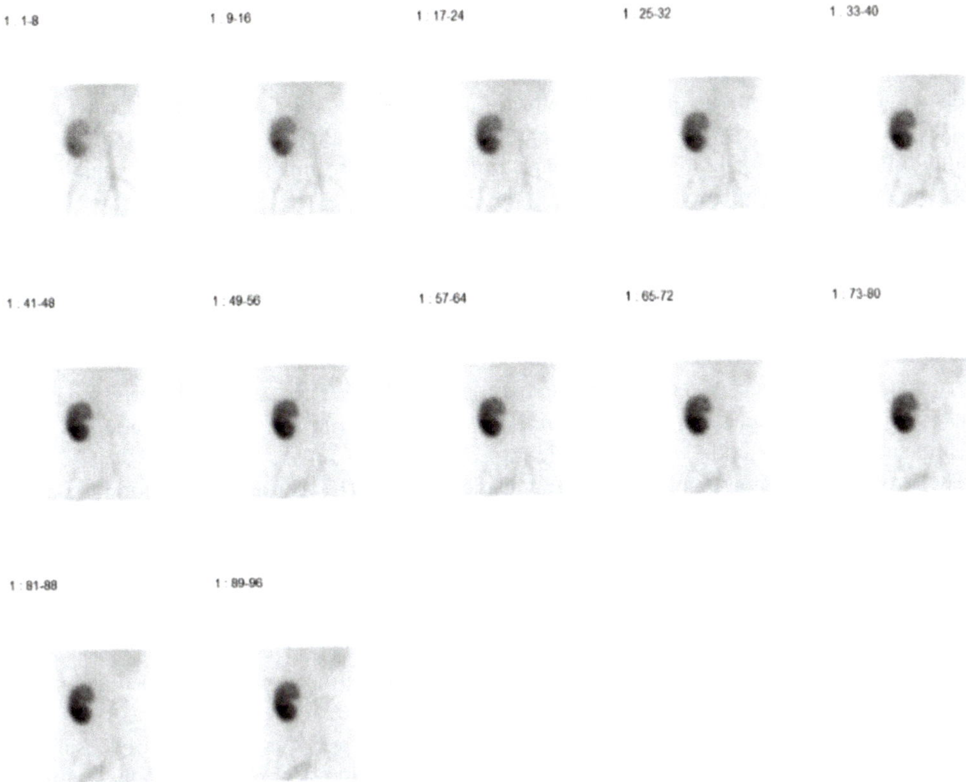

Image 2

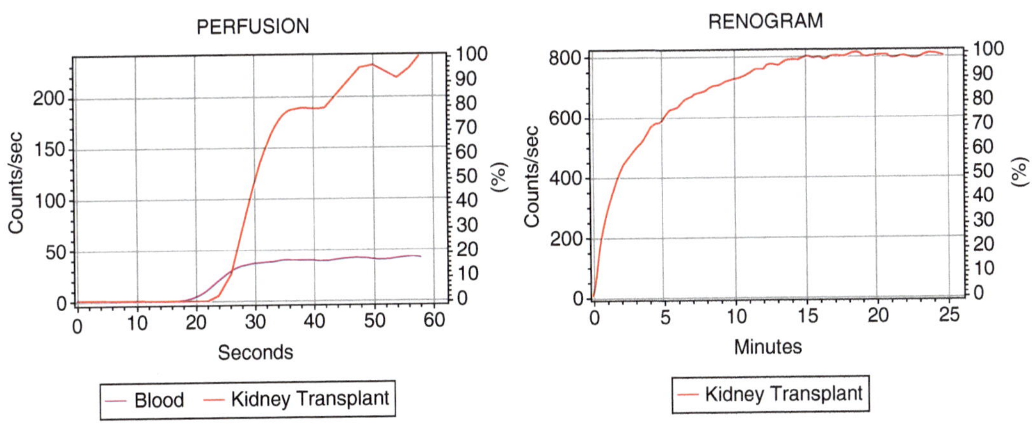

Image 3

Upon evaluation of the above images, you conclude that the patient likely has:

A. Normal function of the transplant kidney
B. Hyperacute rejection
C. Acute rejection
D. Acute tubular necrosis (ATN)

Answer: D

Renal allograft evaluation is performed using the dynamic scintigraphy protocol with Tc-99m MAG3. The camera should be placed anterior to the allograft in the lower pelvis. Because the medical and surgical complications occur at certain time periods after transplant, renal scintigraphy should be interpreted with the age of the transplant and the clinical context in mind. The type of allograft (cadaveric or living related donor transplant) is also important when evaluating the expected level of renal function.

Tc-99m MAG3 renal scintigraphy perfusion images (image 1) demonstrate normal perfusion to the transplanted kidney. Dynamic images (image 2) demonstrate delayed tracer uptake and concentration without evidence of tracer secretion from the cortex to the renal pelvis. Also, there was no tracer washout towards the bladder by the end of the images. The findings are a characteristic of acute tubular necrosis (ATN), which is often reversible.

Time-activity curves (TAC) confirm these findings (image 3). There is normal perfusion of the renal transplant (left) but progressively increasing cortical activity over time with no significant tracer excretion into the collecting system (left).

The two most common complications to consider in the early posttransplant period are acute tubular necrosis (ATN) and acute rejection. Both conditions show decreased function and marked cortical retention. Unlike ATN, acute rejection affects small renal parenchymal vessels and the allograft typically shows significantly diminished blood flow on the perfusion portion of the examination. The classic dynamic imaging pattern of acute rejection is decreased perfusion and then marked cortical retention. ATN, on the other hand, shows normal or near-normal perfusion but poor function with delayed cortical clearance and decreased urine formation immediately after surgery.

Hyperacute rejection usually occurs minutes to hours, with the common scenario of the surgeon witnessing the kidney turning blue in the operating room after performing the vascular anastomosis.

9. A Tc-99m MAG3 renal scintigraphy was repeated 5 weeks after renal transplantation. Images are shown below.

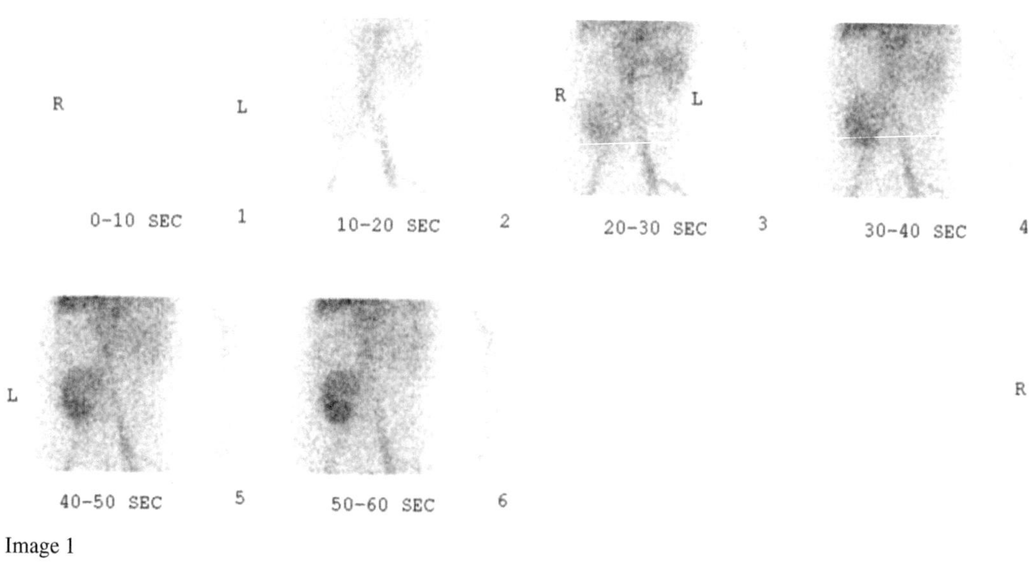

Image 1

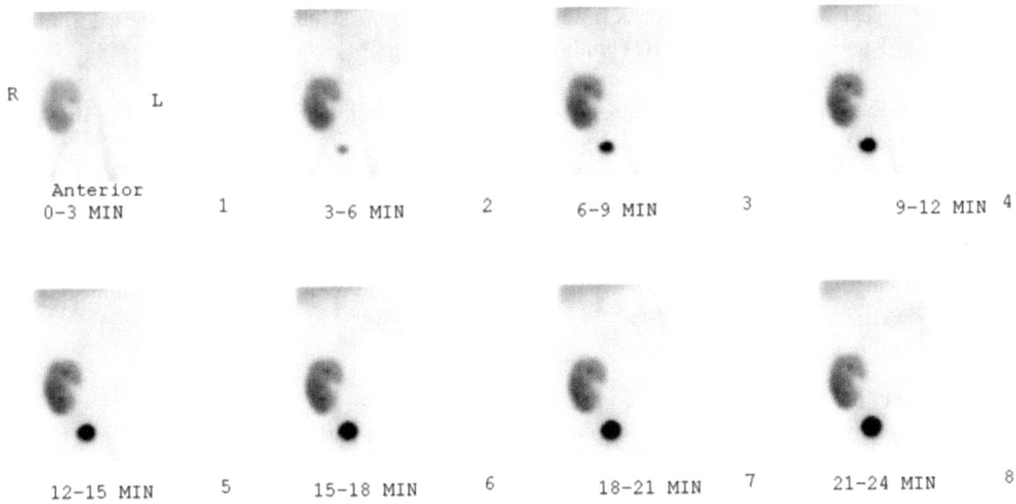

Image 2

Upon evaluation of above images, you conclude that the patient is likely having:

A. Normal renal scintigraphy scan
B. Acute rejection
C. Recovering ATN
D. Chronic rejection

Answer: C
Tc-99m MAG3 renal scintigraphy perfusion images (image 1) demonstrate normal perfusion to the transplanted kidney with the transplant seen a few seconds after the visualization of the iliac artery. Dynamic images (image 2) still demonstrate cortical retention, but now there is evidence of tracer excretion into the collecting system, consistent with recovering acute tubular necrosis.

Chronic rejection is a delayed and irreversible process that is usually seen months to years after renal transplant. Vascular constriction, chronic fibrosis, tubular atrophy, and glomerulosclerosis from immunological or nonhumoral causes may occur, resulting in cortical loss and decreased function. Dilation of the collecting system may occur as the cortex thins.

10. The most common cause of renal artery stenosis is:

A. Atherosclerotic disease
B. Fibromuscular dysplasia
C. Takayasu arteritis
D. A mycotic aortitis extending to the ostia of the renal arteries

Answer: A
Atherosclerotic disease accounts for up to about 66% of renal artery stenosis. Fibromuscular dysplasia is the second most frequent etiology, which usually occurs in younger patients, particularly women. Takayasu affects more frequently the thoracic vessels and a mycotic process weakens the arterial wall resulting in aneurysm formation.

• Soulez G, Oliva VL, Turpin S, et al. Imaging of renovascular hypertension: respective values of renal scintigraphy, renal Doppler US, and MR angiography. RadioGraphics 2000;20:1355–68.

Case 4 A 67-year-old female patient is referred for a renogram evaluation due to the CT findings shown in Fig. 1 and subsequent PET/CT findings shown in Fig. 2.

11. Which of the following statements better explains the suspicious findings that lead to the renogram request?

    a. Increased cortical retention in the right kidney, suggesting functional obstruction.
    b. Left hydronephrosis and lack of excreted activity in the left renal collecting system suggesting obstruction.
    c. Right-sided nephrolithiasis with the absence of activity in the right collecting system suggesting obstruction
    d. Bilateral mechanical obstruction

    Answer: b

    Figure 1 demonstrates a severe left-sided hydronephrosis and cortical atrophy. The PET/CT study (Figure 2) demonstrates decreased cortical uptake of radiotracer and absent excretion into the dilated collecting system. The findings are suspicious for obstruction. The right renal cortex appears to demonstrate preserved cortical enhancement (Figure 1). The level of increased cortical activity seen in the right renal cortex on the PET/CT study can be expected in normal functioning kidneys.

12. The patient is currently on chronic diuretic therapy with oral furosemide. For how long should the patient be instructed to hold the furosemide in preparation for the examination?

    a. The patient should hold furosemide the morning of the study.
    b. The patient should hold furosemide for 24 h
    c. The patient should hold furosemide for 48 h
    d. The patient should hold furosemide for 72 h

    Answer: a

    Since the main indication for the renal scintigram is to evaluate for obstruction in addition to determine the differential renal function, furosemide will be used during the study. The patient should be instructed to hold his regular dose of furosemide at least the morning of the study.

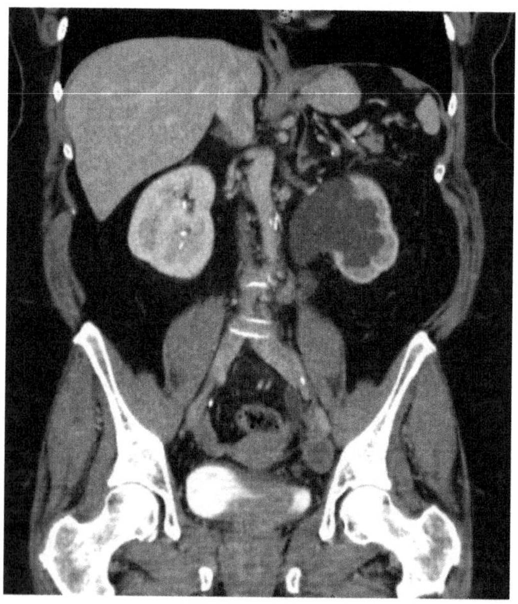

Fig. 1 Coronal CT

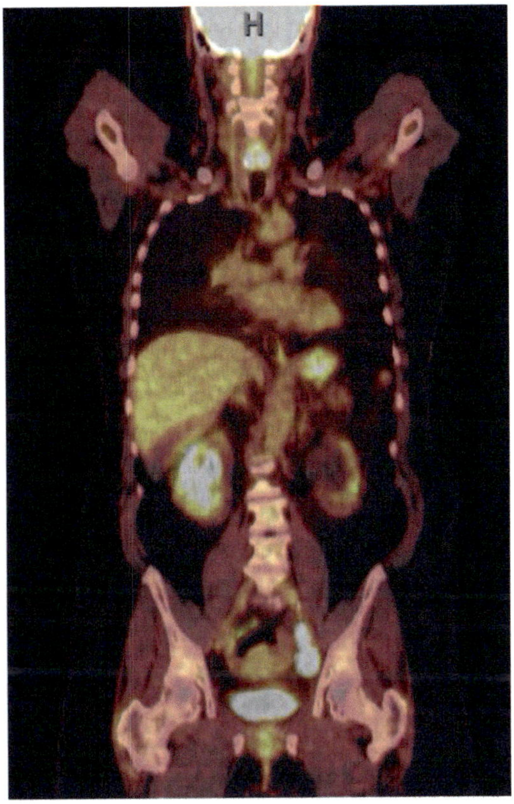

Fig. 2 Fused coronal 18F-FDG PET/CT

13. The patient has a creatinine level of 3.0 mg/dL. How much furosemide should be used during the study?

    a. 30 mg
    b. 40 mg
    c. 60 mg
    d. 80 mg

    Answer: d

    An elevated creatinine level is an indication of impaired renal function. An insufficient dose of furosemide in patients with elevated creatinine levels, even in the absence of obstruction, may prevent adequate diuretic response. The recommendations for the customized diuretic dose based on the creatinine levels are particularly emphasized in the most recent procedure standard recommendations of the Society of Nuclear Medicine and Molecular Imaging. At least 80 mg of furosemide should be administered to patients with an elevated creatinine level equal to or greater than 3 mg/dL.

14. One milligram of Bumetanide is equivalent to approximately how many milligrams of furosemide?

    a. 10 mg
    b. 20 mg
    c. 30 mg
    d. 40 mg

Answer: d

Bumetanide is an acceptable diuretic agent if furosemide is unavailable. It is 40 times more potent than Lasix.

References

1. Taylor AT, et al. SNMMI procedure standard/EANM practice guideline for diuretic renal scintigraphy in adults with suspected upper urinary tract obstruction 1.0. Semin Nucl Med. 2018.
2. Perez-Brayfield, M.R., et al. A prospective study comparing ultrasound, nuclear scintigraphy and dynamic contrast enhanced magnetic resonance imaging in the evaluation of hydronephrosis. J Urol. 2003;170:1330–4.
3. Taylor AT, et al. Society of nuclear medicine procedure guideline for diagnosis of renovascular hypertension. Version 3.0, 2003.

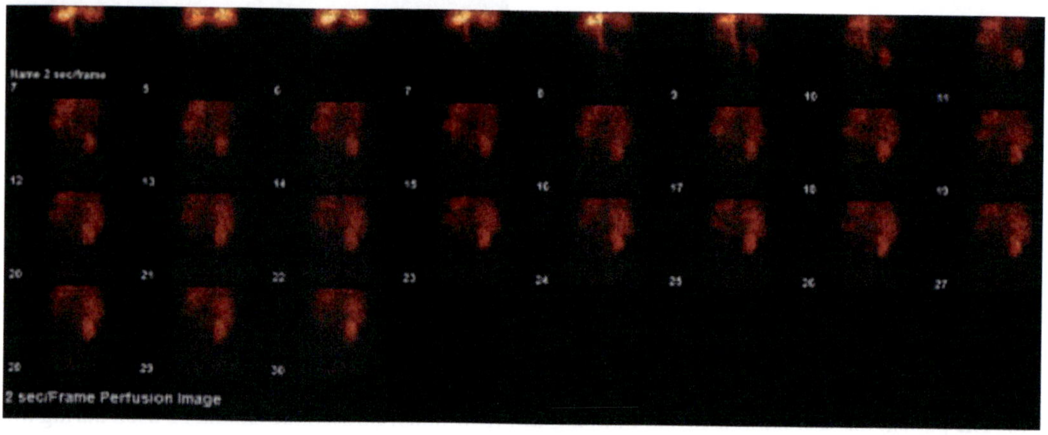

Fig. 3 Perfusion phase

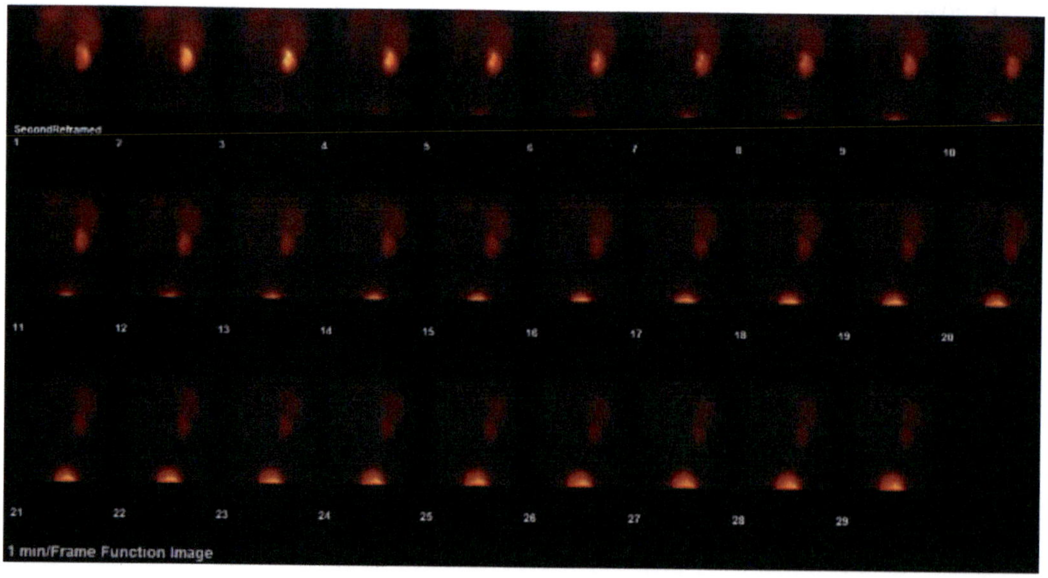

Fig. 4 Functional phase

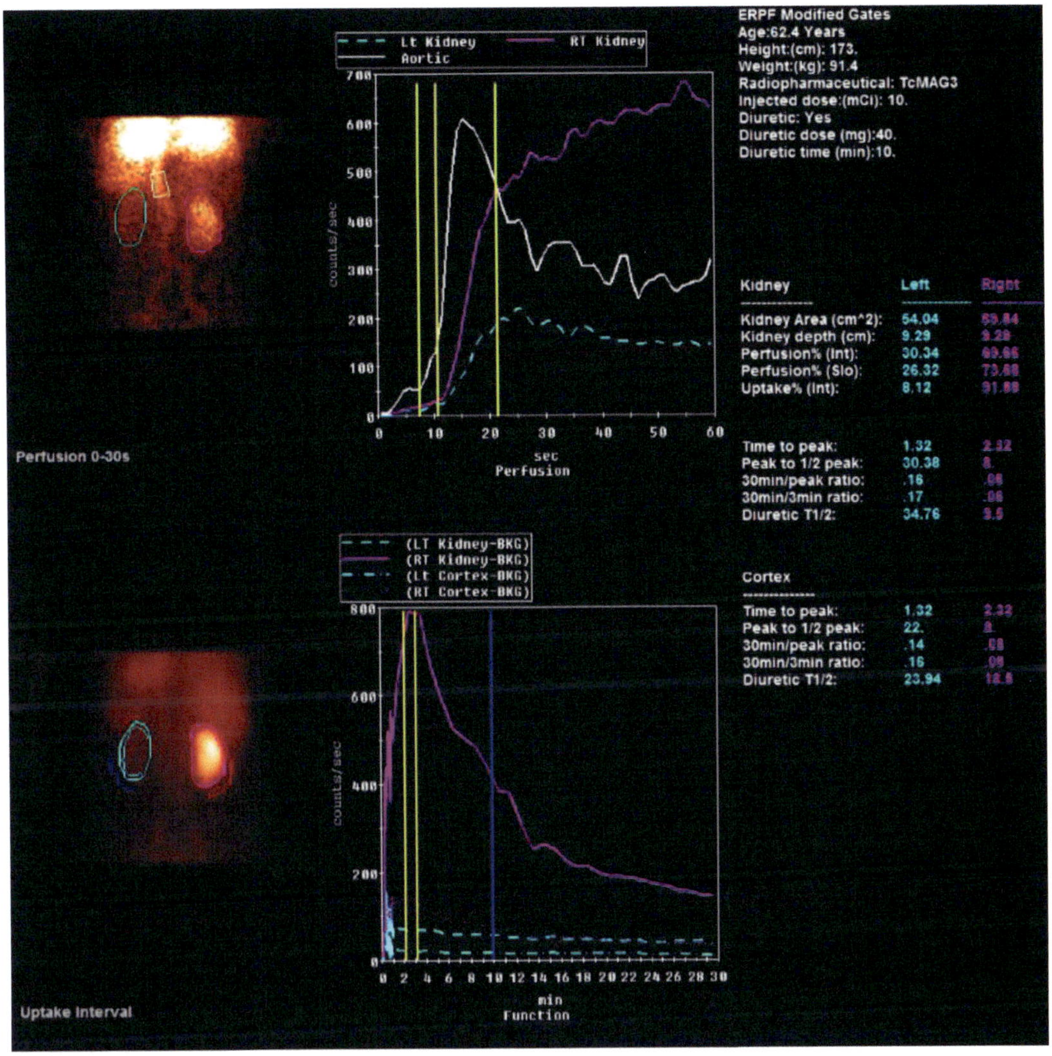

Fig. 5 Renal Quantitation

The renal scintigraphic study shown above is from the above mentioned 67-year-old patient with a newly diagnosed malignant pelvic mass on physical examination, which was subsequently confirmed in a contrast-enhanced CT of the abdomen and pelvis.

15. The Tc-99m MAG3 study shown above demonstrates,

    a. Delayed perfusion to the left kidney with normal function
    b. Negligible perfusion to the left kidney with no significant function
    c. Delayed perfusion to the right kidney with normal function
    d. Negligible perfusion to the right kidney with no significant function

    Answer: b

    There is no significant perfusion or cortical uptake in the left kidney (Figures 3–5). The loss of function of the left kidney is likely secondary to chronic left-sided obstruction as a result of the adnexal mass. The right kidney demonstrates normal perfusion, function, and excretion.

16. Which of the following statements best defines a high-grade renal obstruction?

    a. Acute onset with decreased perfusion and delayed parenchymal uptake
    b. Acute onset with persistent parenchymal uptake but absent washout
    c. Chronic onset with preserved perfusion and gradual loss of function
    d. Chronic onset with decreased perfusion and gradual loss of function

    Answer: b
    The acute nature of the high-grade obstructions allows for visualization of parenchymal uptake given relative preservation of perfusion. If left untreated, the delayed scintigraphic findings will be similar to those of intermediate or low-grade obstruction. The low- and intermediate-grade obstructions typically result in a gradual and progressive loss of function.

17. What instructions should be given to a breastfeeding mother scheduled for a Tc-99m MAG3 renal scan?

    a. The patient should be instructed to pump and dump for 24 h after the study.
    b. The patient should be instructed to pump and dump for 48 h after the study
    c. The patient should be instructed to pump and dump for 72 h after the study.
    d. The patient should be reassured that interruption of breastfeeding is not required

    Answer: d
    Studies showed that the radioactivity concentration in the breastmilk after a renal scan is minimal and it is not essential to interrupt breastfeeding.

References

1. Taylor AT et al. SNMMI procedure standard/EANM practice guideline for diuretic renal scintigraphy in adults with suspected upper urinary tract obstruction 1.0. Semin Nucl Med. 2018.
2. Perez-Brayfield MR, et al. A prospective study comparing ultrasound, nuclear scintigraphy and dynamic contrast enhanced magnetic resonance imaging in the evaluation of hydronephrosis. J Urol. 2003;170:1330–4.
3. Taylor AT, et al. Society of nuclear medicine procedure guideline for diagnosis of renovascular hypertension. Version 3.0. 2003.

Central Nervous System

Medhat Sam Gabriel, Bital Savir-Baruch, Robert H. Wagner, Bruch J. Barron, Davide Bova, Aamna Hassan, Stephanie Rodriguez-Santiago, and Yekaterina Kucerova

Central Nervous System (CNS 2020)

1. F-18 florbetapir study is used to determine the presence of which of the following?

   A. Moyamoya disease
   B. Beta-Amyloid
   C. Normal pressure hydrocephalus
   D. Parkinson's Disease

   Answer: B

   F-18 florbetapir (Amyvid—Eli Lilly) is a PET radiopharmaceutical that is used to determine the presence of beta-amyloid in the brain. Beta-amyloid deposition in the brain is associated with Alzheimer's disease.

   Chronic vascular insufficiency (such as Moyamoya disease) is better assessed using Tc99m HMPAO in combination with acetazolamide. Normal pressure hydrocephalus is evaluated with intrathecal In^{111} DTPA with the evaluation of the ventricular activity. Parkinson's disease is best evaluated using I-123 ioflupane (DaTScan—GE Healthcare).

M. S. Gabriel (✉) · B. Savir-Baruch · R. H. Wagner · D. Bova
Division of Nuclear Medicine, Department of Radiology, Loyola University Medical Center, Maywood, IL, USA
e-mail: sgabriel@lumc.edu; Bital.savir-Baruch@lumc.edu; rwagner@lumc.edu; dbova@lumc.edu

B. J. Barron
Division of Nuclear Medicine and Molecular Imaging, Department of Radiology and Imaging Sciences, Atlanta, GA, USA
e-mail: bjbarro@emory.edu

A. Hassan
Department of Nuclear Medicine, Shaukat Khanum Memorial Cancer Hospital and Research Center, Lahore, Punjab, Pakistan

S. Rodriguez-Santiago
Department of Nuclear Medicine, C.W. Bill Young VA Medical Center, Bay Pines, FL, USA
e-mail: stephanie.rodriguezsantiago@va.gov

Y. Kucerova
Division of Nuclear Medicine, Department of Radiology, Loyola University Medical Center, Maywood, IL, USA

Department of Radiology, Cook County Healthy, Chicago, IL, USA
e-mail: yekaterina.kucerova@cookcountyhealth.org

M. S. Gabriel, B. Savir-Baruch (eds.), *RadTool Nuclear Medicine MCQs*,
https://doi.org/10.1007/978-3-030-69281-0_10

2. The defining characteristic of a positive study using F-18 florbetapir is the presence of:

   A. Decreased activity in the putamen
   B. Persistent activity in the ventricles
   C. Gyri with minimal uptake of tracer
   D. Loss of the gray-white matter differentiation

   Answer: D

   In normal patients, the gyri have little accumulation of tracer with significantly greater tracer present in the white matter.

   A positive F-18 florbetapir study is characterized by loss of the gray-white matter differentiation on scan. This is compared to the gray-white matter differentiation that is seen in the cerebellum which is not affected by Alzheimer's disease. This loss of the gray-white matter differentiation must be seen in two or more regions of the brain. Bilaterally affected areas of the brain such as the frontal or parietal regions count as a single region.

   Decreased activity in the putamen is characteristic of Parkinson's disease when imaged using I-123 Ioflupane.

3. A positive study using F-18 florbetapir:

   A. Is diagnostic of Alzheimer's disease
   B. Is diagnostic of frontotemporal dementia
   C. May be seen in the normal elderly
   D. Should be interpreted in conjunction with the clinical findings

   Answer: C

   Image interpretation should be performed independently of the patient's clinical information.

   The loss of the gray-white differentiation that can be seen with Amyloid deposition is consistent with the diagnosis of Alzheimer's disease, though it is not diagnostic. Elderly patients who do not have symptoms of dementia may still demonstrate loss of gray-white differentiation. It is not clear if this is predictive of later development of Alzheimer's dementia or not.

   A negative F-18 florbetapir study in the setting of dementia should direct the physician to other (and possibly treatable) etiologies of dementia.

4. I-123 Ioflupane has a high affinity to:

   A. The pre-synaptic dopamine transporter system
   B. The pre-synaptic norepinephrine transporter system
   C. The post-synaptic dopamine transporter system
   D. The post-synaptic norepinephrine transporter system

   Answer: A

   I-123 ioflupane has a high affinity for the pre-synaptic dopamine transporter system and can be used to help differentiate essential tremor from Parkinsonian syndrome. Patients receiving I-123 ioflupane should receive potassium iodide oral solution equivalent to 100 mg of iodide in an effort to radioprotect the thyroid.

5. A 49-year-old male with a history of abdominal mass underwent exploratory laparotomy. Following the procedure, the patient developed acute respiratory distress with multiorgan failure. He was found

to be unresponsive on clinical examination, suspicious of brain death. To support the diagnosis of brain death, cerebral blood flow study with Technetium 99m pertechnetate was performed.

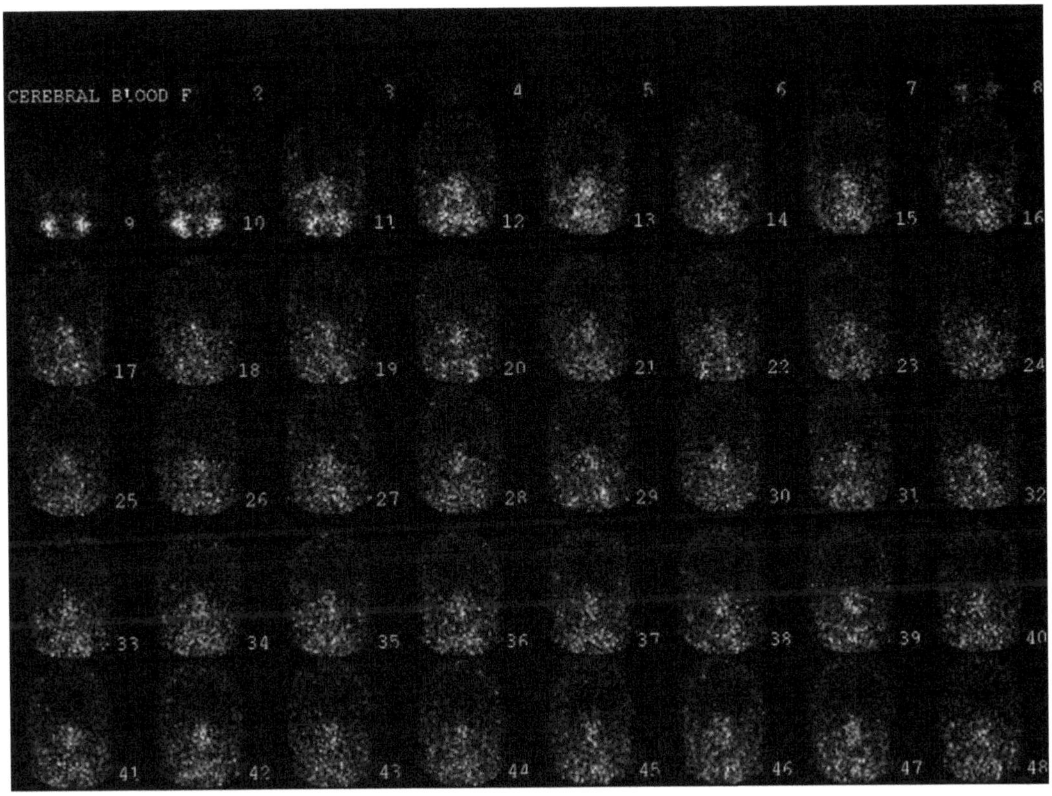

Image 1

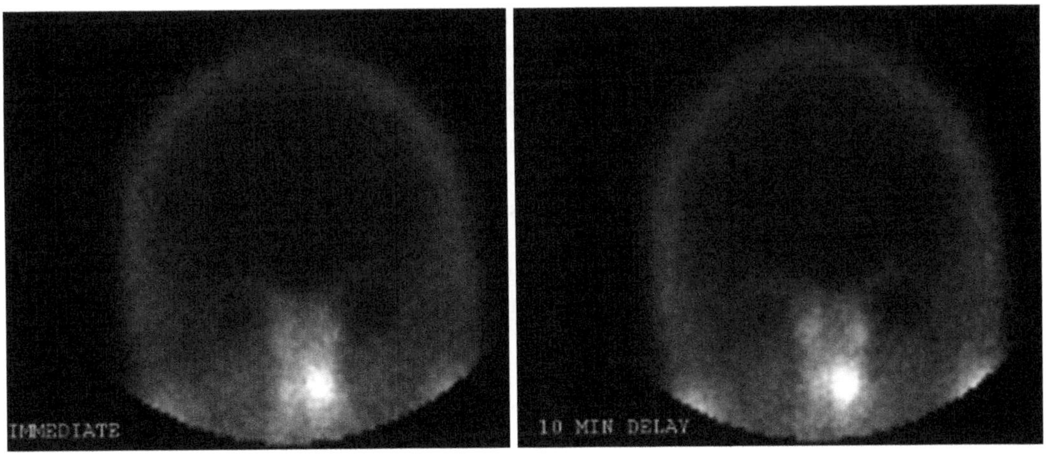

Image 2

Upon evaluation of images you conclude:

A. Repeat the study because of an inadequate bolus injection of the radiotracer
B. The tracer is lipophobic and normally does not cross blood-brain barrier. Therefore, the findings are expected in individuals with normal cerebral perfusion
C. There is cutoff of blood flow at the level of extracranial carotid arteries, consistent with bilateral arterial occlusion
D. The scan is consistent with the literature description for cerebral death

Answer: D

Tc-99m Cerebral Perfusion Scan (or brain death scan) is performed to support the clinical diagnosis of brain death. Brain-specific or lipophilic tracers which cross the blood-brain barrier such as Tc-99m HMPAO and Tc-99m ECD are preferred over nonspecific or lipophobic tracers such as Tc-99m DTPA, which are considered angiographic radionuclides.

The basic protocol is flow (or angiographic) dynamic images followed by immediate and delayed (obtained 5–10 min after injection of tracer) static images. **Flow imaging normally** demonstrates prompt blood flow to the extracranial carotid systems (whether the tracer is brain-specific like Tc-99m HMPAO, Tc-99m ECD or is angiographic like Tc-99m DTPA, Tc-99m pertechnetate). This confirms an adequate bolus was given. On the flow images of angiographic radiotracers, radioactivity does not normally lie within the brain itself because of the integrity of the blood-brain barrier, but rather it is located in the overlying scalp soft tissues, calvarium, and subarachnoid spaces that outline the cerebral hemispheres. Activity is also seen in the larger blood pool accumulations, such as the sagittal and transverse sinuses. B is incorrect.

The normal distribution of lipophilic brain perfusion agents is proportional to regional blood flow, with significantly greater activity seen in the cortical gray matter (activity is symmetric and greatest in the strip of cortex along the convexity of the frontal, parietal, temporal, and occipital lobes, as well as subcortical gray matter, including the basal ganglia and the thalamus).

In the delayed phase (15–20 min after tracer injection), the brain-specific tracer will be **normally** seen in the cerebrum, basal ganglia, thalami, and cerebellum. The angiographic tracer will be seen in the dural venous sinuses.

Positive for brain death study. The flow imaging will demonstrate tracer flowing through the carotid arteries and into the base of the skull (where the flow of tracer stops). There will be the absence of intracerebral perfusion. This is due to increased intracranial pressure. Adequate bolus administration must be ensured, and distinct activity in the carotid arteries is used as a confirmatory sign (C is incorrect). **In the immediate and delayed phases**, the brain-specific tracer will be absent in brain tissue. The angiographic tracer will not be seen in the dural venous sinuses. This combination of findings is interpreted as positive for brain death (because of absent intracerebral perfusion).

The above scenario is an example of a scan positive for brain death using an angiographic tracer Tc 99m-pertechnetate. In the flow dynamic phase (image 1), tracer is seen in the common carotid arteries, confirming an adequate bolus (A is incorrect). No significant intracranial arterial perfusion is noted. Immediate and delayed static images (image 2) also do not demonstrate any significant sagittal sinus activity. This study is abnormal. Findings are consistent with the literature description for cerebral death.

*Ziessman, Harvey A., MD; O'Malley, Janis P., MD; Thrall, James H., MD. Nuclear Medicine: The Requisites.
*Fred A. Mettler and Milton J. Guiberteau. Essentials of Nuclear Medicine Imaging.
*http://www.nucmedresource.com/neurology-nm.html

6. Which radiotracer is used for the study?

A. In-111 DTPA
B. Tc-99m Sulfur colloid
C. Tc-99m tilmanocept
D. Ytterbium-70

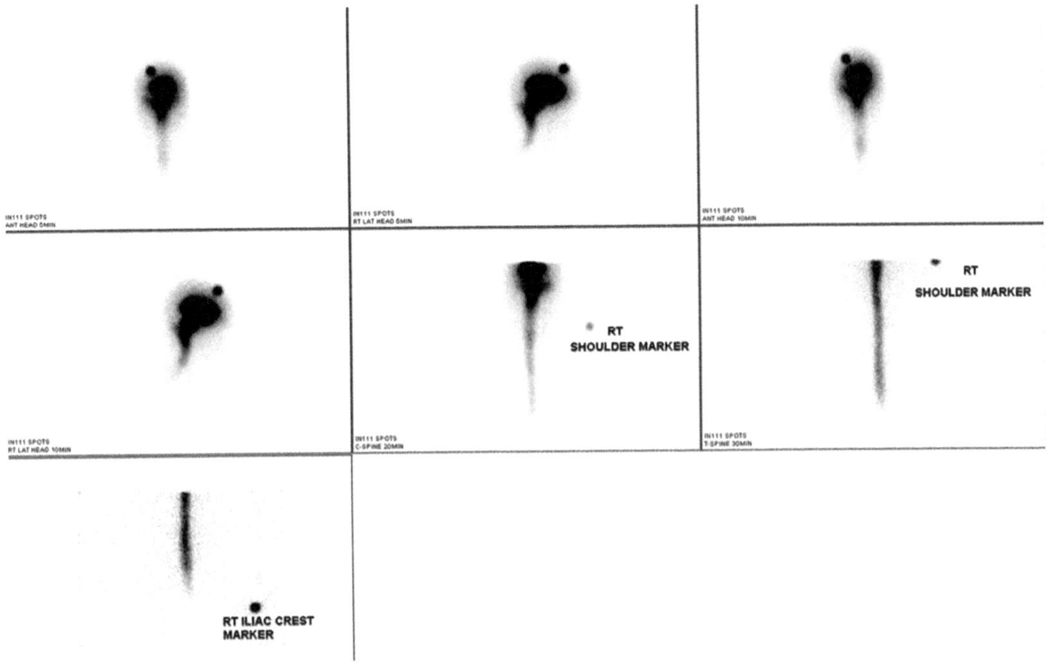

Answer: A

Compounds attached to a DTPA moiety are excreted by the kidneys after absorption. In-111-DTPA is commonly used for cisternography and shunt patency studies. Tc-99m-DTPA is sometimes used in children and adults. This tracer has been linked to aseptic meningitis. Special filtered forms of Tc-99m-DTPA may eliminate the problem. Ytterbium is a beta emitter. It was first used in cisternography. However, in cases of obstruction, the radiation dose became extremely high. Tc-99m Sulfur colloid and Tc-99m tilmanocept (Lymphoseek) are used for lymphoscintigraphy.

7. Which statement is true about this study?

A. The distal limb is obstructed
B. The patient has hydrocephalus
C. The catheter is in the proper location for chemotherapy
D. There is reflux into the ventricles

Answer: C

This is an Ommaya reservoir which is used to deliver chemotherapy agents via the ventricles into the CSF. This patient had DLBCL with malignant cells in the CSF. This was discovered after the patient complained of headaches and numbness in her legs. CSF involvement in diffuse large b-cell lymphoma occurs in between 1–10% of patients. Some patients at risk will receive prophylactic therapy.

8. What is the diagnosis of this FDG PET/CT study on patient with cognitive impairment?

A. Classic Alzheimer's disease
B. Epilepsy
C. Frontotemporal dementia
D. HIV dementia

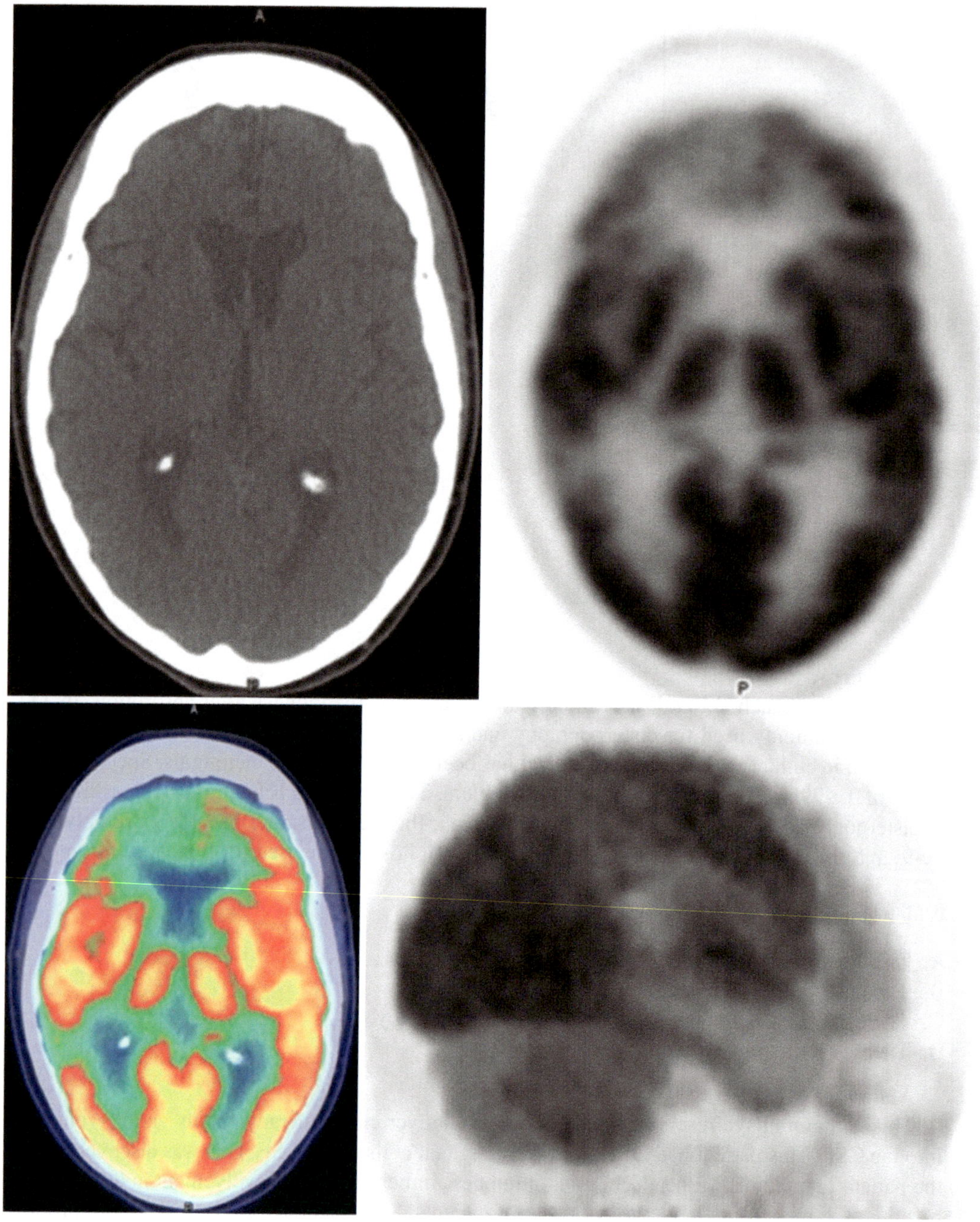

Answer: C

There are several disease entities that cause frontotemporal degeneration. They generally involve either the tau protein or a protein called TDP43. This disorder is called Pick disease. In behavior variant frontotemporal dementia, there are prominent changes in personality and interpersonal relationships and occur in the 50s and 60s. A second type, primary progressive aphasia, affects language skills and speaking. This disorder is more commonly seen in midlife. Classic Alzheimer's disease will generally affect the temporal and parietal lobes before frontal lobe involvement. Diffuse frontal lobe hypometabolism is not usually seen in patients with epilepsy. HIV-induced dementia generally causes global hypometabolism on FDG PET/CT imaging.

Radionuclide Therapy

Medhat Sam Gabriel, Bital Savir-Baruch, Robert H. Wagner, Bruch J. Barron, Davide Bova, Aamna Hassan, Stephanie Rodriguez-Santiago, and Yekaterina Kucerova

Case 1 Transarterial Radioembolization with Yttrium-90 microspheres.

The images shown above are from a 57-year-old male patient recently diagnosed with hepatocellular carcinoma of the right hepatic lobe (Fig. 1).

1. What is the most likely radiotracer used for the pre-Yttrium-90 therapy planning scan shown above?

   A. Gallium-67
   B. Indium-111
   C. Tc-99 m labeled RBCs
   D. Tc-99 m MAA

M. S. Gabriel (✉) · B. Savir-Baruch · R. H. Wagner · D. Bova
Division of Nuclear Medicine, Department of Radiology, Loyola University Medical Center, Maywood, IL, USA
e-mail: sgabriel@lumc.edu; Bital.savir-Baruch@lumc.edu; rwagner@lumc.edu; dbova@lumc.edu

B. J. Barron
Division of Nuclear Medicine and Molecular Imaging, Department of Radiology and Imaging Sciences, Atlanta, GA, USA
e-mail: bjbarro@emory.edu

A. Hassan
Department of Nuclear Medicine, Shaukat Khanum Memorial Cancer Hospital and Research Center, Lahore, Punjab, Pakistan

S. Rodriguez-Santiago
Department of Nuclear Medicine, C.W. Bill Young VA Medical Center, Bay Pines, FL, USA
e-mail: stephanie.rodriguezsantiago@va.gov

Y. Kucerova
Division of Nuclear Medicine, Department of Radiology, Loyola University Medical Center, Maywood, IL, USA

Department of Radiology, Cook County Healthy, Chicago, IL, USA
e-mail: yekaterina.kucerova@cookcountyhealth.org

M. S. Gabriel, B. Savir-Baruch (eds.), *RadTool Nuclear Medicine MCQs*,
https://doi.org/10.1007/978-3-030-69281-0_11

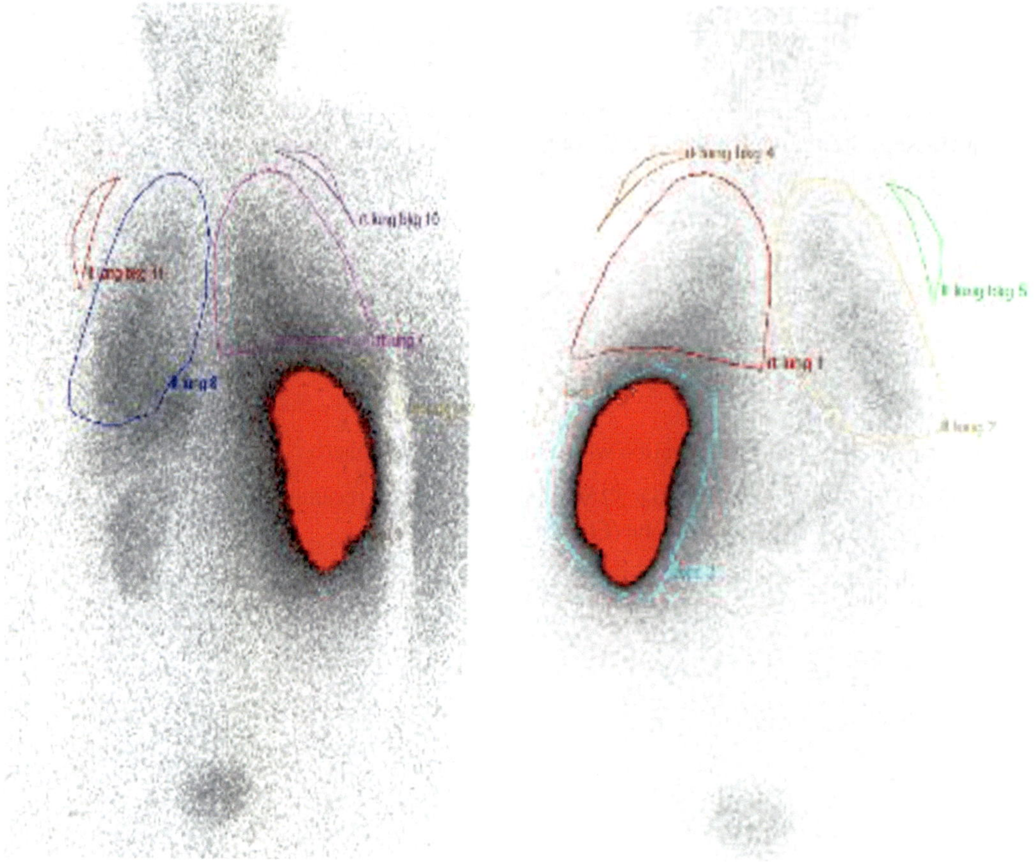

Fig. 1 ^{99m}Tc-MAA planar images with Regions Of Interest (ROIs) over the lungs and the liver

Answer: D

Prior to Y-90 transarterial radioembolization (TARE), a planning hepatic perfusion scan is done. A dose of 3–5 mCi of ^{99m}Tc labeled macroaggregated albumin (^{99m}Tc-MAA) is commonly used for this purpose. The MAA particle diameter ranges from 30–90 μm, comparable to the 20–60 μm and the 20–30 μm of the commercially available resin and glass microspheres, respectively. The radiopharmaceutical is administered intraarterially following optimal access of the hepatic artery supplying the tumor. This is followed by planar and SPECT/CT images in order to determine the percentage of activity deposited in the lungs (lung shunt) and to evaluate for additional unexpected extrahepatic deposition of the radiopharmaceutical. After ^{99m}Tc-MAA is administered intraarterially, hepatic biodegradation takes place liberating free ^{99m}Tc-pertechnetate, which may be seen in the thyroid gland, kidneys, and stomach. This uptake increases with time; hence the importance of early imaging promptly following the injection. Expected activity in these organs should be recognized and should not be included in the calculations of the lung shunt [1, 2].

2. What is the mode of decay of Yttrium-90?

   A. Alpha emission
   B. Beta (−) emission
   C. Beta (+) emission
   D. Electron Capture

Answer: B

Yttrium-90 is a decay product of Strontium-90 and is commonly produced by chemical separation from its parent isotope and less commonly by neutron irradiation of stable Yttrium. Yttrium-90 has a half-life of 64.2 h and decays to stable Zirconium-90 predominantly by beta emission with a maximum energy of 2.25 MeV. It is also a weak positron emitter. Interactions of the high energy (−) betta emission with matter can lead to Bremsstrahlung radiation. These two characteristics make possible posttreatment imaging, including SPECT/CT and PET/CT, respectively, for quality control purposes with evolving implications in radiation dosimetry [1].

3. At what time posttreatment is more than 90% of the radiation from a 3 GBq dose delivered to the tumor tissue?

   A. In the first 24 h posttreatment.
   B. In the first 5 days posttreatment.
   C. In the first 7 days posttreatment.
   D. In the first 11 days posttreatment.

   Answer: D

   The half-life of Y-90 is 64.2 h and most of the energy of a treatment dose is deposited in the target tissue in the first 11 days posttreatment [4].

4. What is the half-life of Yttrium-90?

   A. 46 h
   B. 64 h
   C. 11 days
   D. 50 days

   Answer: B

   Yttrium-90 has a half-life of 64.2 h decaying to stable Zirconium-90 predominantly by beta emission [3]. Other palliative agents include Samarium-153 (half-life of 46 h), Radium-223 (half-life of 11 days), and Stransium-89 (half-life of 50 days).

5. When comparing the only two commercially available devices for transarterial radioembolization, what is true about the approximate activity per microsphere?

   A. SIR-Spheres have lower activity per microspheres, thus requiring a greater number of particles per dose with a moderate embolic effect.
   B. SIR-Spheres have a higher activity per microspheres, thus requiring a lower number of particles per dose with a mild embolic effect.
   C. TheraSpheres have a lower activity per microspheres, thus requiring a greater number of particles per dose with a mild embolic effect.
   D. TheraSpheres have a higher activity per microspheres, thus requiring a higher number of particles with a mild embolic effect

   Answer: A

   Yttrium-90 is the radioisotope present in both commercially available microspheres for transarterial radioembolization (TARE). In the production process, Y-90 is attached to the surface of the resin spheres (SIR-Spheres) and incorporated into the matrix of the glass spheres (TheraSpheres). Because of the difference in the method of attachment, the activity per resin microspheres is significantly lower compared to the glass microspheres, requiring the administration of a larger number of particles per treatment dose. However, even with the larger number of particles administered, the embolization effect is only moderate, and its principal way for target tissue destruction is tumor radionecrosis and not embolization [4, 5].

References

1. Uliel L, et al. From the angio suite to the G-camera: vascular mapping and 99mTc-MAA hepatic perfusion imaging before liver radioembolization—A comprehensive pictorial review. J Nucl Med. 2012;53:1736–47.
2. Hamami ME, et al., SPECT/CT with 99mTc-MAA in radioembolization with 90Y microspheres in patients with hepatocellular cancer. J Nucl Med. 2009;50:688–92.
3. TheraSphere Yttrium-90Glass-Microspheres Package Insert, BTG
4. Kallini JR, et al., Transarterial radioembolization with Yttrium-90 for the treatment of hepatocellular carcinoma. Adv Ther. 2016;33:699–714
5. Salem R, et al., Y90 radioembolization significantly prolongs time to progression compared with chemoembolization in patients with hepatocellular carcinoma. Gastroenterology. 2016;151: 1155–63.
6. Garin E, et al., Personalized dosimetry with intensification using 90Y-loaded glass microsphere radioembolization induces prolonged overall survival in hepatocellular carcinoma patients with portal vein thrombosis. Soc Nucl Med Molecul Imag. 2015.

Case 2 Y-90 Radioembolization (RAE); Pre-therapy planning scan.

6. A 63-year-old male patient is referred for a ^{99m}Tc-MAA planning scan, in preparation for his Y-90 therapy with glass microspheres. The lung shunt was calculated at 7.9% with an expected delivery dose to the lungs of approximately 3.5 mCi. Based on the imaging and semiquantitative findings illustrated above, this patient is at an elevated risk for?

   A. Radiation pneumonitis
   B. Liver failure
   C. Extrahepatic radionecrosis
   D. Hypothyroidism.

   Answer: C

   Increased deposition of activity in the lungs following intraarterial administration of ^{99m}Tc-MAA occurs due to the characteristic angiopathic pattern of neovascularization of primary and metastatic hepatic tumors. These tumors tend to create arteriovenous shunts while developing their vascular supply [1]. The probability of undesired effects of radiation pneumonitis increases as the fraction of activity delivered to the lung increases. When the pre-therapy analysis in preparation for TARE with glass microspheres reveals that a patient might receive greater than 16 mCi of Y-90 to the lungs, or that the total delivered dose to the lungs might exceed 30 Gy in a single treatment, the therapy is contraindicated [2]. A different dose modification approach is recommended for resin microspheres, involving reduction of the treatment dose based on the percentage of lung shunt [3].

   The expected physiologic distribution ^{99m}Tc-MAA increases in intensity the latter the images are taken from the time of intraarterial administration. It is important to recognize that in vivo biodegradation of ^{99m}Tc-MAA and the presence of free ^{99m}Tc, increase the activity seen in the thyroid gland, kidneys, salivary glands, stomach, and other common locations [1, 4]. The intensity of the desired distribution of activity within the liver tumor has not been related to a potential risk for liver failure.

   The anterior planar image shown in Fig. 2 demonstrates irregular curvilinear activity originating inferior to the location of the highest liver activity. The fused SPECT/CT images shown in Fig. 4 localize the activity to collateral vasculature in the abdominal wall. This patient had portal hypertension at the time of presentation. The presence of portosystemic collateral veins is common in patients with portal hypertension, and it can be a potentially difficult problem to solve, particularly if significant deposition of activity to these collaterals is evidenced in the planning scan. If no safe alternatives for pre-therapy embolization exist, the potential risk for significant extrahepatic radionecrosis might prevent Y-90 treatment [6, 7]. Hence, the importance of the planning scan and of the careful examination of the fused images.

7. What is a potential absolute contraindication for Yttrium-90 radioembolization therapy for this patient, based on the findings shown in Figs. 2, 3, and 4?

   A. Portal Vein Thrombosis
   B. Involvement of both hepatic lobes
   C. Tumor volume > 60%
   D. Uncorrectable extrahepatic deposition of ^{99m}Tc-MAA

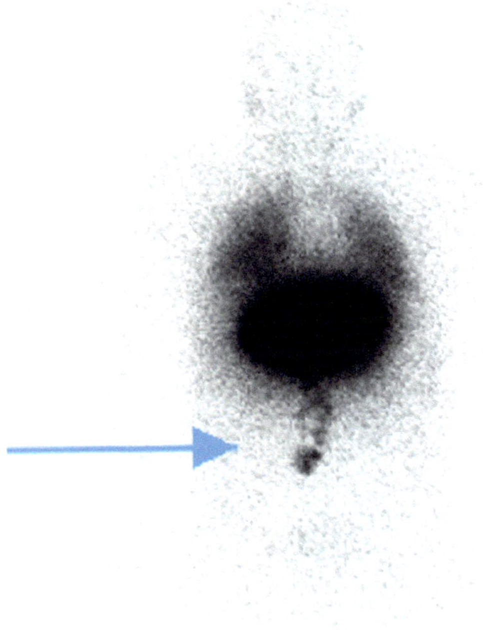

Fig. 2 Anterior view

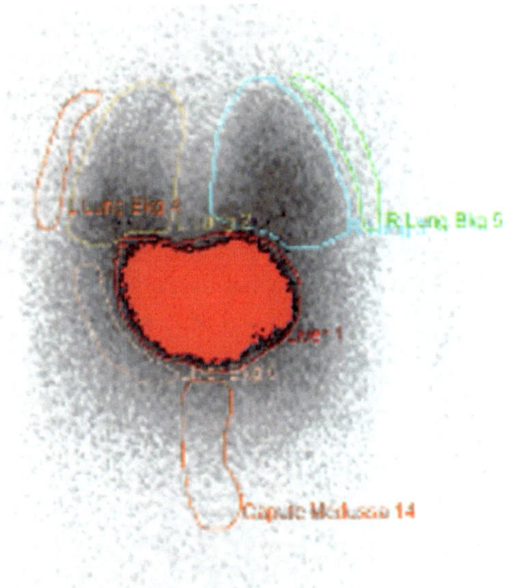

Fig. 3 Posterior view with regions of interest (ROI) for lung shunt determination. Lung Shunt = 7.9%

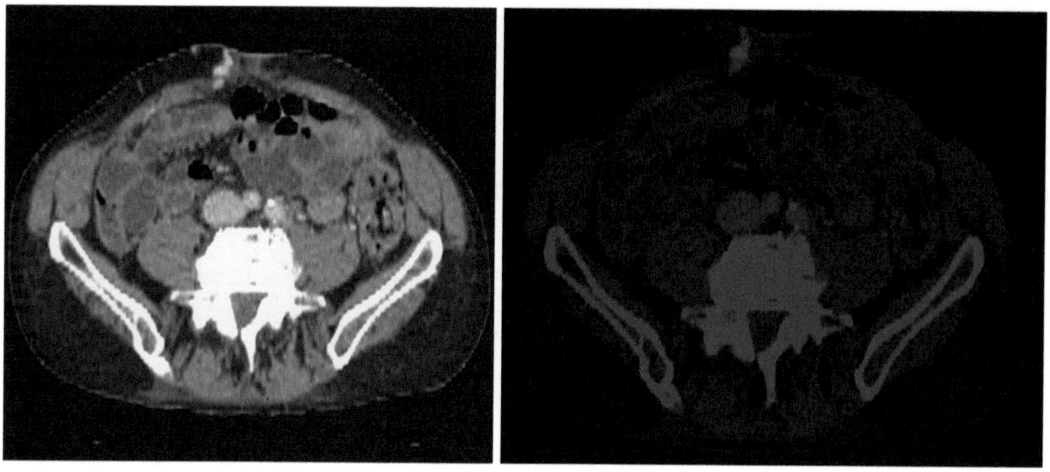

Fig. 4 CT (Left) and Tc-99 m MAA SPECT/CT (Right)

Answer: D

Several studies have demonstrated a survival benefit for patients that originally were not considered candidates for transarterial chemoembolization, some due to the presence of portal vein thrombosis (PVT), and that were subsequently treated with Y-90 RAE. Resolution of PVT has been demonstrated following Y-90 RAE [6].

Involvement of both hepatic lobes by tumor is not an absolute contraindication. However, additional important pre-Y-90 treatment high-risk factors to consider, particularly with the use of glass microspheres include a large tumor volume involving more than 70% of the liver parenchyma, infiltrative tumor type, elevated AST and ALT above five times the upper limit of normal, bilirubin levels above 2 mg/dL, and low albumin levels below 3 g/dL (particularly for tumor volumes involving greater than 50% of the liver) [2].

An absolute contraindication for treatment can be encountered if extrahepatic deposition of ^{99m}Tc-MAA is evidenced in the planning scan, and no safe alternatives for pre-therapy embolization exist, with an elevated risk for undesired extrahepatic radionecrosis [7].

8. What is the recommended time at which the patient should return to the Nuclear Medicine department following a planning session with transarterial deposition of ^{99m}Tc-MAA?

   A. 30 min after radiopharmaceutical administration
   B. 2–4 h after the radiopharmaceutical administration
   C. 3–6 h after radiopharmaceutical administration.
   D. 24 h after radiopharmaceutical administration

Answer: A

The expected physiologic distribution ^{99m}Tc-MAA increases in intensity the latter the images are taken from the time of intraarterial administration. It is important to recognize that in vivo biodegradation of ^{99m}Tc-MAA and the presence of free ^{99m}Tc, increase the activity seen in the thyroid gland, kidneys, salivary glands, stomach, and other common locations. The activity seen in these expected locations should not be included in the calculation of the lung shunt [1, 4].

9. What is a common absolute contraindication for transarterial chemoembolization although it is only a relative contraindication to transarterial radioembolization?

   A. High albumin levels
   B. Low bilirubin levels
   C. Portal vein thrombosis
   D. Tumor volume > 30%

   Answer: C
   Studies have shown that patients that were deemed not appropriate candidates for transarterial chemoembolization due to the presence of portal vein thrombosis (PVT) may be candidates for radioembolization with Y-90 microspheres. Resolution of PVT has been reported following Y-90 RAE [6]. Additional important pre-Y-90 treatment high-risk factors to consider are large tumor volume involving more than 70% of the liver parenchyma, infiltrative tumor type, elevated AST and ALT above five times the upper limit of normal, bilirubin levels above 2 mg/dL, and low albumin levels below 3 g/dL (particularly for tumor volumes involving greater than 50% of the liver) [2].

10. Following treatment Y-90 microspheres, radiation precautions should include?

   A. Contact isolation for 3 days to reduce radiation exposure to the general population.
   B. Contact isolation for the first week following treatment.
   C. Contact isolation for the first 24 h following treatment
   D. No contact isolation required

   Answer: D
   Given the short travel distance of the beta particles, with the majority of the energy deposited in the first 5 mm of tissue in the first 11 days following treatment, no strict contact isolation restrictions are needed [8]. The patient should be released with the recommended general instructions as per the nuclear regulatory commission recommendations regarding patients receiving unsealed byproduct material or implants containing byproduct material if the total effective dose equivalent to any other individual from exposure to the released individual is not likely to exceed 5 mSv [9].

References

1. Uliel L et al. From the angio suite to the g-camera: vascular mapping and 99mTc-MAA hepatic perfusion imaging before liver radioembolization—a comprehensive pictorial review. J Nucl Med. 2012;53:1736–47.
2. TheraSphere Yttrium-90Glass-Microspheres *Package Insert*. BTG Rev. 14.
3. Sir-Spheres Microspheres. Sirtex Package Insert. CE mark issued 2002.
4. Kallini JR et al. Transarterial radioembolization with Yttrium-90 for the treatment of hepatocellular carcinoma. Adv Ther. 2016;33:699–714.
5. Salem R et al. Y90 Radioembolization significantly prolongs time to progression compared with chemoembolization in patients with hepatocellular carcinoma. Gastroenterology. 2016;151: 1155–63.
6. Garin E et al. Personalized dosimetry with intensification using 90Y-loaded glass microsphere radioembolization induces prolonged overall survival in hepatocellular carcinoma patients with portal vein thrombosis. Soc Nucl Med Molecul Imag. 2015.
7. Hamami ME et al. SPECT/CT with 99mTc-MAA in radioembolization with 90Y microspheres in patients with hepatocellular cancer. J Nucl Med. 2009;50:688–92.

8. Gulec SA, Siegel JA, Posttherapy radiation safety considerations in radiomicrosphere treatment with 90Y-microspheres. J Nucl Med. 2007;48:2080–6.
9. NRC Regulations (10 CFR). § 35.75 Release of individuals containing unsealed byproduct material or implants containing byproduct material. Reviewed August 29, 2017.

Case 3 Bone ablation

11. Which of the following radiopharmaceuticals used for palliative therapy for metastatic osseous disease is an alpha emitter?

    A. Samarium-153
    B. Radium-223
    C. Strontium-89
    D. Rhenium-186

    Answer: B

    Patients who have extensive painful osseous metastases (from various primary lesions that are not amenable to external beam radiotherapy or that are unresponsive to chemotherapy), may benefit from radionuclide palliative therapy. It is directed to reduce pain (not to cure), decreased need for opiates, and improved quality of life. The commonly used radiopharmaceuticals are Samarium-153, Radium-223, Strontium-89, Rhenium-186. Among them, only Radium-223 is an alpha emitter. Others are predominantly beta emitters. Rhenium-186 has not been approved for use in the United States but is widely used in Europe.

12. Which of the following radiopharmaceuticals used for palliative therapy for metastatic osseous disease can be proficiently imaged with a gamma camera?

    A. Radium-223
    B. Samarium-153
    C. Strontium-89
    D. Phosphorus-32

    Answer: B

    Among above all mentioned radiopharmaceuticals, only Samarium-153 is effectively suitable for imaging. Samarium-153 is a mixed beta and gamma emitter. In practice, it is possible to image 103 KeV gamma emissions (has 28% abundance). Imaging is usually done at 6 h after administration.

    Radium-223 is an alpha emitter with np gamma radiation to image.

    Strontium-89 and Phosphorus-32 are pure beta emitters. They also produce Bremsstrahlung (BS) radiation, which is difficult to image.

    Rhenium-186 is another radiopharmaceutical used for palliative therapy for metastatic osseous disease. It is a gamma emitter with an energy of 187 KeV that is suitable for imaging. Rhenium-186 has not been approved for use in the United States.

    *StatDx. https://my.statdx.com/.

    *Ziessman HA, O'Malley JP, Thrall JH, Nuclear medicine: the requisites.

13. Radium-223 dichloride is approved by the FDA to be used in which of the following cancer?

    A. Breast cancer
    B. Lung cancer
    C. Prostate cancer
    D. Renal cell carcinoma

Answer: C

The FDA approved the use of Radium-223 (Xofigo) in May 2013 only in men with castration-resistant prostate cancer and painful bone metastasis. Exclusion criteria include soft tissue metastasis. The treatment consists of a single monthly intravenous injection for a total of six doses. Xofigo is not approved for any other malignancies. Common side effects are nausea, vomiting, diarrhea, and bone marrow suppression. Therefore, blood counts should be performed a few days before each injection. Also, the dose depends on body weight, which should be updated routinely.

Miscellaneous

Medhat Sam Gabriel, Bital Savir-Baruch, Robert H. Wagner, Bruch J. Barron, Davide Bova, Aamna Hassan, Stephanie Rodriguez-Santiago, and Yekaterina Kucerova

Miscellaneous 2020

Case 1

1. A 62-year-old female with PMH of recurrent gastric MALT treated with chemotherapy and currently in remission for the past 15 years. Surveillance CT demonstrates mild increase in the size of axillary lymph nodes, prompting further evaluation with F18-FDG PET/CT (select images below). In addition to the hypermetabolic right axillary lymph nodes, what additional information can be provided?

M. S. Gabriel (✉) · B. Savir-Baruch · R. H. Wagner · D. Bova
Division of Nuclear Medicine, Department of Radiology, Loyola University Medical Center, Maywood, IL, USA
e-mail: sgabriel@lumc.edu; Bital.savir-Baruch@lumc.edu; rwagner@lumc.edu; dbova@lumc.edu

B. J. Barron
Division of Nuclear Medicine and Molecular Imaging, Department of Radiology and Imaging Sciences, Atlanta, GA, USA
e-mail: bjbarro@emory.edu

A. Hassan
Department of Nuclear Medicine, Shaukat Khanum Memorial Cancer Hospital and Research Center, Lahore, Punjab, Pakistan

S. Rodriguez-Santiago
Department of Nuclear Medicine, C.W. Bill Young VA Medical Center, Bay Pines, FL, USA
e-mail: stephanie.rodriguezsantiago@va.gov

Y. Kucerova
Division of Nuclear Medicine, Department of Radiology, Loyola University Medical Center, Maywood, IL, USA

Department of Radiology, Cook County Healthy, Chicago, IL, USA
e-mail: yekaterina.kucerova@cookcountyhealth.org

M. S. Gabriel, B. Savir-Baruch (eds.), *RadTool Nuclear Medicine MCQs*,
https://doi.org/10.1007/978-3-030-69281-0_12

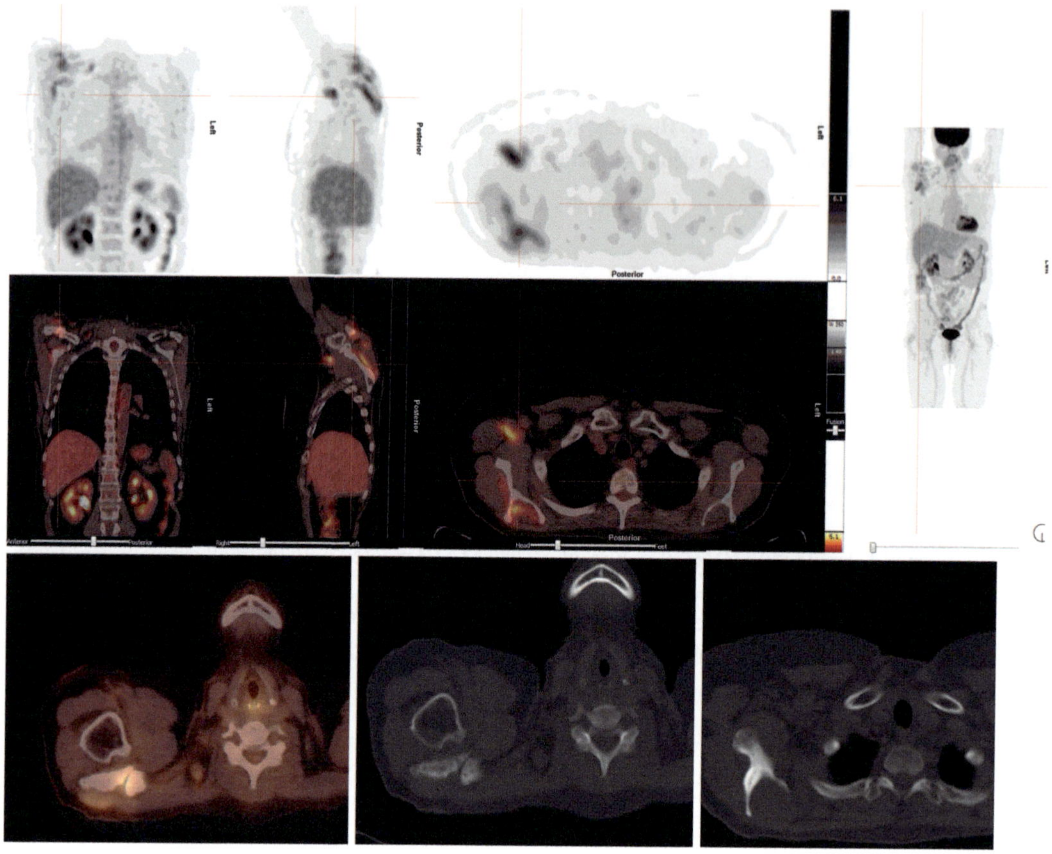

A. Physiological muscular uptake secondary to muscle strain
B. Inflammatory arthropathy of the AC joint, with misregistration due to patient motion.
C. Scapular lymphoma
D. The abnormal metabolic activity is located in the muscle

Answer: D

July 2017 PET-CT study showed several hypermetabolic lymph nodes in the right supraclavicular and right axillary regions, consistent with lymphoma. Plaque-like areas of increased FDG uptake in the muscles overlying the right scapula and the right shoulder (supraspinatus, deltoid, etc.) are also noted. However, associated non-enhanced CT demonstrates no focal abnormality of the affected muscular structures.

2. What is the next diagnostic test to further evaluate the nature of this uptake?

A. CT with contrast
B. WBBS with SPECT-CT of the upper thorax
C. Radiographs of the right shoulder
D. MRI of the shoulder

Answer: C

Standards radiographs are almost invariably the best first-line diagnostic test for musculoskeletal disease, because of its low cost and wide field of view. It provides first assessment of the bony architecture and possible associated abnormalities, offering a global framework of differential diagnoses. Also, it highlights abnormalities for targeted cross-sectional evaluation. It can occasionally provide a visual target for tissue diagnosis under fluoroscopic guidance. Fluoroscopy and ultrasonography allow for faster and therefore better tolerated invasive procedures, in light of their real-time guidance.

3. Radiographs of the right shoulder are shown. What would you do next?

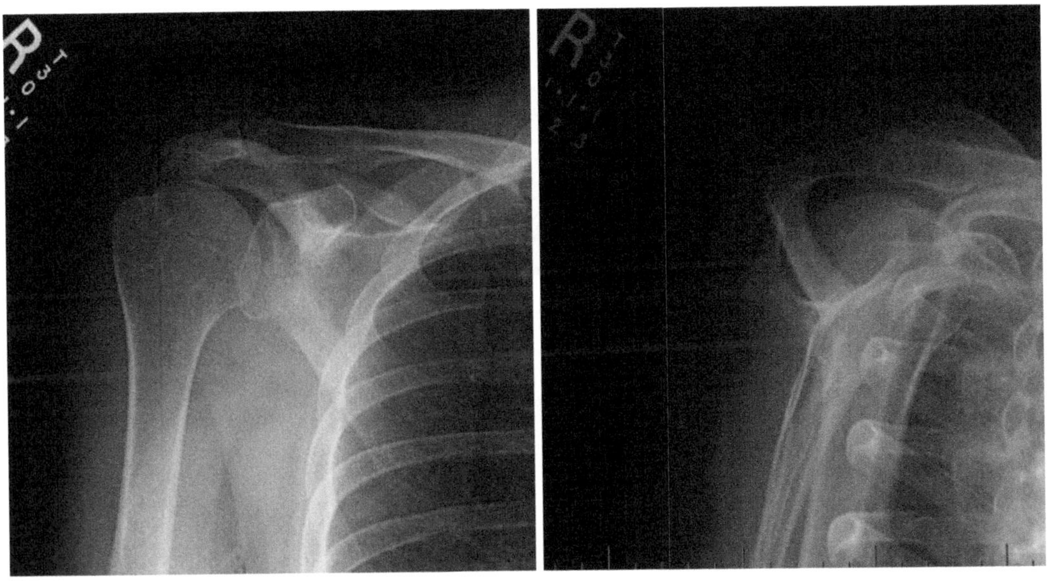

A. CT with contrast
B. WBBS with SPECT-CT of the upper thorax
C. Rest and NSAIDS.
D. MRI of the shoulder with and without contrast

Answer: D

There are mixed sclerotic and cystic changes of the subchondral aspect of the acromion clavicular joint, with the appearance of degenerative-inflammatory changes, corresponding to mild hypermetabolic activity. Non-enhanced CT (from the PET-CT study) shows mild sclerosis of the scapular body without associated hypermetabolism. None of these findings account for the markedly increased muscular metabolic activity. Prior CT with contrast demonstrating enlarging axillary nodes did not show any focal or asymmetric enhancement within these muscular structures. Bone scanning agents are going to electively precipitate in areas of new bone matrix formation and heterotopic calcification/ossification. Therefore, WBBS with or without SPECT would be of little use in evaluating soft tissue processes.

4. MRI of the shoulder T1 sagittal image obtained after IV administration of Gadolinium and oblique coronal T2 weighted image are shown below.

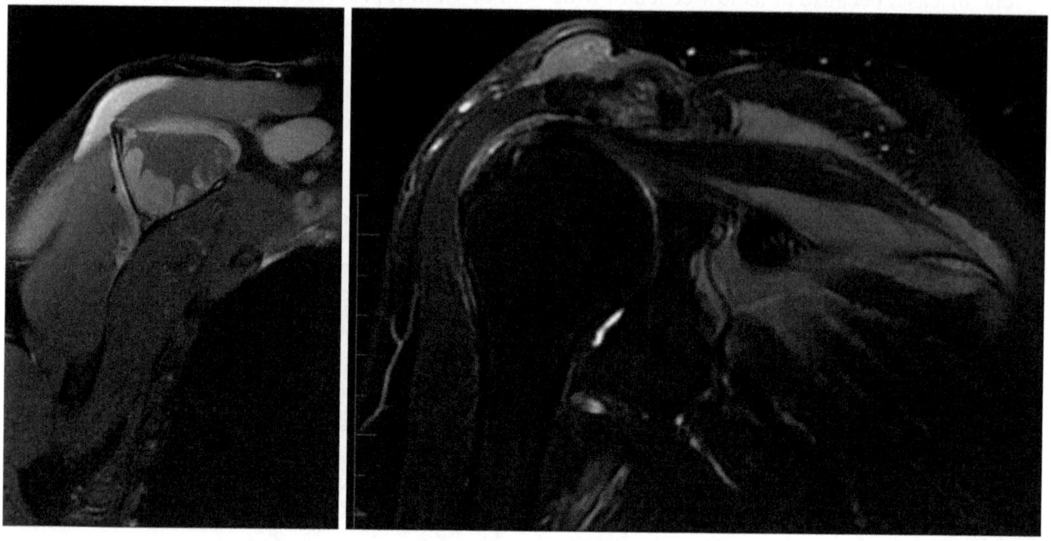

What is your diagnosis?

A. Muscular edema secondary to muscle strain
B. Full-thickness tear of the supraspinatus tendon (part of the rotator cuff
C. Scapular lymphoma
D. Muscular lymphoma

Answer: D

Provided images demonstrate a diffuse infiltrative soft tissue process affecting the substance of the supraspinatus and infraspinatus muscles, along with a soft tissue mass overlying the deltoid, acromion, and trapezius, as well as axillary and suprascapular lymphadenopathy are most suggestive of lymphoma.

Disease process demonstrates marked diffuse and homogeneous contrast enhancement (1: Sag. T1 with Contrast) and markedly as well as homogeneously increased T2 signal (2: coronal fat-suppressed T2 image).

On axial post-contrast at-suppressed T1 images (see below, images 3 and 4) there is mild heterogeneous marrow enhancement of the subchondral portion of the right clavicle, due to inflammatory changes. The sclerotic portion of the scapular body demonstrates neither increased T2 signal nor appreciable enhancement, therefore a benign process, probably reactive to the overlying soft tissue neoplasm.

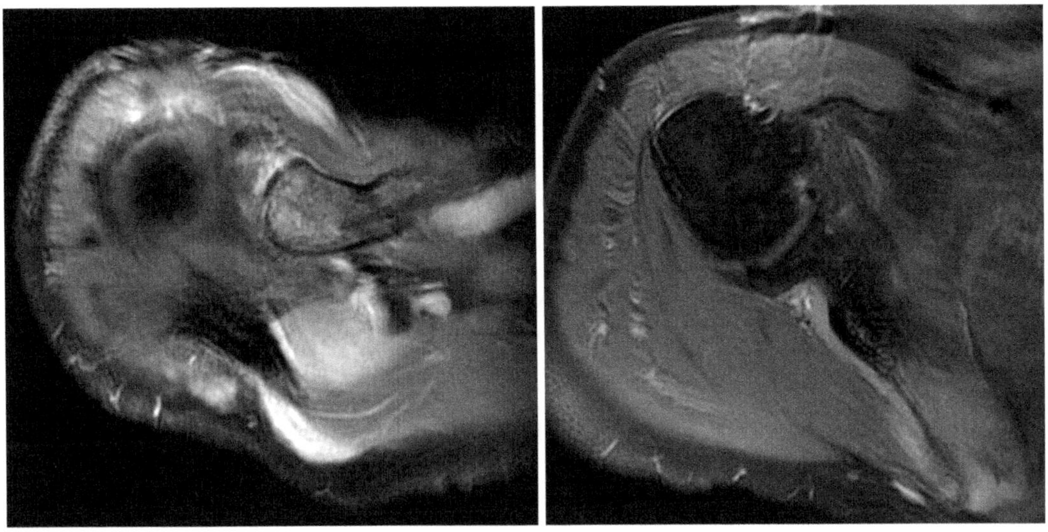

Percutaneous biopsy of a right axillary lymph node demonstrated low-grade Marginal Zone B cell lymphoma

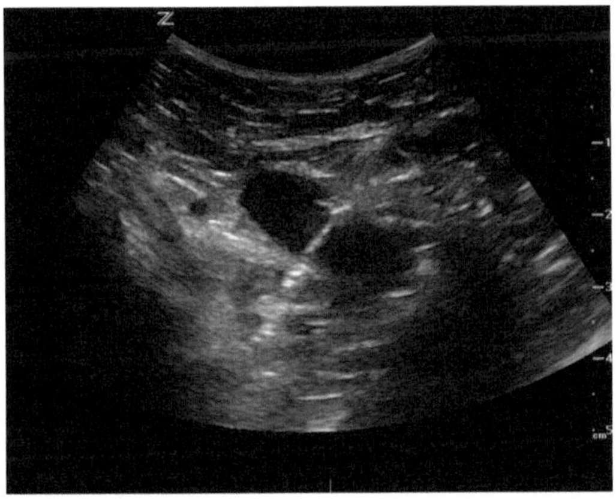

5. Which of the following statements about muscle lymphoma is true?

   A. It frequently demonstrates large areas of central necrosis
   B. It characteristically displays heterogeneous enhancement
   C. It preferentially grows along the length of the muscular belly
   D. It should be controlled with local external beam radiation.

   Answer: C

   Lymphoma more frequently presents with longitudinal often peripheral growth along the muscle than with discrete round or oval masses. It typically displays homogeneous signal intensity and enhancement. Rare necrosis occurs when it is large. It is primarily treated with systemic chemotherapy. Radiation therapy is most often reserved for local symptomatic/palliative control in case of systemic disease.

6. Which of the following statements about muscle lymphoma is true?
   A. It is most frequent in Hodgkin's disease.
   B. Extranodal lymphoma is an exceedingly rare occurrence
   C. CT is the modality of choice for diagnosing muscular lymphoma
   D. Both B cell and T cell NHL are associated with muscular lymphoma.

   Answer: D
   Extranodal lymphoma is a relatively frequent occurrence, particularly in more aggressive subtypes of both T cell and B cell non-Hodgkin lymphomas. They can be exceedingly difficult to detect on CT because of their frequent isodensity to surrounding or adjacent normal muscle. They may show just enlargement of the affected muscles with relatively low and uniform enhancement. MRI and PET-CT are therefore more sensitive imaging modalities.

*Chun CW, Jee WH, Park HJ, et al., MRI features of skeletal muscle lymphoma. AJR. 2010;195:1355–60.
*Lim CY, Ong KO. Imaging of musculoskeletal lymphoma. Cancer Imaging. 2013;13(4): 448–57.

Case 2

7. A 70-year-old woman presents for staging studies after a recent diagnosis of right breast infiltrating ductal carcinoma grade 2. Contrast-enhanced CT of the chest and abdomen demonstrates likely:

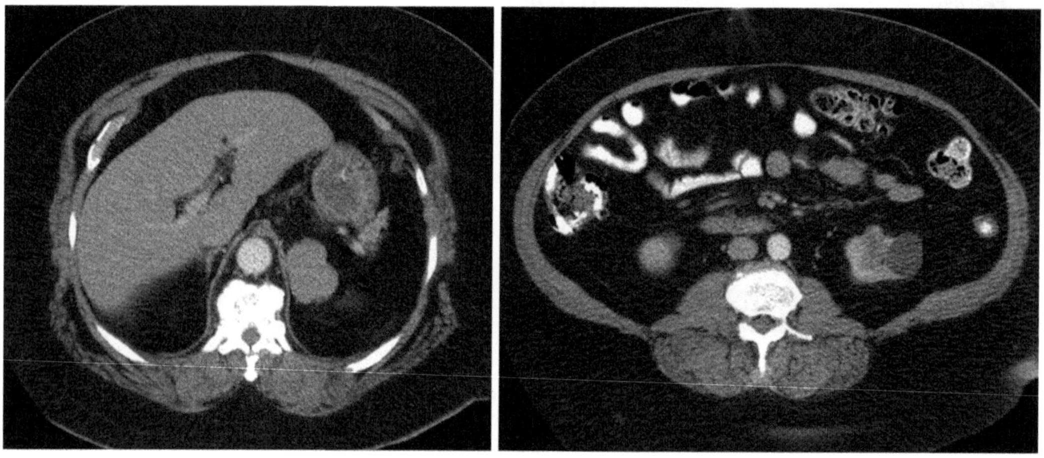

   A. Metastatic breast carcinoma to the left adrenal gland and to the left renal cortex
   B. Non-hyperfunctioning left adrenal adenoma with simple and complex cysts of the lower pole of the left kidney
   C. Left renal cell carcinoma with left adrenal metastasis
   D. Indeterminate left adrenal mass and suspected left renal solid mass adjacent to a simple cyst

   Answer: D
   The 4.5 cm left adrenal mass measures greater than 10 HU and is therefore indeterminate because it does not meet the criteria for an adrenal adenoma. The anterior deformity in the lower pole of the left kidney is also indeterminate because it appears solid and heterogeneously enhancing.

8. What is the next best diagnostic test to further characterize the CT findings?

   1. MRI before and after intravenous Gadolinium
   2. Renal and adrenal Ultrasound
   3. CT of the adrenal gland and CT Urogram at a later date
   4. F18-FDG PET-CT

   Answer: A

   MRI of the abdomen, augmented by chemical shift imaging (better known as in-phase and out-of-phase or opposite-phase sequences), is well suited to detect intracellular lipids with the adrenal adenomas, often obviating the need for Gadolinium-based IV contrast or needled biopsies. In addition, it would provide further evaluation of the left renal mass, by means of dynamic post-contrast acquisition and T2 signal appearance. Ultrasound has a limited role in characterizing solid masses, especially in patients with large body habitus causing severe beam degradation. CT has already been performed and would not allow the same number of dynamic acquisitions of MR because of the exposure to ionizing radiations. FDG-PET demonstrates poor diagnostic accuracy for malignant renal masses.

9. Patient failed attempt to MRI of the abdomen due to large body habitus. Below are selected FDG-PET images.

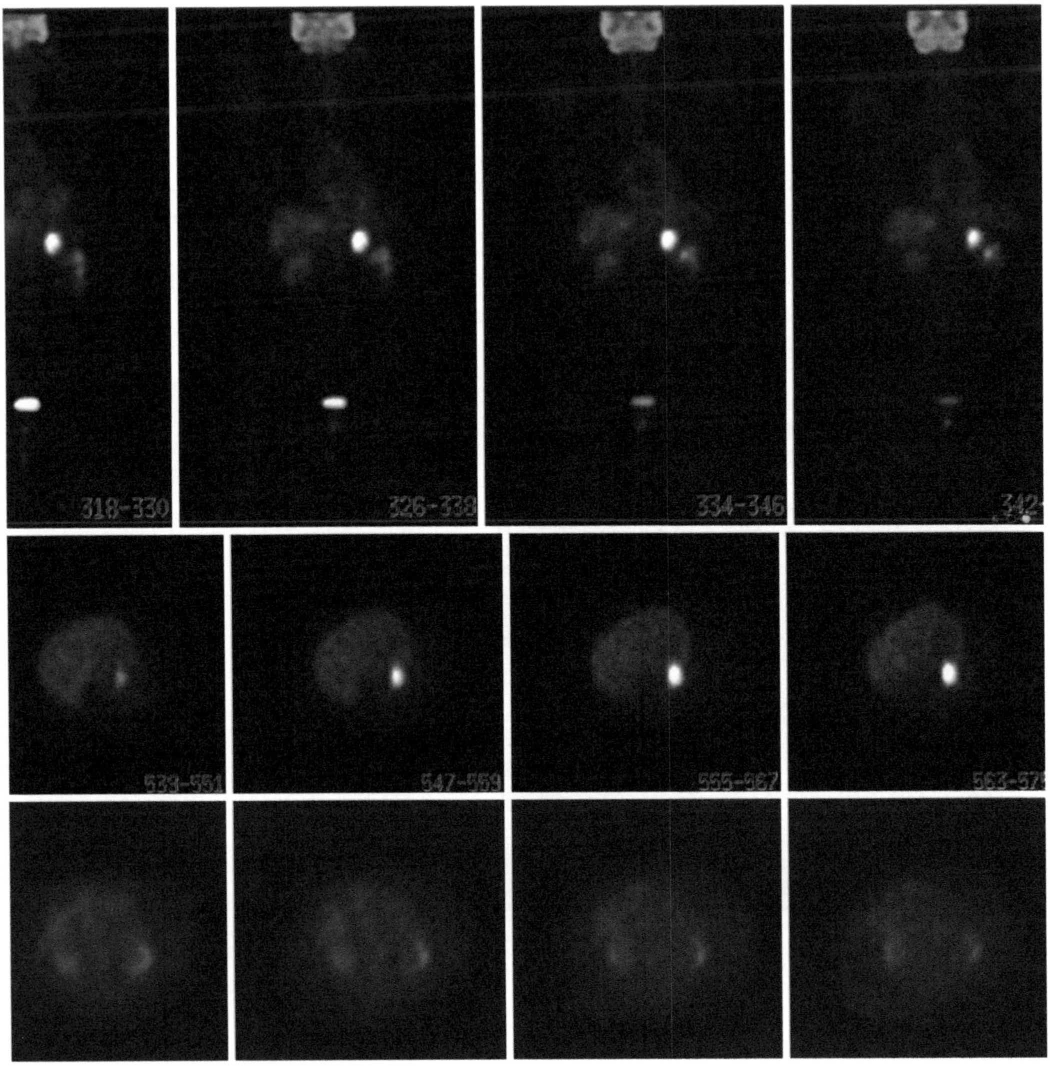

What would your diagnosis be, in light of a SUV left adrenal is 19.8, and a SUV not measurable in the left renal anterior lesion?

A. Negative for renal cell carcinoma and left adrenal metastatic breast carcinoma
B. Left renal cell carcinoma and left adrenal metastatic breast carcinoma
C. Indeterminate left adrenal mass and possible left renal cell carcinoma
D. Left adrenal adenoma but negative for renal cell carcinoma

Answer: C

High SUV in the left adrenal mass remains suspicious for, but not diagnostic of, metastatic disease, especially in the absence of other metastatic lesions. Lack of appreciable metabolic activity in the left solid renal lesion should not be viewed as an indication of its benign nature, because renal cell carcinomas frequently display low metabolic activity and are difficult to separate from the adjacent urinary activity, especially when small such as this one.

10. Which of the following statements is true about adrenal masses?

A. Adrenal adenomas can be reliably diagnosed by MRI if they demonstrate signal drop-out on fat-suppressed T1 AND T2 sequences.
B. Percutaneous biopsies of the adrenal glands cannot be performed because of their unacceptably high incidence of pneumothorax and renal AV-fistulas
C. Pheochromocytomas are most frequently incidental findings at cross-sectional imaging of the abdomen performed for other indications.
D. Adrenal adenomas can rarely be markedly hypermetabolic

Answer: D

Adenomas demonstrate signal drop-out at chemical shift imaging because their lipid content is exclusively intracellular. They can display greater uptake than the liver in approximately 3% of cases, particularly if they are secreting adenomas. Pheochromocytomas are incidentally diagnosed in only about 10% of cases and can be quite hypermetabolic, especially if they are malignant. Pneumothoraces can be avoided during percutaneous CT-guided biopsies by using an oblique caudo-cranial approach and decubitus to minimize chances of traversing the pleura (see below).

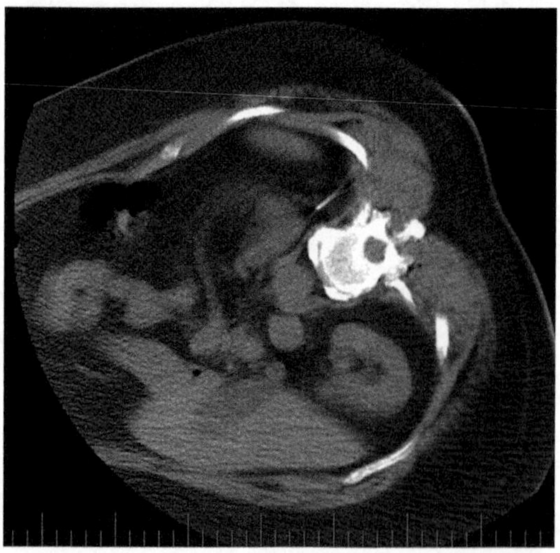

Patient underwent robotic-assisted left partial nephrectomy and left adrenalectomy. Pathology revealed "LEFT ADRENALECTOMY: (SPECIMEN A) ADRENAL CORTICAL ADENOMA, SIZE 5.0 CM.—LEFT KIDNEY PARTIAL NEPHRECTOMY: (SPECIMEN B) RENAL CELL CARCINOMA, CLEAR CELL TYPE, FUHRMAN GRADE 3, MEASURING 2.3 CM IN GREATEST DIMENSION."

11. If during initial staging for breast carcinoma, the adrenal lesion below is identified at contrast-enhanced CT:

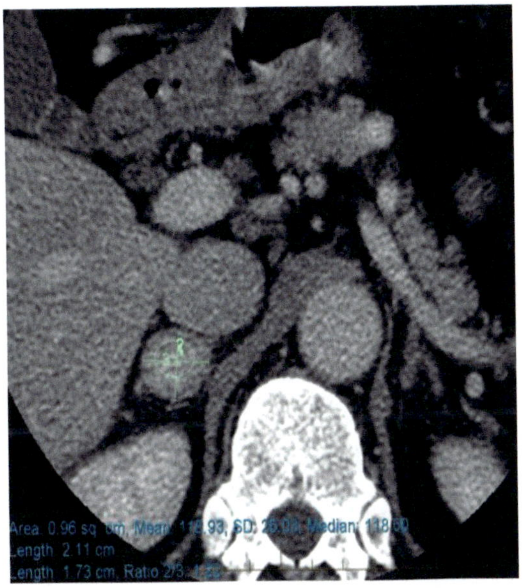

What would the next best diagnostic test be?

A. Follow up in 12 months for interval stability
B. Biochemical assays
C. MRI abdomen
D. PET/CT

Answer: C

Even lipid-rich adenomas can display marked enhancement after contrast medium administration, because of their frequent internal rich vascularity. Any adrenal lesion displaying greater than 10 H.U. at CT is therefore indeterminate. MRI of the abdomen with chemical shift imaging (better known as in-phase and out-of-phase or opposite-phase sequences) is sensitive to intracellular lipids, causing signal drop-out on opposite-phase sequences. Sub-voxel mixed lipid and fluid, such as the intracellular lipid content of adrenal adenomas or hepatic steatosis, cause this "signal degradation" and is therefore very specific, often obviating the need for Gadolinium IV contrast administration or tissue diagnosis. Of note, adrenal adenomas and hepatic steatosis will NOT display signal loss on Fat-Suppressed MR sequences; this is characteristic of lesions containing "macroscopic" fat, such as adrenal myelolipomas, renal angiomyolipomas, or ovarian dermoids.

*Dong A, Cui Y, Wang Y, Zuo C, Bai Y, 18F-FDG PET/CT of adrenal lesions. AJR. 2014;203: 245–52.

Case 3

12. A 70-year-old woman presents for initial staging of right lung carcinoma. Select PET-CT images are shown below.

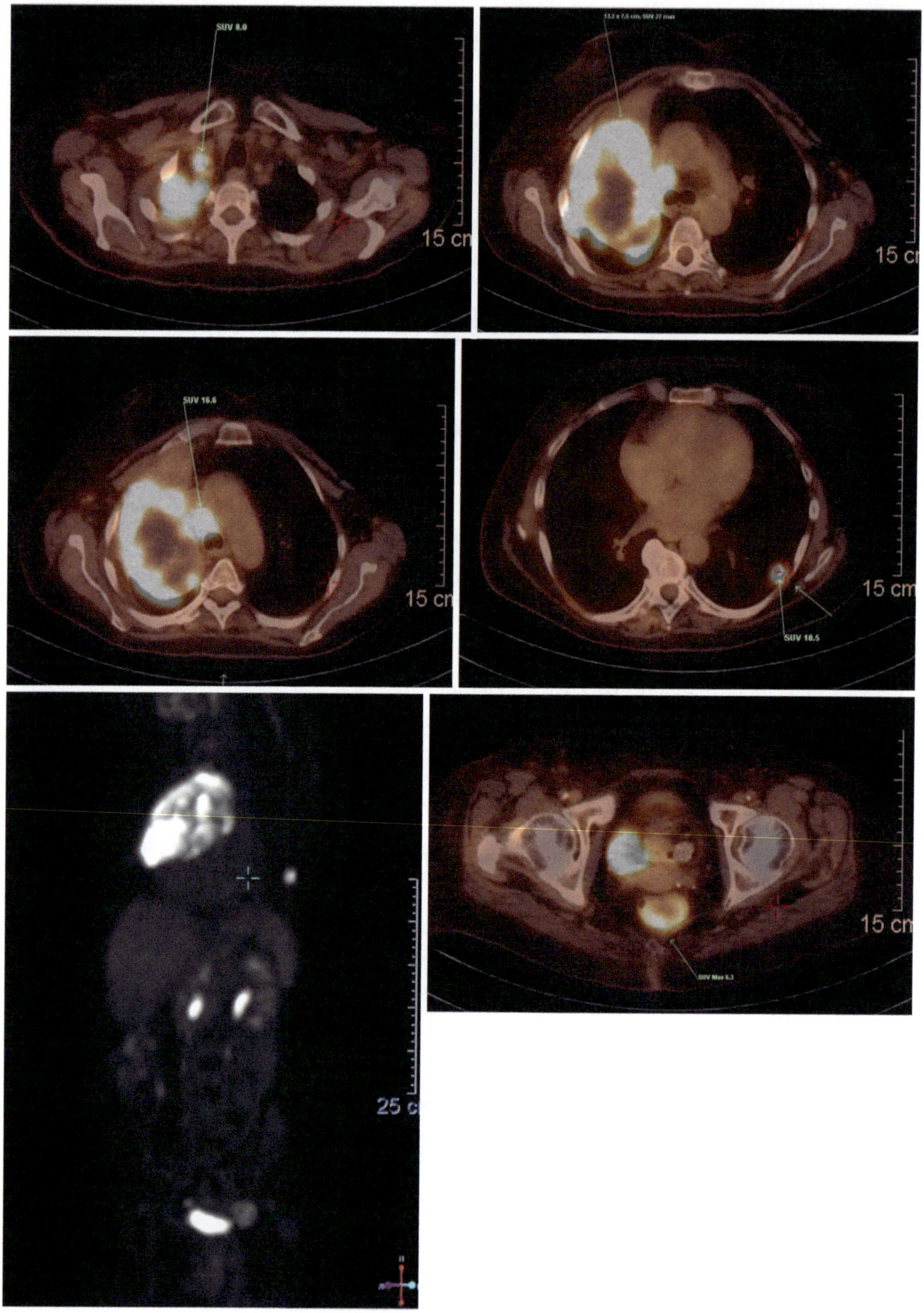

Which of the following statements is true?

A. T2 N2 M0 lung cancer and recto-sigmoid carcinoma
B. Stage IV lung cancer with metastases to the colon
C. Stage IV lung cancer and primary lesion of the recto-sigmoid colon
D. Bilateral synchronous lung carcinomas with distant metastasis to the colon

Answer: C
Primary lung mass with ipsilateral metastatic mediastinal nodes and contralateral lung metastasis (M1a). The hypermetabolic area in the recto-sigmoid is suspicious for a second malignancy.

13. What is the most appropriate next step?

A. Colonoscopy
B. CT Abdomen and Pelvis with contrast
C. CT colonography
D. Endoscopic Ultrasound

Answer: A
Standard CT has poor sensitivity for GI lesions because of the variable distention of the bowel lumen. CT colonography is indicated as an alternate screening tool or in patients unable to tolerate colonoscopy. Endoscopic ultrasound is useful, in expert hands, to assess depth of invasion for colonic carcinoma. Colonoscopy has high diagnostic yield for intraluminal lesions and allows concurrent resection and/or tissue diagnosis of colonic lesions.

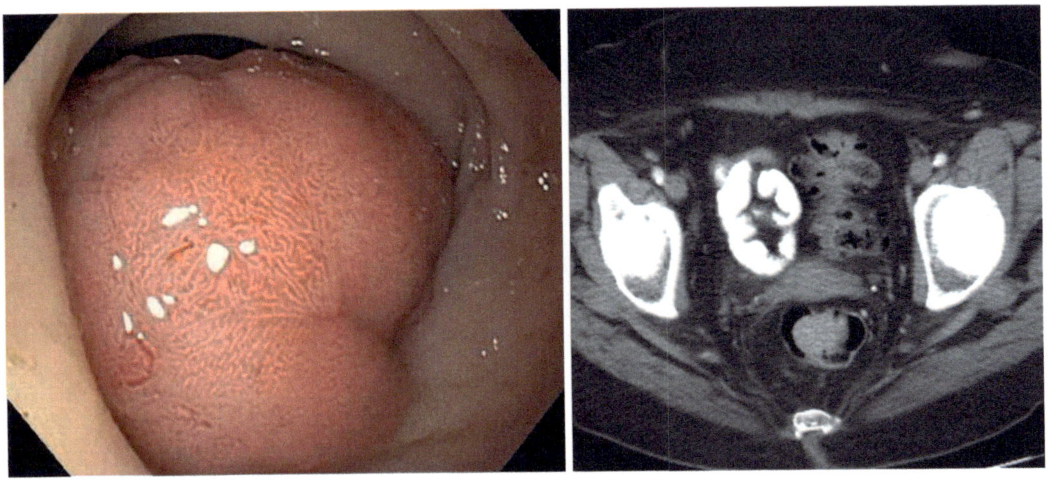

14. Which of the following statements is true regarding FDG uptake in the gastrointestinal tract?

A. It is a frequent incidental finding and of no clinical significance
B. Diffuse colonic uptake is suspicious for familial polyposis
C. Focal FDG uptake in the colon is associated with skip lesions from Crohn disease
D. Focal FDG uptake of the colon is frequently associated with malignant and premalignant lesions

Answer: D

Diffuse uptake is not suspicious for malignancy and can be seen with inflammatory bowel disease or associated with certain medications, such as metformin. Colonoscopy reveals either malignant lesions or villous/tubule-villous adenomas in most cases referred for incidentally noted focal FDG uptake of the colon. At pathology, the lesion above was proven to represent a "tubulo-villous adenoma with focal high-grade dysplasia."

*Kei PL, Vikram R, Yeung HWD, Stroehlein JR, Macapinlac HA, Incidental finding of focal FDG uptake in the bowel during PET/CT: CT features and correlation with histopathologic results. AJR. 2010;194:W401–6.

15. A Tc-99 m labeled WBC study is performed on this patient with a history of fevers of unknown origin. The CT study of the chest is unremarkable. The24 h images on this study demonstrate:

    A. A focal area of pneumonia
    B. A technical artifact
    C. Vasculitis
    D. Pulmonary thromboembolism

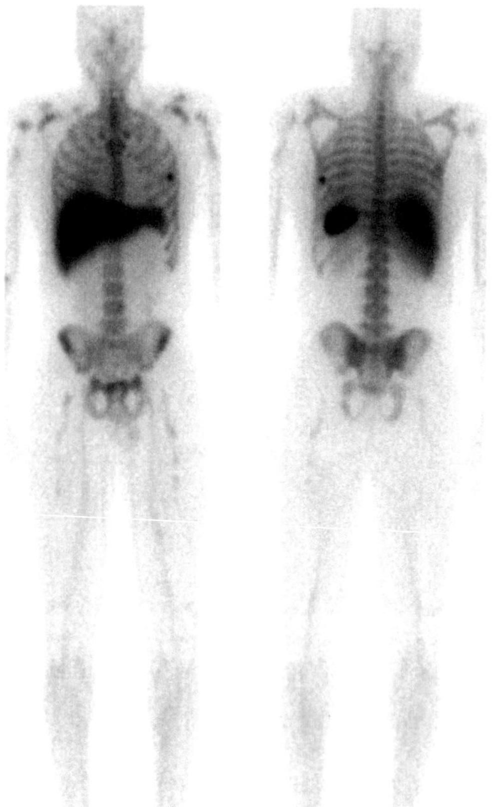

Answer: B

The process of labeling WBC's requires removal of the patient's blood, separation of the white cells, labeling them, and then reinjection. It is not uncommon for some of the white cells to clump together after labeling and upon injection remain within the lung as a "hot clot."

The white cells distribute within the reticuloendothelial system and demonstrate significant uptake in the liver, spleen, and bone marrow. The study (with the exception of the "hot clot" artifact) is unremarkable.

16. A 63-year-old female with liver failure and ascites presents with a history of hypercalcemia that is thought to be secondary to a parathyroid adenoma. The following initial and 2 h delay planar images are obtained. A SPECT/CT study was also performed. Your impression is:

    A. Negative for parathyroid adenoma, further imaging is required
    B. Asymmetric uptake is noted in the right lobe of the thyroid, suggestive of an intrathyroidal parathyroid adenoma
    C. There is decreased uptake in the septal area of the heart, indicating ischemia at rest or scar. Suggest a rest/stress myocardial perfusion study or CT angiogram.
    D. The findings suggest the presence of renal osteodystrophy secondary to parathyroid adenoma.

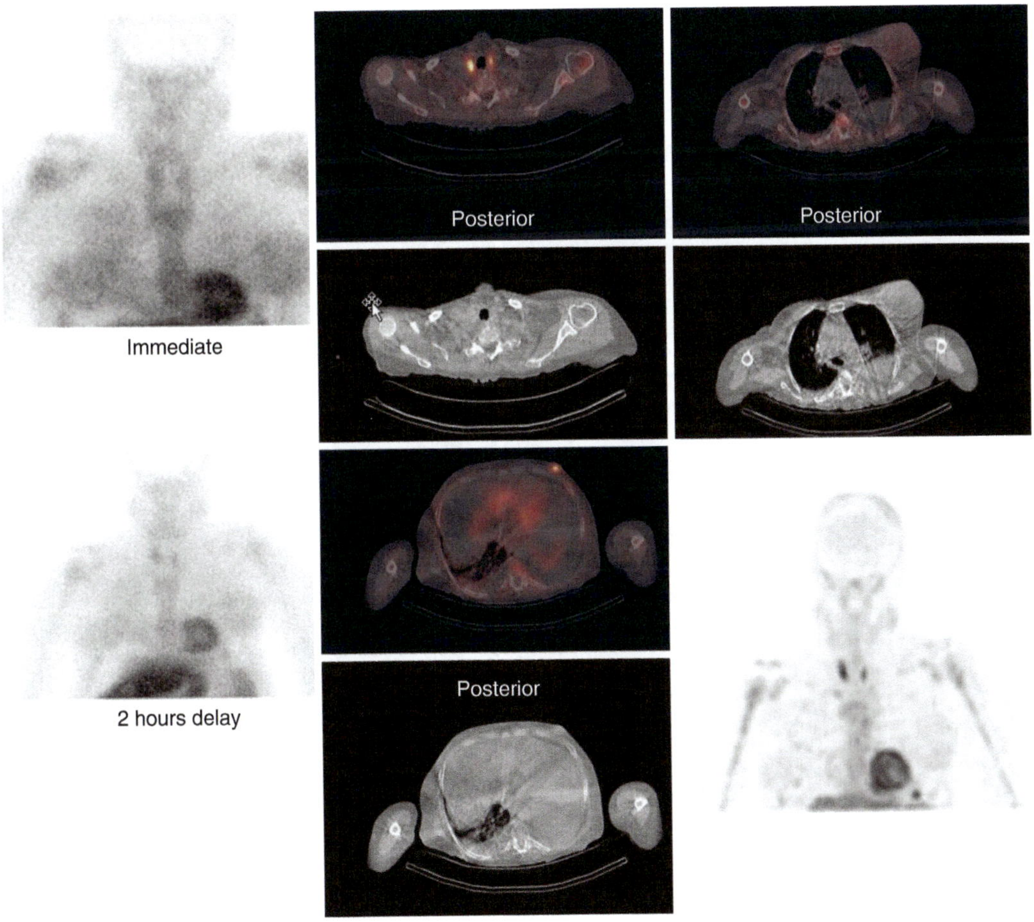

Answer: A

Although there is some asymmetry in the thyroid with slightly greater uptake on the right seen on the SPECT study this is due to asymmetric thyroid tissue, which is commonly seen. No significant asymmetry is seen on the planar images. There is, however, a large asymmetric mass in the left breast that demonstrates low-level uptake. An additional smaller focus of significant uptake is noted inferior to the breast. Further imaging with ultrasound of the breast and biopsy of the large breast revealed a phyllodes tumor. Most are benign but tend to grow quickly. The biopsy of the more intense focus of uptake inferior to the breast in the chest wall revealed a metastatic focus from invasive ductal carcinoma.

The septal portion of the heart cannot be well viewed in the anterior projection. Renal osteodystrophy is best seen on a bone scan.

Case 4

17. A urology colleague explains that he has a 68-year-old patient with a history of splenectomy for prior traumatic rupture and a mass on his kidney believed to be renal cell carcinoma on a CT study. Subdiaphragmatic nodules are seen on the CT study and a MRI study of the abdomen confirms the right lower pole mass. The nodules are most suggestive of splenules. He asks you to confirm that the nodules are not malignant.
    Your first choice of agents to evaluate the nodules is:

    A. F18 FDG/CT imaging
    B. Ga67 citrate
    C. Tc-99 m sestamibi
    D. Heat damaged Tc-99 m RBC's

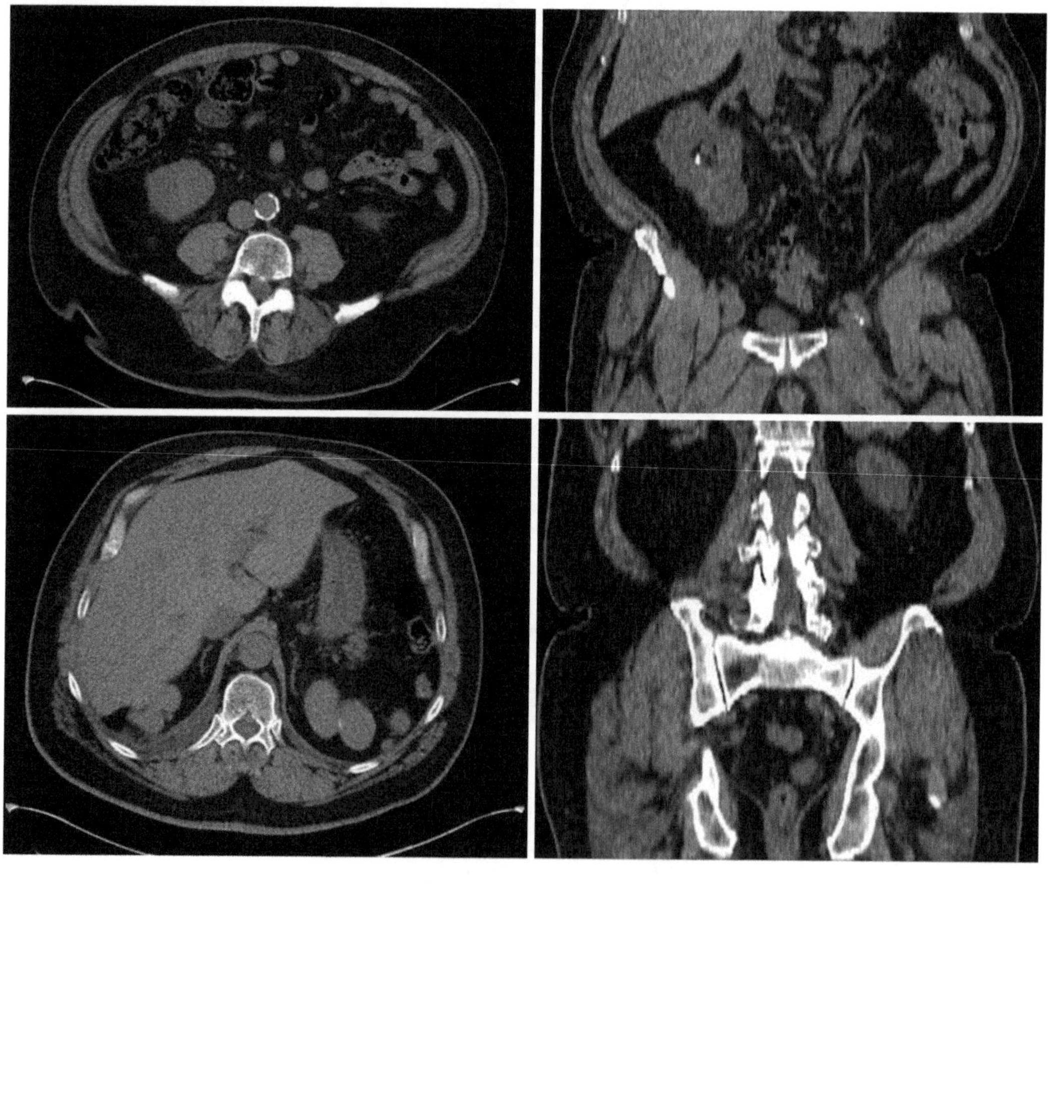

Answer: D

Heat-damaged RBC's are an excellent choice for the determination of splenosis. They are accumulated in the splenules to a great degree and can be helpful to exclude the presence of metastatic foci which would have little to no uptake.

F18 FDG imaging uptake in renal cell carcinoma is variable and not routinely used in initial staging or detection of recurrence.

Gallium 67 citrate and Tc-99 m sestamibi are not routinely used in either renal cell carcinoma imaging or in the detection of splenules.

18. You perform the study from the previous question and obtain planar images and a SPECT/CT. Your impression to the urology colleague is:

    A. The greater metabolic activity present in the nodules is most consistent with metastatic disease.
    B. Cardiac and blood pool activity is seen suggesting a failure of radiopharmaceutical preparation
    C. The nodules can be confidently interpreted as benign
    D. You suggest that the renal mass is an oncocytoma since it does not accumulate sestamibi

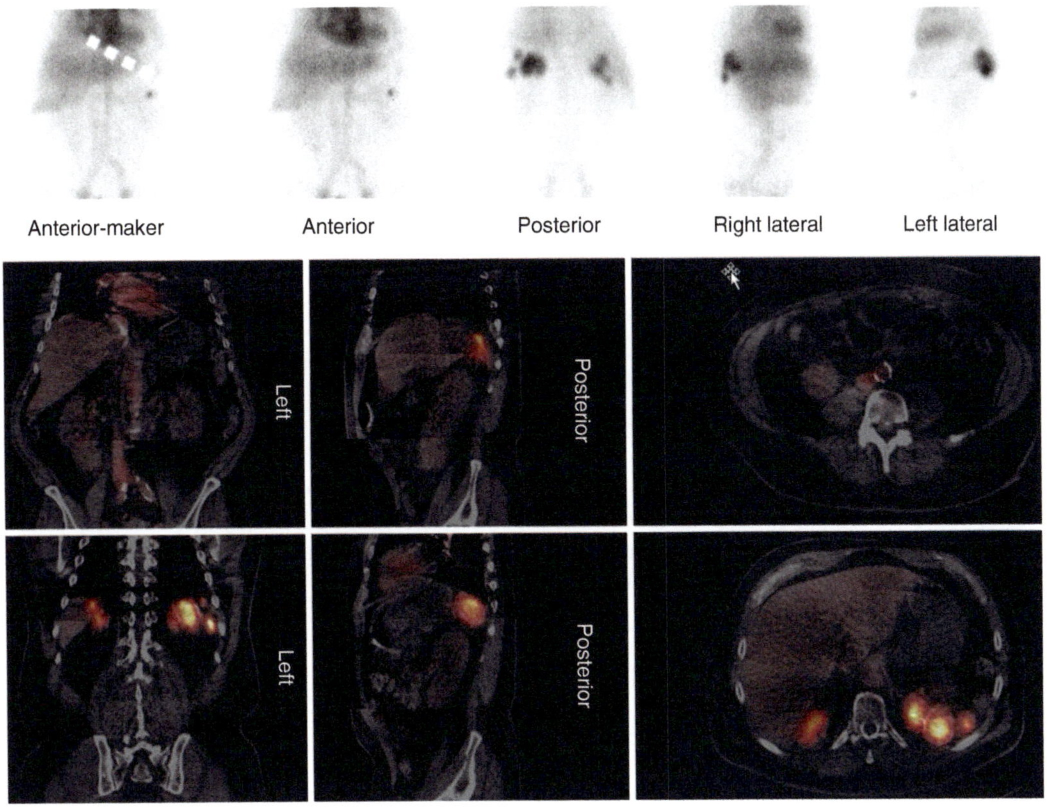

Answer: C

The tracer used for imaging is heat-damaged RBCs.

The images illustrate significant accumulation of the heat-damaged RBCs consistent with splenosis. An additional splenules can be seen anteriorly in the abdomen on the planar images.

Since the agent used is heat-damaged RBCs, activity within the heart and vessels is expected.

Some studies have attempted to use Tc-99 m sestamibi to assist in the differentiation of renal cell carcinoma from oncocytoma. In these studies, it has been found that the oncocytomas accumulate Tc-99 m sestamibi while renal cell carcinomas do not.

19. A study using F-18 florbetapir is used to determine the presence of:

    A. Moyamoya disease
    B. Beta Amyloid
    C. Normal pressure hydrocephalus
    D. Parkinson's Disease

    Answer: B

    F-18 florbetapir (Amyvid—Eli Lilly) is a PET radiopharmaceutical that is used to determine the presence of beta amyloid in the brain. Beta amyloid deposition in the brain is associated with Alzheimer's disease.

    Chronic vascular insufficiency (such as Moyamoya disease) is better assessed using Tc-99 m HMPAO in combination with acetazolamide [1]. Normal pressure hydrocephalus is evaluated with intrathecal In^{111} DTPA with evaluation of ventricular activity. Parkinson's disease is best evaluated using I123 ioflupane (DaTScan—GE Healthcare).

Reference

1. Moyamoya Disease Update edited by Byung-Kyu Cho, Teiji Tominaga. Perfusion imaging in Moyamoya Disease. Jin Chul Paeng and Dong Soo Lee. pp. 181–9.

20. The defining characteristic of a positive study using F-18 florbetapir is the presence of:

    A. Decreased activity in the putamen
    B. Persistent activity in the ventricles
    C. Gyri with minimal uptake of tracer
    D. Loss of the gray-white matter differentiation

    Answer: D

    In normal patients, the gyri have little accumulation of tracer with significantly greater tracer present in the white matter.

    A positive F-18 florbetapir study is characterized by loss of the gray-white matter differentiation on the scan. This is compared to the gray-white matter differentiation that is seen in the cerebellum which is not affected by Alzheimer's disease. This loss of the gray-white matter differentiation must be seen in two or more regions of the brain. Bilaterally affected areas of the brain such as the frontal or parietal regions count as a single region.

    Decreased activity in the putamen is characteristic of Parkinson's disease when imaged using I-123 Ioflupane.

21. A positive study using F-18 florbetapir:

    A. Is diagnostic of Alzheimer's disease
    B. Possible even in the normal elderly
    C. Interpreted in conjunction with the clinical findings
    D. A negative scan precludes the development of dementia

    Answer: B

    Image interpretation should be performed independently of the patient's clinical information [1].

The loss of the gray-white differentiation that can be seen with Amyloid deposition is consistent with the diagnosis of Alzheimer's disease but is not diagnostic of it. Elderly patients who do not have symptoms of dementia may also demonstrate an abnormal gray-white differentiation. It is not clear if this is predictive of later development of Alzheimer's dementia.

The negative scan reflects the findings at a particular time point. It does not preclude the later development of Alzheimer's dementia. A negative F-18 florbetapir study in the setting of dementia should direct the physician to other (and possibly treatable) etiologies of dementia.

22. I-123 Ioflupane is useful for imaging:

    A. The presynaptic dopamine transporter system
    B. The presynaptic norepinephrine transporter system
    C. The postsynaptic dopamine transporter system
    D. The postsynaptic norepinephrine transporter system

    Answer: A

    I-123 ioflupane has a high affinity for the presynaptic dopamine transporter system and can be used to help differentiate essential tremor from Parkinsonian syndrome. Patients receiving I-123 ioflupane should receive potassium iodide oral solution equivalent to 100 mg of iodide in an effort to radioprotect the thyroid.

Case 5 49-year-old male with a history of abdominal mass underwent exploratory laparotomy. Following the procedure, the patient developed acute respiratory distress with multiorgan failure. He was found to be unresponsive on clinical examination, suspicious of brain death. To support the diagnosis of brain death, cerebral blood flow study with Technetium 99 m pertechnetate was performed.

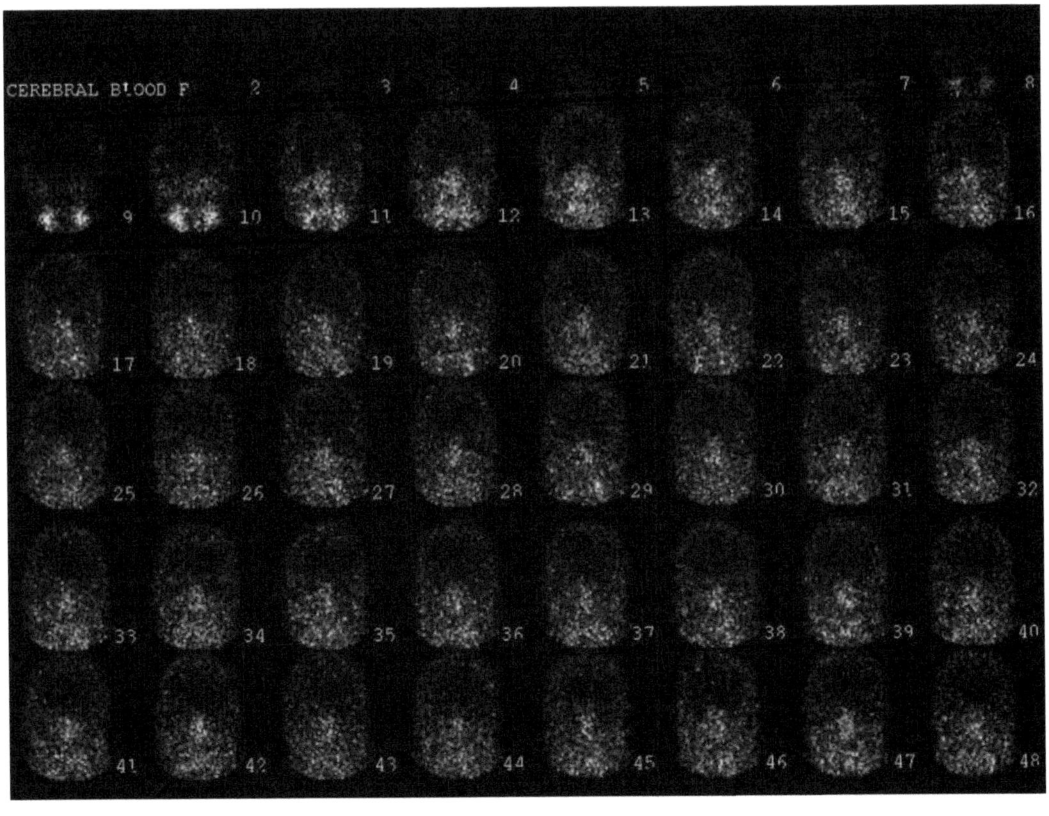

Image 1

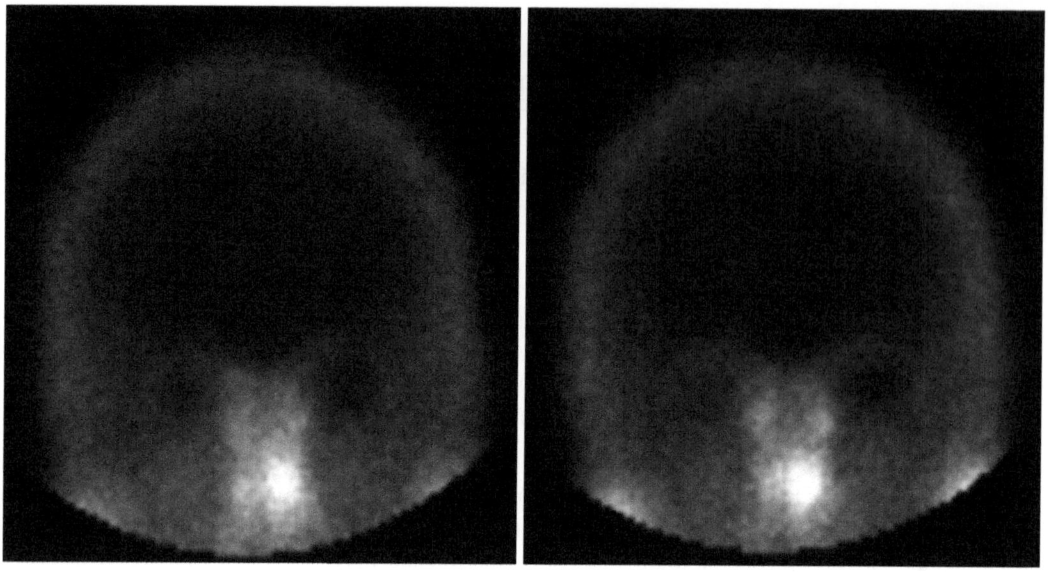

IMMEDIATE 10 MIN DELAY

Image 2

23. Upon evaluation of images you conclude:
    A. Repeat the study because of inadequate bolus injection of the radiotracer
    B. The tracer is lipophobic and normally does not cross blood-brain barrier. Therefore, the findings are expected in individuals with normal cerebral perfusion
    C. There is a cutoff of blood flow at the level of extracranial carotid arteries, consistent with bilateral arterial occlusion
    D. The scan is consistent with the literature description for cerebral death

Answer: D

Tc-99 m Cerebral Perfusion Scan (or brain death scan) is performed to support the clinical diagnosis of brain death. Brain-specific or lipophilic tracers which cross the blood-brain barrier such as Tc-99 m HMPAO and Tc-99 m ECD are preferred over nonspecific or lipophobic tracers such as Tc-99 m DTPA, which are considered angiographic radionuclides.

The basic protocol is a flow (or angiographic) dynamic images followed by an immediate and delayed (obtained 5–10 min after injection of tracer) static images. **Flow imaging normally** demonstrates prompt blood flow to the extracranial carotid systems (whether the tracer is brain-specific like Tc-99 m HMPAO, Tc-99 m ECD, or is angiographic like Tc-99 m DTPA, Tc-99 m pertechnetate). This confirms an adequate bolus was given. On the flow images of angiographic radiotracers, radioactivity does not normally lie within the brain itself because of the integrity of the blood-brain barrier, but rather it is located in the overlying scalp soft tissues, calvarium, and subarachnoid spaces that outline the cerebral hemispheres. Activity is also seen in the larger blood pool accumulations, such as the sagittal and transverse sinuses. B is incorrect.

The normal distribution of lipophilic brain perfusion agents is proportional to regional blood flow, with significantly greater activity seen in the cortical gray matter (activity is symmetric and greatest in the strip of cortex along the convexity of the frontal, parietal, temporal, and occipital lobes, as well as subcortical gray matter, including the basal ganglia and the thalamus).

In the delayed phase (15–20 min after tracer injection), the brain-specific tracer will be **normally** seen in the cerebrum, basal ganglia, thalami, and cerebellum. The angiographic tracer will be seen in the dural venous sinuses.

Positive for brain death study. The flow imaging will demonstrate tracer flowing through the carotid arteries and into the base of the skull (where the flow of tracer stops). There will be the absence of intracerebral perfusion. This is due to increased intracranial pressure. Adequate bolus administration must be ensured, and distinct activity in the carotid arteries is used as a confirmatory sign (C is incorrect). **In the immediate and delayed phases**, the brain-specific tracer will be absent in brain tissue. The angiographic tracer will not be seen in the dural venous sinuses. This combination of findings is interpreted as positive for brain death (because of absent intracerebral perfusion).

The above scenario is an example of a scan positive for brain death using an angiographic tracer Tc-99 m pertechnetate. In the flow dynamic phase (image 1), tracer is seen in the common carotid arteries, confirming an adequate bolus (A is incorrect). No significant intracranial arterial perfusion is noted. Immediate and delayed static images (image 2) also do not demonstrate any significant sagittal sinus activity. This study is abnormal. Findings are consistent with the literature description for cerebral death.

*Ziessman HA, O'Malley JP, Thrall JH, Nuclear medicine: the requisites.
*Mettler FA, Guiberteau MJ, Essentials of nuclear medicine imaging.
*http://www.nucmedresource.com/neurology-nm.html.

Palliative Therapy

24. Which of the following radiopharmaceuticals used for palliative therapy for metastatic osseous disease is an alpha emitter?

    A. Samarium-153
    B. Radium-223
    C. Strontium-89
    D. Rhenium-186

    Answer: B

    Patients who have extensive painful osseous metastases (from various primary lesions that are not amenable to external beam radiotherapy or that are unresponsive to chemotherapy), may benefit from radionuclide palliative therapy. It is directed to reduce pain (not to cure), decreased need for opiates, and improved quality of life. The commonly used radiopharmaceutical are Samarium-153, Radium-223, Strontium-89, Rhenium-186. Among them, only Radium-223 is an alpha emitter. Others are predominantly beta emitters. Rhenium-186 has not been approved for use in the United States but is widely used in Europe.

25. Which of the following radiopharmaceuticals used for palliative therapy for metastatic osseous disease can be proficiently imaged with a gamma camera?

    A. Radium-223
    B. Samarium-153
    C. Strontium-89
    D. Phosphorus-32

Answer: B

Among above all mentioned radiopharmaceuticals, only Samarium-153 is effectively suitable for imaging. Samarium-153 is a mixed beta and gamma emitter. In practice, it is possible to image 103 keV gamma emissions (has 28% abundance). Imaging is usually done at 6 h after administration.

Radium-223 is an alpha emitter.

Strontium-89 and Phosphorus-32 are pure beta emitters. They also produce "secondary radiation", or Bremsstrahlung (BS), as a result of stopping (or slowing) the primary radiation (beta particles). Bremsstrahlung imaging is possible but is not practical. BS radiation is not ideal for diagnostic purposes, because of the continuous energy range, interseptal penetration of high energy photons, and the creation of photons far from the radiation emission site.

Rhenium-186 is another radiopharmaceutical used for palliative therapy for metastatic osseous disease. It is a gamma emitter with an energy of 187 keV that is suitable for imaging. Rhenium-186 has not been approved for use in the United States.

*StatDx. https://my.statdx.com/.
*Ziessman HA, O'Malley JP, Thrall JH, Nuclear medicine: the requisites.

26. Radium-223 dichloride is approved by the FDA to be used in which of the following cancer?

   A. Breast cancer
   B. Lung cancer
   C. Prostate cancer
   D. Renal cell carcinoma

Answer: C

The FDA approved the use of Radium-223 (Xofigo) in May 2013 only in men with castration-resistant prostate cancer and painful bone metastasis. Exclusion criteria include soft tissue metastasis. The treatment consists of a single monthly intravenous injection for a total of 6 doses. Xofigo is not approved for any other malignancies. Common side effects are nausea, vomiting, diarrhea, and bone marrow suppression. Therefore, blood counts should be performed a few days before each injection. Also, the dose depends on body weight, which should be updated routinely.

Case 6 Sarcoid Protocol ^{18}F-FDG PET/CT.

Figure 1 shows a sarcoid protocol cardiac ^{18}F-FDG PET/CT performed in a 58-year-old female patient for further evaluation of suspicious findings described on a recent cardiac MRI. Incidentally, focal ^{18}F-FDG uptake was seen in the inferior pole of the right thyroid lobe.

27. Based on the imaging findings seen in Fig. 1a, b, what is the most appropriate next step in management?

   A. Follow up with thyroid ultrasound and FNA when indicated
   B. Follow up with I-123 whole-body scan to evaluate for possible metastasis
   C. Follow up with I-123 thyroid uptake and scan to assess for cold nodule(s)
   D. Total thyroidectomy followed with radioiodine ablation

Answer: A

Given the high incidence of thyroid cancer for thyroid nodules demonstrating focal FDG uptake, incidental focal FDG thyroid uptake should be followed with a dedicated thyroid ultrasound and fine-needle aspiration if a correlative nodule with suspicious characteristics is identified [1, 2].

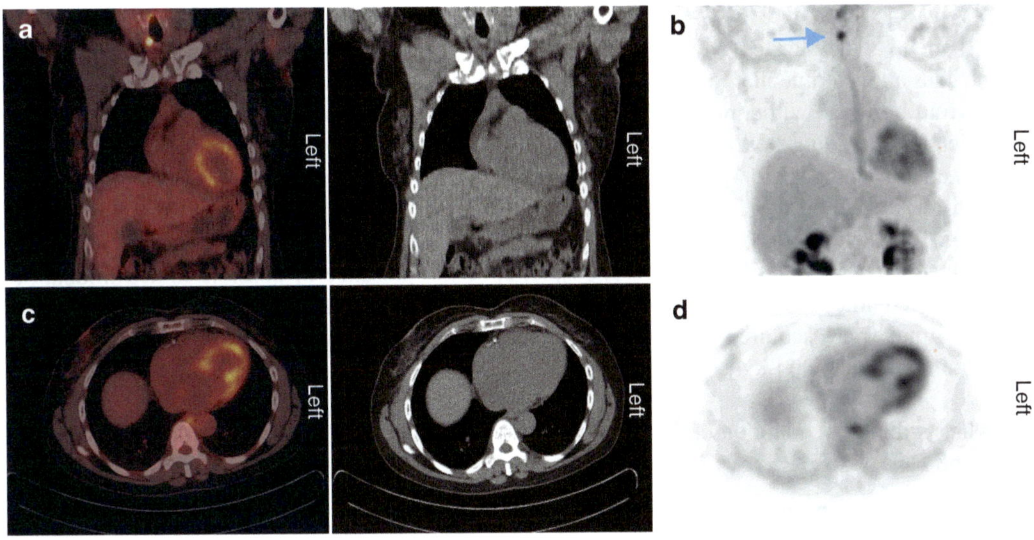

Fig. 1 (**a**) Coronal fused 18F-FDG PET/CT and CT. (**b**) Coronal 18F-FDG PET. (**c**) Axial fused 18F-FDG PET/CT and CT. (**d**) Coronal 18F-FDG PET

28. What is the diagnostic performance of PET/CT for differentiated thyroid cancer?

    A. Poor sensitivity
    B. High sensitivity
    C. High positive predictive value
    D. High negative predictive value

    Answer: A
    Differentiated thyroid cancer has a low glycolytic rate, and usually it is not significantly FDG avid. High-risk patients diagnosed with differentiated thyroid cancer are better evaluated with I-123 whole-body scan [1, 2]

29. What is the reported percentage of malignancy in incidentally identified focal ^{18}F-FDG uptake in the thyroid gland?

    A. Less than 2%
    B. 2–10%
    C. 15–50%
    D. 50–75%

    Answer: C
    The incidence of focal thyroid uptake in ^{18}F-FDG PET/CT done for other indications has been reported as high as 4%. Given the high incidence of thyroid cancer within these "incidentalomas" ranging from 16–50%, the NCCN guidelines recommend further workup for adequate characterization of the focal thyroid uptake, typically by thyroid US and FNA if needed [1, 2].

30. For patients in which Iodine resistant thyroid cancer is suspected, what would be the most appropriate imaging protocol to recommend?

    A. I-123 whole-body scan followed by Thyrogen injections and ^{18}F-FDG PET/CT
    B. Thyrogen injections and I-123 whole-body scan followed by ^{18}F-FDG PET/CT 10 days later
    C. ^{18}F-FDG injection and scan, followed by Thyrogen injection and I-123 scan
    D. Thyrogen injections and I-123 whole-body scan, followed shortly by ^{18}F-FDG PET/CT

Answer: D

Patients with elevated Thyroglobulins will undergo the typical preparation and workup. This includes a low iodine diet for at least 10 days prior to the scans and either Thyrogen injections or Thyroid hormone withdrawal to increase the TSH levels. This preparation is very important for patients undergoing both, I-123 scans and FDG-PET/CT. Ideally, both studies should be performed during the same week, beginning with Thyrogen injections on day 1 and day 2, followed with I-123 dose on day 2 after the Thyrogen injection has been performed. The patient should return for a whole-body I-123 scan on day 3. If the iodine scan is negative in the presence of elevated Thyroglobulins or if there is suspicion of the presence of possible mixed findings of well-differentiated and less differentiated thyroid cancer, FDG dose can be given the same day the I-123 images are acquired (day 3), with subsequent PET/CT images performed approximately 60 min postinjection [3, 4].

31. What is the most common etiology of diffuse low to moderate level of thyroid uptake seen in Ga-68 DOTATATE PET/CT?

    A. Medullary Thyroid Cancer
    B. Follicular Variant Thyroid Cancer
    C. Physiologic uptake
    D. Chronic Thyroiditis

Answer: C

Studies have demonstrated that diffuse uniform ^{18}F-FDG thyroid uptake does not demonstrate the high incidence of cancer seen with the more focal uptake. This presentation is most commonly secondary to chronic or lymphocytic thyroiditis. In contrast, given the presence of somatostatin receptors in the normal thyroid tissue, diffuse thyroid uptake is commonly seen as part of the physiologic distribution of Ga-68 DOTATATE [5].

References

1. Nishimori H et al. Incidental thyroid "PETomas": clinical significance and novel description of the self-resolving variant of focal FDG-PET thyroid uptake. Can J Surg. 2011;54(2):83–8
2. Bae JS et al. Incidental thyroid lesions detected by FDG-PET/CT: prevalence and risk of thyroid cancer. World J Surg Oncol. 2009:63.
3. Chen W et al. Evaluation of thyroid FDG uptake incidentally identified on FDG-PET/CT imaging. Nucl Med Commun. 2009;30(3):240–4.
4. Vaisman F et al. A new appraisal of iodine refractory thyroid cancer. Soc Endocrinol. 2015;22:R301–10
5. Pavel N et al. The rate and clinical significance of incidental thyroid uptake as detected by Gallium-68 DOTATATE positron emission tomography/computed tomography. Thyroid. 2016;26(6):831–5.

Inflame and Infection 2020

Case 7 A 43-year-old male patient is referred for evaluation of episodes of fever and chills of undetermined etiology. Recent radiological imaging described hepatosplenomegaly, retroperitoneal, and inguinal adenopathy, concerning lymphoma. The patient's past medical history is significant for prior L4–L5 laminectomy complicated by paraspinal epidural abscess and drainage. In addition, he has a history of Brucellosis.

32. What radiotracer was administered for the study shown in Fig. 2a?

    (a) Indium-111 WBC
    (b) Tc-99 m HMPAO WBC
    (c) Gallium-67
    (d) Tc-99 m Sestamibi

    Answer: c

    The images shown in Fig. 2a demonstrate increased liver uptake with moderate bone marrow activity and some splenic uptake, which is the expected physiologic distribution of gallium-67. The salivary glands, lacrimal glands, and soft tissues of the breasts show variable levels of uptake depending on the amount of lactoferrin and binding. Gallium-67 demonstrates initial renal clearance and delayed gastrointestinal clearance after 24 h.

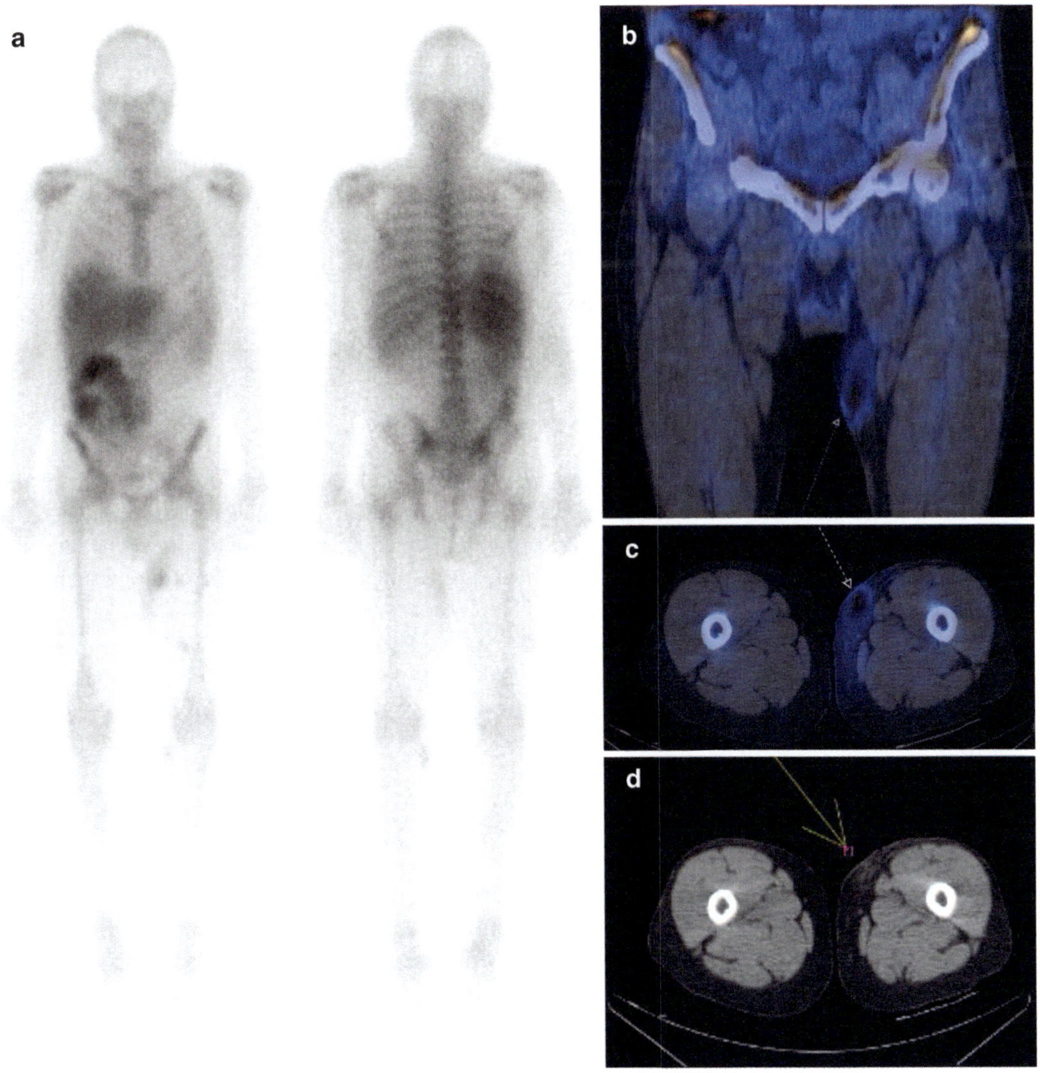

Fig. 2 (a–d)

Indium-111 and Tc-99 m labeled leukocytes demonstrate a similar pattern of tracer uptake in the liver but more intense splenic activity and some lung uptake, particularly on early imaging. In contrast to Indium-111 labeled leukocytes, Tc-99 m labeled leukocyte demonstrates renal and gastrointestinal excretion. Tc-99 m Sestamibi is not used for evaluation of infection and can be easily differentiated from other radiotracers by its prominent myocardial uptake.

33. Based on the tracer distribution shown in Fig. 2a, at what time postinjection the images were acquired?

    (a) 2 h
    (b) 4 h
    (c) 24 h
    (d) 48 h

    Answer: d

    The initial route of excretion is via the kidneys during the first 24 h. After the initial phase of excretion, the major route of excretion is the colon. The images shown in Fig. 2 demonstrate increased bowel activity, indicating that they were taken after the gastrointestinal excretion started. The standard 48-h images provide a better target-to-background ratio when compared to the 24 h images. However, a whole-body study performed 24-h postinjection is frequently diagnostic with no need for additional imaging.

34. What is the most likely explanation for the findings seen in the left lower extremity (Fig. 2b–d)?

    (a) Gallium-67 avid inguinal adenopathy
    (b) Abscess
    (c) Urinary contamination
    (d) Inflammatory process

    Answer: d

    The whole-body images demonstrate a small area of mild activity in the subcutaneous soft tissues of the proximal left lower extremity. The fused SPECT/CT images confirm the activity correlates to a subcutaneous soft tissue stranding in the medial left proximal thigh. The intensity of the uptake is less than the intensity of the liver, suggesting a less aggressive etiology such as an inflammatory process. This was confirmed upon physical examination.

35. How many photopeaks are produced during the decay of Gallium-67?

    (a) 1
    (b) 2
    (c) 3
    (d) 4

    Answer: d

    Gallium-67 is cyclotron produced radiotracer with a physical half-life of 78 h. It decays by electron capture emitting gamma rays with energy peaks of 93, 185, 288, and 394 KeV, many refer to as 100, 200, 300, and 400 KeV. Typically the three lower energy peaks are acquired during imaging. The higher energy peak (394 KeV) is avoided due to the poor detection efficiency for the higher energy photons. Newer nuclear scanners are best performing around Tc-99 m energy (140 KeV).

36. In addition to Gallium-67, what other radiotracer can be used for the evaluation of inflammatory bowel disease/abscess?
    (a) Tc-99 m HMPAO labeled leukocytes
    (b) Ga-68 DOTATATE
    (c) Tc-99 m Sulfur colloid
    (d) Tc-99 m MDP

    Answer: a
    Early images of Tc-99 m labeled WBCs have demonstrated sensitivity comparable to that of the 48 h images following gallium-67 administration for the evaluation of inflammatory bowel disease. The images should be performed early within 2 h of injection, where there is no significant gastrointestinal excretion of radiotracer. Such excretion may incorrectly result in interpreting physiologic activity as infection.

References

1. Palestro C et al. ACR-SNM-SPR practice guideline for the performance of scintigraphy for inflammation and infection. 2009.
2. Society of nuclear medicine procedure guideline for gallium scintigraphy in inflammation version 3.0, approved June 2, 2004.

Case 8 A 57-year-old male patient is referred for evaluation due to persistent fevers of unknown origin. The patient was admitted 7 days prior with leukocytosis and shortness of breath. Pneumonia was suspected. A chest radiograph revealed a large right-sided pleural effusion and a CT revealed the presence of an empyema. A draining tube was placed. The patient is referred for evaluation 2 days after the chest tube was removed because of persistent fever.

37. Based on the uptake shown in Fig. 3, what radiotracer was used?
    (a) Indium-111 labeled leukocytes
    (b) Gallium-67 citrate
    (c) Tc-99 m labeled leukocytes
    (d) ^{18}F-FDG

    Answer: a
    Following reinfusion of ^{111}In-leukocyte radiolabeled leukocytes, physiologic activity is seen in the blood pool, lungs, liver, and spleen in the first 4 h. The blood pool and lung activity decrease by 24 h. If lung infection is not suspected, the presence of higher than usual lung and liver activity at 24 h should suggest poor leukocyte viability. No activity should be seen in the genitourinary or gastrointestinal tract, neither evidence of hepatobiliary excretion.

38. What is the most appropriate interpretation of the findings shown in Fig. 3?
    (a) Right lower chest wall abscess.
    (b) Inflammatory response at the site of previous chest tube.
    (c) Persistent right lower lobe empyema.
    (d) Splenic abscess.

Fig. 3 Indium-111 labeled WBC scan

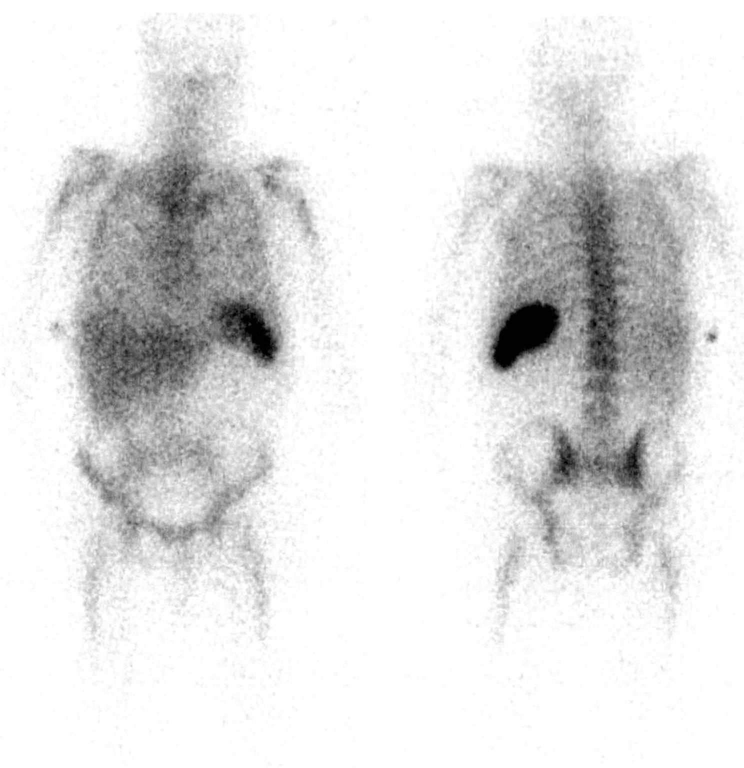

Answer: b

Pertinent information prior to performing a radiolabeled white blood cell study includes but is not limited to surgical history, previous trauma, the presence and location of surgical drains, sites of skin or soft tissue infection, particularly important for patients referred for evaluation of additional sites of infection or abscess. The presence of feeding tubes, ostomies, nasogastric tubes, etc., should be detailed by the referring physician since these sites can demonstrate variable levels of activity and if the history is unclear, these findings may be reported as abnormal (false positive). If the activity in a given concerning focus is equal to or higher than the liver activity, the findings should be described as highly concerning for an abscess. Inflammatory activity typically demonstrates similar or lower intensity than the activity in the bone marrow, as it is seen in the small right lower chest wall focus in this patient.

39. What is the standard imaging time for an Indium-111 labeled WBC study?

(a) 4 and 24 h postinjection
(b) 4, 24, and 48 h postinjection
(c) 24, 48, and 72 h postinjection
(d) 24, 48, and 96 h postinjection

Answer: a

Indium-111 as Gallium-67 is a group III element in the periodic table. It is a cyclotron produced decaying by electron capture and should be ordered the day prior to the desired date of the study. Both gamma rays produced during its decay (173 KeV and 247 KeV) are acquired during imaging. The standard 24-h images have the highest sensitivity; however, early 4-h images should be acquired in the evaluation of inflammatory bowel disease, to ensure adequate localization of the site of inflammation. Imaging past 24 h is usually suboptimal with low counts.

40. What is the minimum volume of blood recommended for appropriate labeling of white blood cells in pediatric patients?

    (a) 10 mL
    (b) 20 mL
    (c) 30 mL
    (d) 40 mL

    Answer: a

    The minimum blood volume obtained for pediatric patients is within 10–15 mL. For adults, the ideal volume from which leukocytes are obtained ranges from 40 to 80 mL of venous blood. To ensure optimal sensitivity the labeled leukocytes should be reinjected as soon as possible after radiolabeling. The reinfusion window should not exceed 1–2 h post labeling to prevent significant loss of viable cells.

References

1. Palestro C et al. ACR-SNM-SPR practice guideline for the performance of scintigraphy for inflammation and infection. 2009.
2. Society of nuclear medicine procedure guideline for gallium scintigraphy in inflammation version 3.0, approved June 2, 2004.

Case 9 The radiographic and scintigraphic images shown above are from a 53-year-old male patient status post total right knee arthroplasty performed a year ago, referred for evaluation due to persistent anterior right knee pain and suspicious radiographic findings concerning loosening. Previous total left knee prosthesis was performed 15 years ago.

41. What significant finding is shown in the right knee radiograph, Fig. 4b (arrow)?

    (a) Linear Sclerosis
    (b) Periprosthetic lucency
    (c) Femoral component malrotation
    (d) Heterotopic ossification

    Answer: b

    On the lateral view (Fig. 4b) the alignment of the femoral and tibial components, as well as the patellar alignment can be assessed. This view also provides a better assessment of the anterior and posterior bone to prosthesis interfaces. The arrow in Fig. 4b is pointing to an area of subtle lucency in the anterior bone to the prosthetic interface of the femoral component. Zones of radiolucency in the interface between the bone and prosthesis is one of the hallmarks for the diagnosis of loosening. However, it is not uncommon to see areas of lucency in the interfaces, even in asymptomatic patients. The zone of radiolucency at the interface should demonstrate progressive widening to be considered representative of loosening. A change in position of the prosthetic components is commonly seen with significant zones of radiolucency.

42. A three-phase bone scan is performed to further evaluate the patient complaints and radiographic findings. How can the angiographic phase shown in Fig. 5a be described?

    (a) Symmetric normal perfusion to both knees
    (b) Symmetric increased perfusion to both knees
    (c) Asymmetric increased perfusion to the right knee
    (d) Asymmetric increased perfusion to the left knee

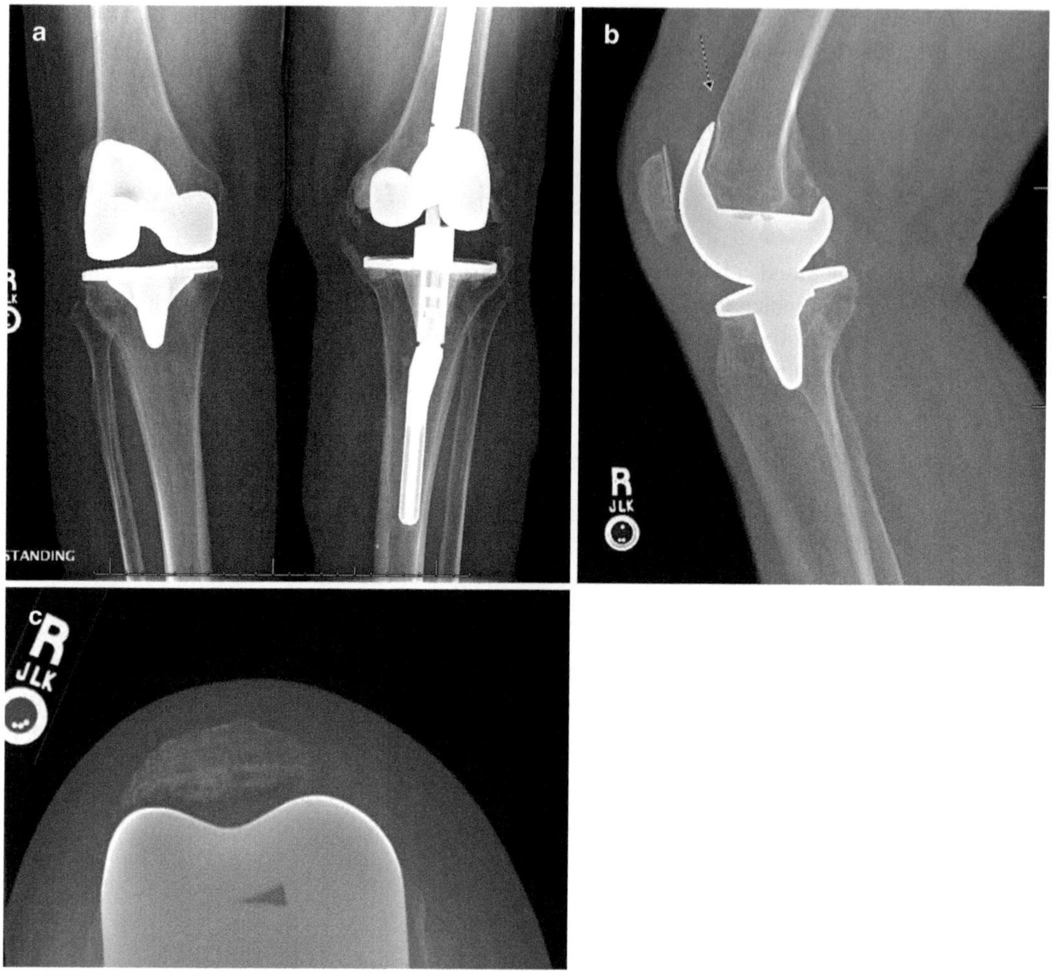

Fig. 4 (**a–c**) Planar radiographs

Answer: a

The anterior and posterior perfusion images are unremarkable and demonstrate symmetric preserved perfusion to both knees. There are no worrisome areas of asymmetric increased perfusion to either knee.

43. The findings in Fig. 5b, c are most consistent with?

(a) Bilateral prosthetic loosening, more prominent in the right knee
(b) Right knee aseptic loosening, more prominent in the femoral component
(c) Right knee aseptic loosening, more prominent in the tibial component
(d) No evidence of prosthetic loosening or infection

Answer: d

The right knee prosthesis was completed only a year ago. Contrary to what is seen after total hip replacement, the osseous remodeling and increased activity at the interfaces of a replaced knee can persist for longer than 2 years. The normal angiographic phase and the only mild blood pool activity are more consistent with expected postsurgical changes, particularly since the right knee activity is not significantly above the uptake seen in the normal and preserved contralateral left knee. In addition, there is no delayed increased activity surrounding the prosthetic stems to suggest loosening. An infected knee characteristically demonstrates increased uptake in all three phases.

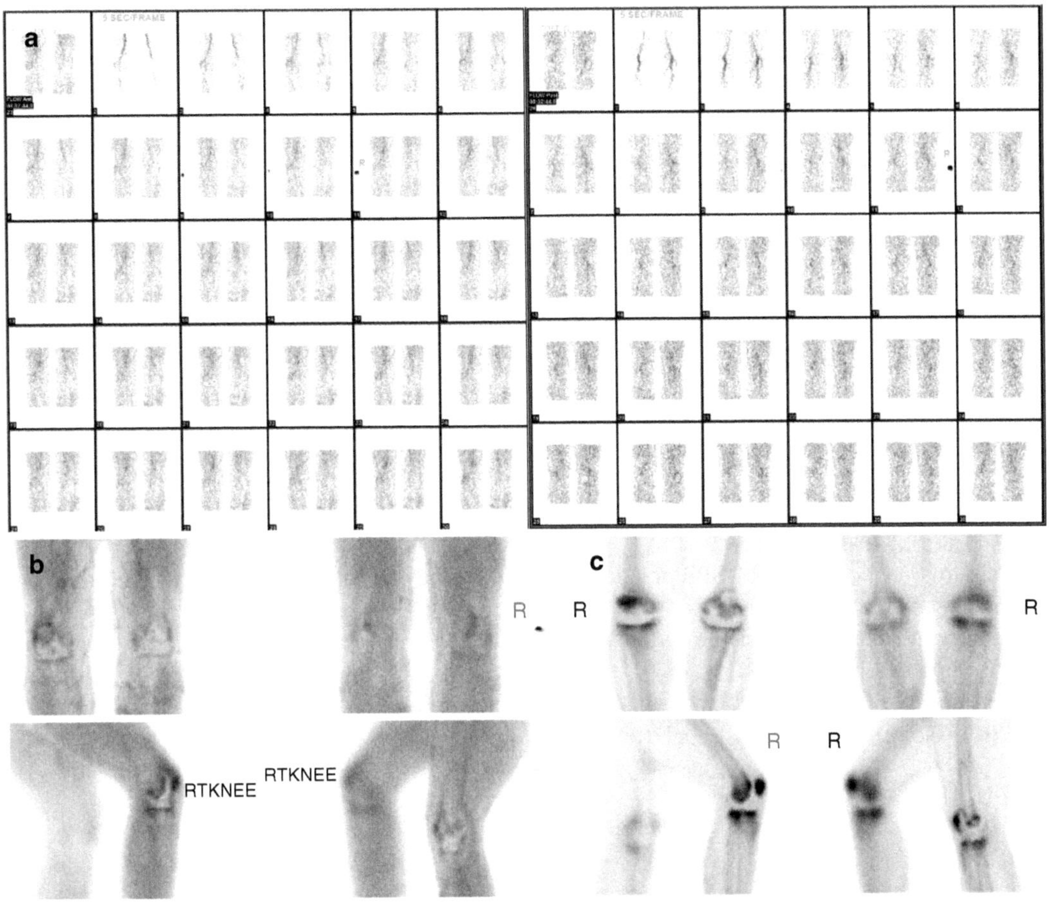

Fig. 5 (**a**) Blood flow. (**b**) Blood pool. (**c**) Delayed images

44. The findings in the delayed images seen in Fig. 5c, suggest what additional diagnosis to be considered in the differential for this patient?

    (a) Patellar fracture
    (b) Progressive osteoarthritic changes of the patella
    (c) Patellar misalignment
    (d) Patellar osteomyelitis

    Answer: b

    Mild to moderate anterior knee pain is the most common complaint following a total knee replacement procedure. The most common complications following this procedure include patellofemoral problems and prosthesis loosening or infection. Following a septic joint treatment, patellofemoral problems are the most common cause leading to the surgical revision of the prosthetic knee. Prominent uptake in the right patella is seen in the delayed images. The Merchant views shown in Fig. 4c provide adequate visualization of the patella and demonstrate degenerative changes without evidence of fracture. There is no significant evidence favoring patellar misalignment in the lateral views shown in Fig. 4b. No increased flow to the patellar region is noted in the angiographic phase to suggest an infectious process.

45. Based on the combination of radiographic and scintigraphic findings shown in Figs. 4 and 5, what should be the next consideration for this patient?

    (a) Correlation of the imaging findings with a radiolabeled leukocyte study.
    (b) Right knee surgical revision
    (c) Prolonged intravenous antibiotic treatment
    (d) Close follow up with radiographs as indicated

    Answer: d

    There is no scintigraphic evidence to favor an infectious process. The benign scintigraphic findings do not suggest that the subtle lucency in the anterior femoral interface seen in the radiographs is the result of early failure. Follow-up with further correlative imaging as needed is an appropriate conservative approach for this patient.

References

1. Math KR, Imaging of total knee arthroplasty. Semin Musculoskel Radiol. 2006;10:1.
2. Villano M et al. Painful knee prosthesis; surgical approach. Clin Cases Miner Bone Metabol. 2011;8(2):26–8.
3. Palestro C et al. ACR-SNM-SPR practice guideline for the performance of scintigraphy for inflammation and infection. 2009.
4. Society of nuclear medicine procedure guideline for gallium scintigraphy in inflammation version 3.0, approved June 2, 2004.
5. Peterson W et al. Anterior knee pain after total knee arthroplasty: a narrative review. Int Orthop (SICOT). 2014;38:319–28.

Case 10 A 55-year-old male patient is referred for evaluation due to painful right knee 2 years post total knee arthroplasty. The main indication was not clear at the time of imaging and since the patient described multiple areas of pain and the order indicated history of prostate adenocarcinoma. A single delayed phase whole-body scan was performed.

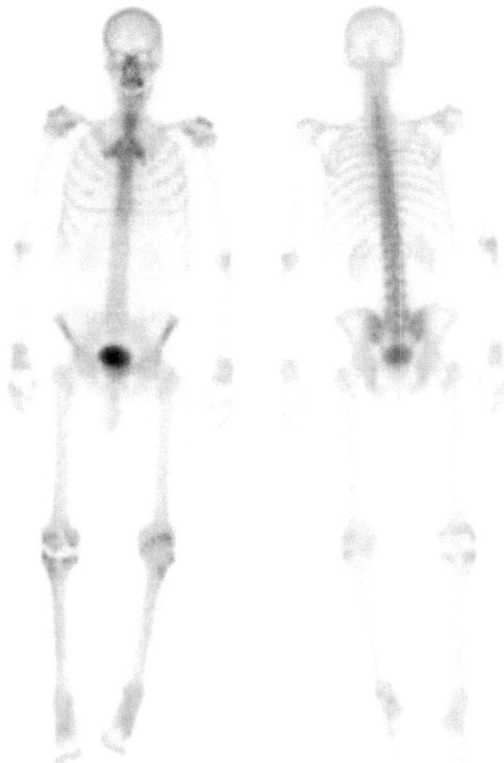

46. How can the findings in the right knee be interpreted?
    (a) Equivocal for infection given the lack of angiographic and blood pool phases
    (b) Suspicious for loosening of the tibial component
    (c) Suspicious for tibial fracture
    (d) No evidence suggestive of loosening or infection

    Answer: d

    The ideal bone scan protocol that will allow optimal evaluation of all the clinical indications in this patient should include a triple-phase bone scan of the knees with a delayed whole-body scan. Despite the lack of angiographic and blood pool images, there are no worrisome areas of activity in the knees in the delayed whole-body images. The delayed whole-body images show no concerning findings to suggest infection or loosening.

47. Based on the scintigraphic findings on the delayed whole-body images, you suggest:

    (a) Reevaluating the knee with a three-phase bone scan
    (b) Correlation with radiolabeled leukocyte study
    (c) Correlation with a Gallium-67 citrate study
    (d) Routine follow-up imaging.

    Answer: d

    The delayed osseous uptake in the right knee is not suspicious for either loosening or infection and further scintigraphic workup is not indicated. Although a triple-phase study was not done, the benign findings on the delayed images are reassuring. Besides mild degenerative changes in the contralateral left knee, there are no additional worrisome scintigraphic findings to suggest blastic metastatic lesions from his known prostate adenocarcinoma. Because the patient is at high risk for developing skeletal metastases, a repeat bone scan is appropriate, particularly if the PSA level is increasing.

48. Which of the following is a common cause of false-positive findings in radiolabeled white blood cell studies?

    (a) Chronic infection
    (b) Bone marrow uptake
    (c) Low neutrophil count
    (d) Areas of bone marrow replaced by fat

    Answer: b

    The sensitivity of radiolabeled leukocyte studies for the detection of chronic infections is debatable with possible false-negative studies. Nevertheless, false-positive scans are also reported. A common reason for false-positive findings, particularly in the postsurgical bones, is the deposition of increased activity in areas of altered marrow distribution.

49. If a false positive WBC study is suspected in the evaluation of a painful prosthetic knee, what additional study should be performed for correlation?

    (a) Tc-99 m MDP triple-phase bone scan
    (b) T-99 m sulfur colloid scan
    (c) T-99 m pertechnetate scan
    (d) T-99 m PYP scan

    Answer: b

    A combined radiolabeled leukocyte study and Tc-99 m Sulfur colloid scan have proven to improve the sensitivity for the evaluation of prosthetic joints when compared to radiolabeled leukocytes and Tc-99 m MDP bone scans. The marrow scan helps discriminate areas of increased leukocyte deposition in altered marrow distribution from areas of infection.

50. Reinjection of the radiolabeled leukocytes should not be performed if the labeling efficiency is below what percentage?

    (a) 80%
    (b) 70%
    (c) 60%
    (d) 50%

Answer: d

Tests to assess sterility and pyrogenicity cannot be performed for radiolabeled leukocytes due to the need for rapid reinjection and to avoid cell damage. However, the labeling efficiency should be measured. The usual labeling efficiency is in the range of 75–90%. A dose with a labeling efficiency of less than 50% should not be reinfused.

References

1. Math KR, Imaging of total knee arthroplasty. Semin Musculoskel Radiol. 2006:10:1.
2. Villano M et al. Painful knee prosthesis; surgical approach. Clin Cases Min Bone Metabol. 2011;8(2): 26–8.
3. Palestro C et al. ACR-SNM-SPR practice guideline for the performance of scintigraphy for inflammation and infection. 2009.
4. Society of nuclear medicine procedure guideline for gallium scintigraphy in inflammation version 3.0, approved June 2, 2004.
5. Peterson W et al. Anterior knee pain after total knee arthroplasty: a narrative review. Int Orthop (SICOT). 2014;38:319–28.

GPSR Compliance

The European Union's (EU) General Product Safety Regulation (GPSR) is a set of rules that requires consumer products to be safe and our obligations to ensure this.

If you have any concerns about our products, you can contact us on ProductSafety@springernature.com

In case Publisher is established outside the EU, the EU authorized representative is:

Springer Nature Customer Service Center GmbH
Europaplatz 3
69115 Heidelberg, Germany

Batch number: 10732817

Printed by Printforce, the Netherlands